MATHEMATICS FOR ELECTRONICS WITH APPLICATIONS

MATHEMATICS FOR ELECTRONICS

with Applications

HENRY M. NODELMAN
Head of Mathematics and Physics Department, RCA Institutes, and Instructor in Mathematics, Polytechnic Institute of Brooklyn

FREDERICK W. SMITH
Television Engineer, National Broadcasting Company

McGRAW-HILL BOOK COMPANY, INC.
New York Toronto London
1956

MATHEMATICS FOR ELECTRONICS
with Applications

Library of Congress Catalog Card Number 55-9549

IV

46850

THE MAPLE PRESS COMPANY, YORK, PA.

Preface

In *Mathematics for Electronics with Applications*, the emphasis has been placed on application rather than on mathematical theory. This approach, which represents a new departure for a mathematics text in this field, is deliberate. It is a direct result of the authors' many years of experience in the teaching of mathematics to students of electronic engineering and in engineering practice. It is our conviction that the wide barrier which exists between mathematics and its applications can best be surmounted by demonstrating the practical uses which industry makes of mathematics. In this way, student motivation is increased considerably, and the student is not confronted with the problem of interrelating two discrete sets of information, one mathematical, the other electronic, as unfortunately has too often been the case.

In order to accomplish this objective, the text offers a complete set of up-to-date problems based on current engineering practice, the result of more than five years of research in the technical literature. All of the leading American and foreign technical periodicals devoted to communications and electronics were carefully examined to determine which mathematical topics were to be included. The mathematical applications are accompanied by worked-out examples and are extensively supported with references so that the reader can supplement his knowledge of the problem by consulting the original article from which the problem was obtained.

Part I of the book describes the methods by which mathematics has been used in electronics and presents a series of electronic engineering problems which involve the use of the calculus. Part II introduces the reader to dimensional analysis as it is applied in electronics and shows the methods which can be used to check the validity of engineering equations and formulas independent of the engineering considerations used to establish such equations. Part III deals with the algebra of determinants. The theory is then applied to the solution of steady-state electronic circuits. Matrix algebra is discussed, and the solution of electronic networks, including those which contain transistors, is undertaken by the matrix method. In Part IV the fundamental properties of series approximations are presented. The use of series to predict the behavior of nonlinear electronic devices is treated; the processes of harmonic generation and intermodulation testing are described. Part V is devoted to solutions of differential equations employing both classical and Laplace transform methods. Part VI gives the elements of Boolean algebra and includes material on switching circuits. In addition, it presents an analysis of the more advanced mathematical topics which logically follow those incorporated in the text. The various parts of the book are independent and may be studied and read as separate units if desired.

The book is intended for those readers who have a background of elementary calculus, physics, and elementary electrical network theory. It is designed for use in industry as well as for use in undergraduate technical institute and engineering college courses in applied mathematics, networks, transients, and nonlinear systems.

Because the technical references incorporated in this book are such an integral part of it, it is obvious that the book could not have become a reality without the generous cooperation of the editors of the various technical publications identified in the footnote references, for which the authors are grateful. We also wish to extend thanks to the industrial organizations elsewhere identified who made illustrations and valuable technical data available.

HENRY M. NODELMAN
FREDERICK W. SMITH

Contents

MATHEMATICS FOR ELECTRONICS WITH APPLICATIONS

Resistance network analog computer provides solutions to problems of temperature distribution in solids, lamellar flow in fluids, and field patterns of electrostatic and magnetic fields. (*National Bureau of Standards.*)

PART 1

The Uses of Mathematics in Electronics

Chapter 1

CASE HISTORIES OF PRACTICAL APPLICATIONS

A crucial question often proposed concerning mathematics is: Of what practical use is it? In the chapters which follow, it is hoped that the reader will discover sufficient evidence of its practical use to agree with J. J. Thomson that "mathematical analysis is the greatest mental labor saving device ever invented."[1] Actually, the practicing engineer nowadays is enormously indebted to the mathematician. Once he is able to express in symbols and equations the physical relations of a problem, he is in a position to forget the original problem for a moment and to solve the equations by the methods of mathematical analysis, which have usually been worked out previously on an abstract basis by the mathematician. How the engineer thus came to be dependent on the mathematician has been described by Van der Pol:[2]

Consider the relation between mathematics on the one hand and physics and technology on the other. History shows us that mathematics has, to a great extent, been developed by pure mathematicians who were under the spell of the beauty of it; who were entranced by its mysterious generality and who spent their life discovering new facts and relations in this very wide and wonderful domain. Whether their results could be applied to astronomy, physics, chemistry, or say, technology, as a rule did not interest them primarily. Their main concern was to make their theorems cover cases of a more and more abstract nature, and generality in their work was their main purpose.

Some mathematicians even went so far as to be proud of the fact that none of their results could be applied in practice. It is said that one great mathema-

[1] J. R. Carson, Mathematics and Electrical Communication, *Bell Labs. Record*, **14** (8): 397–399, August, 1936.

[2] B. Van der Pol, Radio Technology and the Theory of Numbers. *J. Franklin Inst.*, **255** (6): 475–495, June, 1953.

tician (Cayley) remarked that "Bessel functions are beautiful functions in spite of their many applications."

Now, physics and technology were developed to a certain extent independently of pure mathematics. It is, therefore, natural that at a given moment the physicists or technologists discovered, to their surprise, that precisely the mathematical tools and methods they needed to solve their problems had already been fully developed by pure mathematicians who, at the time, did not have the slightest notion of, or even interest in, a possible practicable application of their results.

An adaptation of mathematics to the purposes of the engineer is aptly portrayed in the history of the complex number.[1] In 1572, Raphael Bombelli pointed out that in order to obtain a solution for an equation of the general form $x^2 + a = 0$, one must either accept the concept of the imaginary number $\sqrt{-a}$ or conclude that the equation is meaningless. Although a dim view was taken of this suggestion by Bombelli's mathematical contemporaries, the concept of the imaginary number flourished. Later, the mathematician Gauss in extending the work of a Norwegian surveyor, Wessel, and a Parisian bookkeeper, Argand, developed the idea of a complex plane wherein points represented by pure imaginaries jy or complex numbers[2] of the form $x + jy$ could be graphically displayed[3] as in Fig. 1-1. These developments held little significance for the electrical engineer until the late nineties, when C. P. Steinmetz of the General Electric Company and others observed that the multiplication of a directed line in the complex plane by $\sqrt{-1}$ (or j) represents a rotation of 90°, and thereby established quite ingeniously a connection between the complex number, which had been up to that time a mathematical curio, and alternating-current theory. This result, amazing at the time, was then further applied by showing that if the inductances and capacitances in electric networks were represented by impedances $j\omega L$ and $1/j\omega C$, respectively, the voltages and currents in the networks could then be calcu-

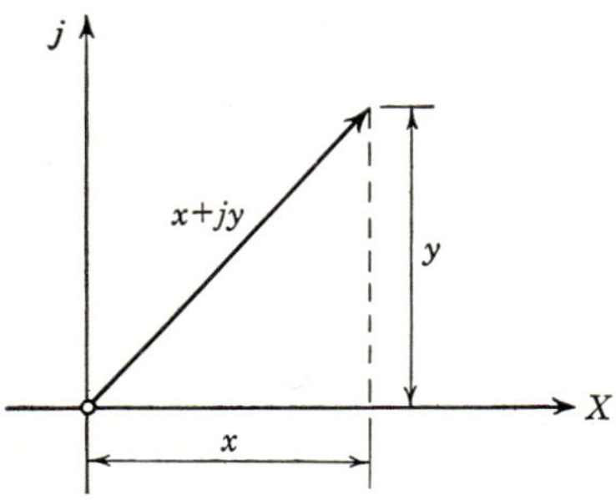

FIG. 1-1 The Argand diagram.

[1] See E. Kasner and J. Newman, "Mathematics and the Imagination," Simon and Schuster, Inc., New York, 1941, for stimulating essays on this and other mathematical topics.

[2] Euler originally represented the imaginary $\sqrt{-1}$ by the symbol i, and mathematicians still use this notation. However, electrical engineers prefer the symbol $\sqrt{-1} = j$ to prevent confusion with the standardized symbol for electric current i.

[3] Such a graph is still known as an Argand diagram because of its origin. See A. T. Starr, "Electric Circuits and Wave Filters," p. 19, Sir Isaac Pitman & Sons, Ltd., London.

lated by the simple algebraic methods previously devised for d-c circuits. Thus, through the keen insight of an engineer, an abstract mathematical idea was converted to a practical laborsaving method for the solution of a-c circuit problems which had heretofore required continual recourse to the drawing board for the graphical construction of phasor diagrams.[1]

Sometimes, of course, the mathematical theory evolves from physical discovery and is then confirmed by further experiment, as in the development of electromagnetic wave theory, as described by Van der Pol.[2]

> One will remember how Faraday, working entirely alone in the Royal Institute in Albemarle Street in London, developed the fundamentals of the science of electricity, his researches being crowned with the famous law of induction. It is most remarkable that he set to work absolutely experimentally and that all his results, visualized in three-dimensional space, were represented without any formulae. Then Maxwell arrived on the scene and translated Faraday's ideas into the language of mathematics; at first, however, without employing the vector notation, which was not introduced into the field of electricity until sometime later by Heaviside and Lorentz.
>
> Thus Maxwell (by solving the famous wave equation) was able to arrive at the conclusion that electromagnetic waves exist; to achieve this result without mathematical symbols would have been beyond human possibility. It was not until 1886 that Heinrich Hertz, stimulated partly by a problem set by Von Helmholtz, based on the Maxwell theory, demonstrated the existence of electromagnetic waves in the laboratory.

In the modern research laboratory, the application of the mathematical method is no less important. Consider the engineering procedure reported by Gilman.[3]

> To evaluate a proposed development [the engineer] usually starts with an exploratory study. Here, whenever possible, he takes advantage of mathematics and physical theory which have frequently demonstrated their capacity to save him time and money. There are notable instances in which mathematical studies have effectively disclosed in quantitative terms basic controlling factors of complex problems without any practical experimentation at all. Mathematical analysis has proved a powerful tool, for example, in clarifying the philosophy of the transmission of information and in the application of statistical theory to telephone trunking.

[1] So-called "vector" diagrams of the phase and amplitude relationships of voltages and currents are properly described as phasor diagrams; and their usage has been so standardized by the Institute of Radio Engineers. See N. R. LePage, Symbolic Nomenclature in Sinusoids, *Elec. Eng.*, **68** (7), July, 1949. Although complex quantities add vectorially in the complex plane, they are not vectors, because they do not possess the vector and scalar products.

[2] B. Van der Pol, Mathematics and Radio Problems, *Philips Research Repts.*, **3**: 174–190, 1948.

[3] G. W. Gilman, Systems Engineering in Bell Telephone Laboratories, *Bell Labs. Record*, **31** (1), January, 1953.

The level of the mathematics employed need not necessarily be high, either. As Fry[1] says,

> There need be no apology associated with the statement that such simple processes as algebra, trigonometry and the elements of calculus are the most common and the most productive in modern industrial research. They frequently lead to results of the greatest practical importance. The single sideband system of carrier transmission, for example, was a mathematical invention. It virtually doubled the number of long distance calls that could be handled simultaneously over a given line. Yet the only mathematics involved in its development was a single trigonometric equation, the formula for the sine of the sum of two angles.

Of course, where blind reliance is placed on it, the application of mathematics can sometimes produce negative results, too. Occasionally, a theoretical result obtained mathematically will prove to be in conflict with the physical facts as determined by experiment because the analyst made some incorrect assumptions, ignored certain factors which in practice may not be neglected, or overgeneralized for all cases from the particular case under study.

Thus, on the basis of an insufficiently general theoretical analysis, J. R. Carson[2] concluded in 1928 that " . . . static, like the poor, will always be with us." Posterity, through the benefits of hindsight, now recognizes that the riddle of static elimination can be solved by means of many of the recently developed modulation schemes, but it is interesting to speculate on the extent to which research into modulation systems other than amplitude was delayed by such a discouraging prophecy so early in the development of radio.

It is often quite surprising to discover how essential the use of mathematics is, particularly for the solution of problems which are apparently very simple. Problems, for instance, which involve solutions in terms of integers only can prove to be quite complicated, and the reader is invited to test his skill on the four which follow.[3]

Problems

Prob. 1-1 One of 27 precision resistors has been improperly marked and is known to have a value which differs considerably from the remainder of the lot. A Wheatstone bridge is available; consequently the

[1] T. C. Fry, Industrial Mathematics, *Bell System Tech. J.*, **20** (3): 267, July, 1941.

[2] J. R. Carson, The Reduction of Atmospheric Disturbances, *Proc. I.R.E.*, **16** (7): 966, July, 1928.

[3] For information on the solution of these see, for example, P. S. Herwitz, The Theory of Numbers, *Sci. American*, **185** (1): 52–55, July, 1951; O. Ore, "Number Theory and Its History," McGraw-Hill Book Company, Inc., New York, 1948.

resistors may be checked against one another by placing them in series in opposite arms of the bridge. What is the minimum number of checks which must be made with the bridge in order to locate the culprit?

Ans. 3 if unknown is known to be high or low; otherwise, 4

Prob. 1-2 A radar observer errs in reporting the distance to a target as read on a radar range indicator which can be read to an accuracy of $\frac{1}{100}$ mile. He accidently reverses his readings and reports in miles what should have been hundredths of a mile and vice versa. The radar plotter notes that if the erroneous range is decreased by 0.62 mile, it is about twice what it should be. What is the correct range of the target?

Ans. 12.25 miles

Prob. 1-3 The over-all height above ground of a television superturnstile antenna and the steel tower on which it is mounted is 364 ft excluding the code beacon at the top. Since such a structure constitutes a hazard to air navigation, it must be painted in alternate uniform bands of international orange and white, the orange bands to be exactly twice the height of the white bands. If each white section must be from 6 to 9 per cent of the over-all tower height, and the base and the top of the tower must be painted orange, lay out a tower painting plan.

Prob. 1-4 In the central stockroom of a television broadcasting network, a shipment of monitor kinescopes is received. The standard procedure is for such a shipment to be divided into three equal parts which are dispatched to the eastern, central, and western divisions of the network, respectively. Before the division occurs, however, one of the tubes is always placed in reserve stock. On this particular occasion the clerk responsible for shipments to the eastern division is alone when the shipment arrives. After consignment of one tube to reserve stock, he forthwith divides the remainder into three equal parts and ships one part to the eastern division.

Subsequently the second clerk, who is responsible for the central division, arrives to find the stockroom unattended. He divides the remaining kinescopes into three groups, after first dispatching one to reserve stock. He then forwards one group to the central division.

Next the clerk for the western division arrives and, in the absence of the other clerks, repeats the previous process of dividing the remaining tubes into three lots, after earmarking and sending one tube to reserve stock. He then sends one lot to the western division.

Finally, during lunch relief, the stockroom supervisor appears and, assuming that the kinescope shipment has not yet been processed, sends one to reserve stock and ships out the three groups remaining to the various divisions of the network. Before the invoice is checked for payment, the minimum number of kinescopes in the original shipment must be known. *Ans.* 79

Chapter 2

ELECTRONIC APPLICATIONS OF CALCULUS

2-1 Applications of Differentiation

EXAMPLE 1.* For a thermistor used as a detector in microwave power measurements,

$$R = J\epsilon^{B/(K+CP)}$$

where R is the thermistor resistance; B, J, and C are constants; P is the applied power; and K is the absolute ambient temperature. The detector sensitivity of the thermistor is defined as the change in resistance with respect to the change in applied power, and an expression for this quantity must be derived.

In the limiting case,

$$\text{Detector sensitivity} = \frac{dR}{dP}$$

and

$$\ln R = \ln J + \frac{B}{K + CP}$$

By differentiation,

$$\frac{1}{R}\frac{dR}{dP} = -\frac{BC}{(K + CP)^2} \qquad \text{and} \qquad \frac{dR}{dP} = -\frac{BCR}{(K + CP)^2}$$

But recall from above that

$$\frac{B}{K + CP} = \ln\frac{R}{J}$$

so that

$$\left(\frac{1}{K + CP}\right)^2 = \frac{1}{B^2}\left[\ln\left(\frac{R}{J}\right)\right]^2$$

* C. G. Montgomery, "Technique of Microwave Measurements," Radiation Laboratory Series, vol. 11, p. 104, McGraw-Hill Book Company, Inc., New York, 1947.

Substitution of this result in the expression for dR/dP yields the detector sensitivity as

$$\frac{dR}{dP} = -\frac{CR}{B}\left[\ln\left(\frac{R}{J}\right)\right]^2$$

which is the desired result.

EXAMPLE 2. In the design of nonlinear precision potentiometers, nonlinearity of resistance with rotation is obtained by suitably shaping the card on which the resistance wire is wound. It can be shown that the card shape is proportional to the derivative of the resistance as a function of angular rotation. For example, a potentiometer card which consists of a true triangular section will give a variation of resistance with shaft rotation conforming to a true square-law function.

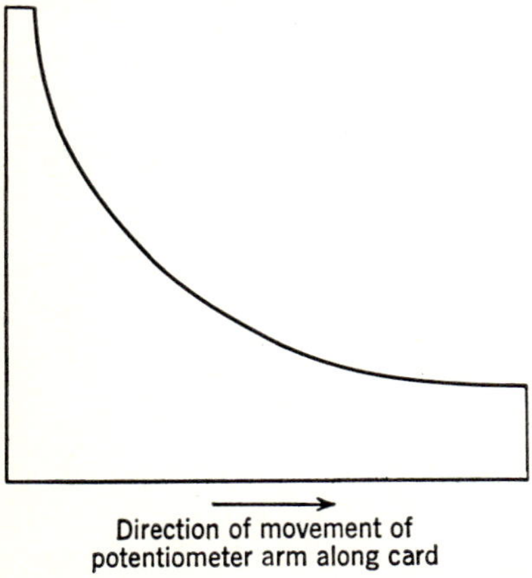

FIG. 2-1 Potentiometer card shape used for approximating logarithmic variation of resistance with rotation of potentiometer.

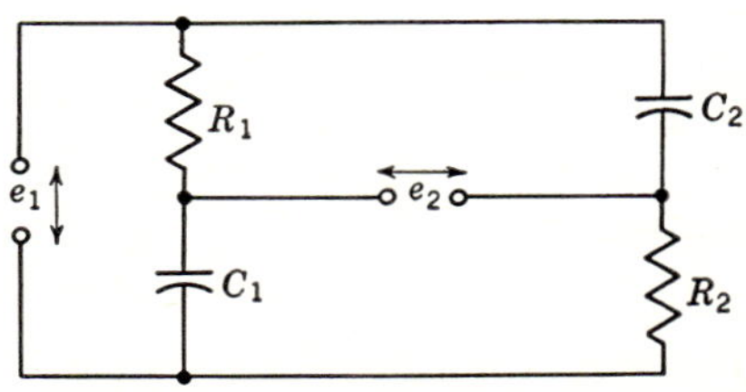

FIG. 2-2 Ninety-degree-phase-shift network.

Suppose it is desired to obtain a logarithmic variation of resistance with potentiometer shaft rotation such that $R = k \ln \theta$, where k is a constant.

According to this method for designing a potentiometer card shape,

$$\text{Potentiometer card shape} = \frac{dR}{d\theta} = \frac{k}{\theta}$$

The potentiometer card shape is therefore hyperbolic, as shown in Fig. 2-1.

EXAMPLE 3.* The ratio of the output to the input voltage of the 90°-phase-shift network, shown in Fig. 2-2, is

$$\frac{e_2}{e_1} = \frac{1/j\omega C_1}{R_1 + 1/j\omega C_1} - \frac{R_2}{R_2 + 1/j\omega C_2}$$

* I. S. Greenwood, J. V. Holdam, and D. MacRae, "Electronic Instruments," Radiation Laboratory Series, vol. 21, pp. 149–150, McGraw-Hill Book Company, Inc., New York, 1948.

or
$$\frac{e_2}{e_1} = \frac{1 + \omega^2 R_1 R_2 C_1 C_2}{1 - \omega^2 R_1 R_2 C_1 C_2 + j\omega(R_1C_1 + R_2C_2)}$$

and the phase shift is 90° if

$$\omega^2 = \omega_0{}^2 = \frac{1}{R_1R_2C_1C_2}$$

Under this condition the expression for e_2/e_1 may be written as

$$\frac{e_2}{e_1} = \frac{1 + (\omega/\omega_0)^2}{1 - (\omega/\omega_0)^2 + j\omega(R_1C_1 + R_2C_2)}$$

If the fraction is multiplied in both numerator and denominator by the complex conjugate of the denominator, it becomes

$$\frac{e_2}{e_1} = \frac{[1 + (\omega/\omega_0)^2][1 - (\omega/\omega_0)^2 - j\omega(R_1C_1 + R_2C_2)]}{[1 - (\omega/\omega_0)^2]^2 + \omega^2(R_1C_1 + R_2C_2)^2}$$

and
$$\varphi = \tan^{-1}\left[\frac{-\omega(R_1C_1 + R_2C_2)}{1 - (\omega/\omega_0)^2}\right]$$

where φ is the phase angle of e_2 with respect to e_1.

It is desired to find the sensitivity of this phase shift with respect to changes in ω. This is given by

$$\frac{d\varphi}{d\omega} = \frac{-(R_1C_1 + R_2C_2)\dfrac{1 + (\omega/\omega_0)^2}{[1 - (\omega/\omega_0)^2]^2}}{1 + \dfrac{\omega^2(R_1C_1 + R_2C_2)^2}{[1 - (\omega/\omega_0)^2]^2}}$$
$$= \frac{-(R_1C_1 + R_2C_2)[1 + (\omega/\omega_0)^2]}{[1 - (\omega/\omega_0)^2]^2 + \omega^2(R_1C_1 + R_2C_2)^2}$$

and at $\omega = \omega_0$,

$$\left(\frac{d\varphi}{d\omega}\right)_{\omega_0} = \frac{-2(R_1C_1 + R_2C_2)}{\omega_0{}^2(R_1C_1 + R_2C_2)^2} = \frac{-2}{\omega_0{}^2(R_1C_1 + R_2C_2)}$$

Substituting $\omega_0{}^2 = 1/(R_1R_2C_1C_2)$,

$$\left(\frac{d\varphi}{d\omega}\right)_{\omega_0} = \frac{-2}{(R_1C_1 + R_2C_2)/R_1R_2C_1C_2} = \frac{-2}{1/R_1C_1 + 1/R_2C_2}$$

If the two time constants R_1C_1 and R_2C_2 are equalized, then

$$\left(\frac{d\varphi}{d\omega}\right)_{\omega_0} = -R_1C_1$$

which is the desired result.

If $R_1 = R_2 = 10^5$ ohms and $C_1 = C_2 = 0.01 \times 10^{-6}$ farad, then, at

$\omega = \omega_0$, $d\varphi/d\omega = 10^{-3}$ radian/cycle, or 0.573° shift per cycle, in the phase of the source e_1.

EXAMPLE 4. Two sine waves of equal amplitude which differ in phase by φ,

$$y = A \sin \omega t$$
$$x = A \sin (\omega t + \varphi)$$

are applied to the horizontal (x) and vertical (y) inputs of an oscilloscope in order to determine the extent of the phase shift between them.

These are parametric equations of the curve or figure which will appear on the face of the scope. If the quantity ωt is eliminated in these equations, the rectangular equation of the curve is obtained.

From the first equation, $\sin \omega t = y/A$, and $\cos \omega t = \sqrt{1 - y^2/A^2}$. From the formula for the sine of the sum of two angles,

$$x = A \sin \omega t \cos \varphi + A \cos \omega t \sin \varphi$$

Substituting for $\sin \omega t$ and $\cos \omega t$,

$$x = y \cos \varphi + A(\sin \varphi) \sqrt{1 - y^2/A^2}$$
$$= y \cos \varphi + \sqrt{A^2 - y^2} \sin \varphi$$

from which $$x^2 - 2xy \cos \varphi + y^2 = A^2 \sin^2 \varphi$$

This is the equation of an ellipse centered on the origin.

It is now desired to find the axes of the ellipse, which by definition intersect the ellipse at right angles. This can be accomplished by finding the slope of the ellipse y'. Differentiating the expression for the ellipse,

$$y' = \frac{y \cos \varphi - x}{y - x \cos \varphi}$$

When $y' = -1$, $y = x$, so that the slope of the ellipse is normal to the straight line $y = x$. Also when $y' = +1$, $y = -x$, so that the slope of the ellipse is also normal to $y = -x$, and the two axes of the ellipse are $y = x$ and $y = -x$.

If the phase angle $\varphi = 0$, so that $\cos \varphi = 1$ and $\sin \varphi = 0$, then

$$x^2 - 2xy - y^2 = 0 \qquad \text{or} \qquad x = y$$

so that the ellipse degenerates into a straight line at 45° to the x or y axis. If $\varphi = 90°$, so that $\cos \varphi = 0$ and $\sin \varphi = 1$, then

$$x^2 + y^2 = A^2$$

so that the ellipse becomes a circle of radius A.

2-2 Maximum-Minimum Problems

EXAMPLE 1. Consider the series resonant circuit shown in Fig. 2-3, which is the basic arrangement used in the Q meter.

If an emf e is applied to the input terminals, the current in the circuit is

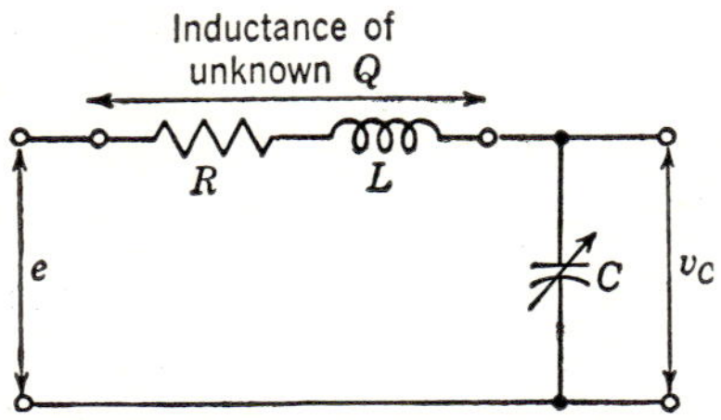

FIG. 2-3 Series resonant circuit used in Q meter.

$$i = \frac{e}{\sqrt{R^2 + (\omega L - 1/\omega C)^2}}$$

and the voltage across the variable element C is

$$v_C = \frac{e}{\omega C \sqrt{R^2 + (\omega L - 1/\omega C)^2}}$$

Under normal conditions of operation, the variable capacitor of the Q meter is adjusted with an input voltage of constant frequency to obtain a maximum value of v_C, which is read by a vacuum-tube-voltmeter circuit. To determine the value of C for which v_C is a maximum, the derivative of v_C with respect to C is equated to zero, or

$$\frac{dv_C}{dC} = e\frac{d}{dC}\left[\frac{1}{\omega C \sqrt{R^2 + (\omega L - 1/\omega C)^2}}\right] = 0$$

$$-\frac{1}{\omega C^2}\left[R^2 + \left(\omega L - \frac{1}{\omega C}\right)^2\right]^{-1/2} - \frac{1}{2\omega C}\left[R^2 + \left(\omega L - \frac{1}{\omega C}\right)^2\right]^{-3/2} \times 2\left(\omega L - \frac{1}{\omega C}\right)\left(\frac{1}{\omega C^2}\right) = 0$$

$$R^2 + \left(\omega L - \frac{1}{\omega C}\right)^2 + \frac{1}{\omega C}\left(\omega L - \frac{1}{\omega C}\right) = 0$$

from which

$$C = \frac{1}{\omega^2 L + R^2/L}$$

If $Q = \omega L/R$, the capacitance at which v_C is a maximum is

$$C = \frac{1}{\omega^2 L(1 + 1/Q^2)}$$

Substituting this value of C in the original expression for v_C yields

$$v_C = \frac{e\omega L(1 + 1/Q^2)}{\sqrt{R^2 + \omega^2 L^2/Q^4}} = eQ\sqrt{(1 + 1/Q^2)}$$

If $Q \geq 10$,

$$v_C \cong eQ$$

so that the maximum value of v_C is directly proportional to the Q of the inductance provided that Q is sufficiently high.

EXAMPLE 2. Suppose it is necessary to determine the optimum characteristic impedance of an air- or nitrogen-filled coaxial transmission line which will best withstand voltage breakdown for a given applied power,

The power flowing in such a line may be expressed as

$$P = \frac{V^2}{2Z_0}$$

where V is the peak voltage between the inner and outer conductor, and Z_0 is the characteristic impedance of the line.

$$Z_0 = 60 \ln \frac{b}{a}$$

where b is the inside radius of the outer conductor and a is the outside radius of the inner conductor.

The peak voltage V may be expressed in terms of the electric field strength E which exists between the inner and outer conductor as

$$V = \int_a^b \frac{Ea}{r}\, dr = Ea \ln \frac{b}{a}$$

so that, by substituting the new values for Z_0 and V,

$$P = \frac{E^2a^2 \ln (b/a)}{120}$$

Since it is the electric field strength or dielectric stress E which determines the voltage breakdown point of the air dielectric, this expression may be solved for E^2, and

$$E^2 = \frac{120P}{a^2 \ln (b/a)}$$

Voltage breakdown for a fixed power P will be least likely when E is at a minimum. Taking the derivative of E with respect to a, obtain

$$\frac{2E}{120P}\frac{dE}{da} = \frac{1 - 2 \ln (b/a)}{a^3[\ln (b/a)]^2}$$

For the minimum value of E,

$$\ln (b/a) = \tfrac{1}{2} \qquad \text{or} \qquad b/a = 1.65$$

and the most desirable characteristic impedance from the standpoint of voltage breakdown is $Z_0 = 60 \cdot \frac{1}{2} = 30$ ohms.

2-3 Applications of Differentials

Example 1. Consider a simple resonant circuit, the Q of which is sufficiently high so that the resonant frequency f_r is given by

$$f_r = \frac{1}{2\pi \sqrt{LC}}$$

If it is desired to find the change in resonant frequency which occurs with variations in either of the tuning elements, then either $\partial f_r/\partial C$ or

$\partial f_r/\partial L$ will be required, since the resonant frequency is a function of both L and C. In the case of a shift in the capacitor value, then

$$\frac{\partial f_r}{\partial C} = \frac{-1}{4\pi\sqrt{LC^3}} = \frac{-f_r}{2C}$$

Thus, for a tank circuit with a resonant frequency of 1 mc which is tuned by a 100-$\mu\mu$f capacitor, the change in frequency with capacitance is $-5{,}000$ cycles/$\mu\mu$f. In other words, if the equipment which contains this circuit is to shift frequency from 1,000 to 995 kc, a change of 1$\mu\mu$f in the tuning capacitor will be necessary.

EXAMPLE 2.* A quantitative relationship between the operating frequency of an f-m resistance-capacitance oscillator and the change in grid potential of the resistance tube can be derived in the following manner.

The frequency of oscillation f is represented approximately by the relation

$$f = a\sqrt{G_m}$$

where

$$a = \frac{1}{2\pi\sqrt{C_1C_2R_1}}$$

If the plate current of the resistance tubes I_p is

$$I_p = k\left(E_g + \frac{E_{sg}}{\mu_{sg}}\right)^{3/2}$$

and G_m is defined as

$$G_m = \frac{\partial I_p}{\partial E_g} = \frac{\partial}{\partial E_g}\left[k\left(E_g + \frac{E_{sg}}{\mu_{sg}}\right)\right]^{3/2}$$

then

$$G_m = \frac{3}{2}k\left(E_g + \frac{E_{sg}}{\mu_{sg}}\right)^{1/2}$$

if μ_{sg} remains constant when E_g is permitted to vary.

By substitution,

$$f = a\left(\frac{3k}{2}\right)^{1/2}\left(E_g + \frac{E_{sg}}{\mu_{sg}}\right)^{1/4}$$

The frequency deviation with respect to the grid potential is then

$$\frac{\partial f}{\partial E_g} = \frac{a}{4}\left(\frac{3k}{2}\right)^{1/2}\left(E_g + \frac{E_{sg}}{\mu_{sg}}\right)^{-3/4}$$

If two resistance tubes are used in the circuit, then

$$f = bG_m$$

where

$$b = \frac{1}{2\pi\sqrt{C_1C_2}}$$

* C. M. K. Chang, A Frequency-modulated Resistance-capacitance Oscillator, *Proc. I.R.E.*, **31** (1): 22–25, January, 1943.

In this case

$$f = \frac{3}{2} bk \left(E_g + \frac{E_{sg}}{\mu_{sg}} \right)^{1/2}$$

The rate of change of frequency with respect to grid voltage is then

$$\frac{\partial f}{\partial E_g} = \frac{3}{4} bk \left(E_g + \frac{E_{sg}}{\mu_{sg}} \right)^{-1/2}$$

EXAMPLE 3. Suppose that in the tuned circuit analyzed in Example 1 a fixed capacitor is to be used for fixed tuning of the circuit. If the capacitor is manufactured with a tolerance of 2 per cent, what is the possible error in f_r?

The relationship established previously was

$$\frac{\partial f_r}{\partial C} = -\frac{f_r}{2C}$$

Hence, the approximate error in f_r is

$$\Delta f_r = -\frac{f_r}{2C} \Delta C$$

and the relative error is approximately

$$\frac{\Delta f_r}{f_r} = -\frac{\Delta C}{2C}$$

In the case cited, $\Delta C/C$ is 2 per cent, and, therefore, the relative error in f_r will be 1 per cent.

EXAMPLE 4.* The frequency of oscillation of the phase-shift oscillator shown in Fig. 2-4 is determined not only by the phase-shifting network

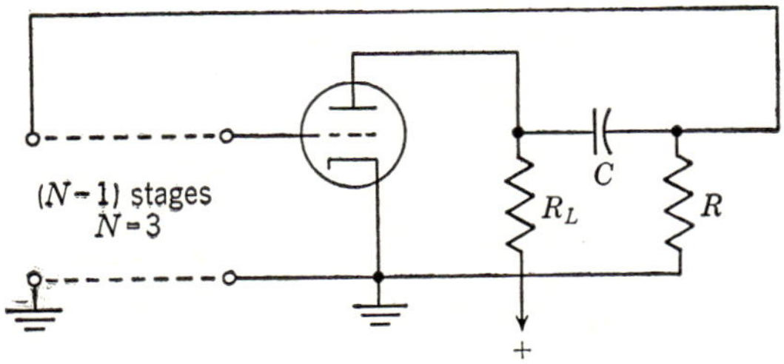

FIG. 2-4 Three-phase RC oscillator circuit (coupling-network phase shift is 60°).

but also by the value of R_1, which consists of the parallel combination of the plate load resistance and the plate resistance of the tube. R_1 therefore will vary with the operating conditions of the oscillator.

The greatest variation of R_1 is caused by the change in plate resistance

* R. M. Barrett, "N"-Phase Resistance-Capacitance Oscillators, *Proc. I.R.E.*, **33** (8): 541–545, August, 1945.

R_p of the tube, due to plate voltage variations. Therefore, in order to compute the frequency stability with reference to plate resistance variations, let df be a small change in the frequency f due to the change in the resistance dR_1 resulting from a plate voltage variation.

Frequency stability is defined as

$$\frac{\text{Incremental change in oscillator frequency}}{\text{frequency of the oscillator}} = \frac{df}{f}$$

The operating frequency of a three-phase oscillator is

$$f = \frac{1}{2\pi C\sqrt{3}\,(R + R_1)}$$

where $R_1 = R_L R_P/(R_L + R_P)$, R_L is the load resistance, R_p is the plate resistance of the tube, and R is the grid resistor of the next stage as shown in Fig. 2-4.

Differentiating the expression for the operating frequency with respect to R_1, obtain

$$df = \frac{-dR_1}{2\pi C\sqrt{3}\,(R_1 + R)^2}$$

and, substituting for R_1, in the numerator only,

$$df = \frac{-1}{2\pi C\sqrt{3}}\,\frac{R_L^2\,dR_p}{(R_1 + R)^2(1 + R_L/R_p)^2 R_p^2}$$

The frequency stability is

$$\frac{df}{f} = \frac{-R_L^2\,dR_p/R_p}{(R_L + R_p)[R_L + R(1 + R_L/R_p)]}$$

For a typical design, $R_L = 10{,}000$ ohms, $R_p = 1$ megohm (pentode), $R = 1$ megohm, and $f = 1{,}000$ cps. Substituting these values, the frequency stability is

$$\frac{df}{f} = -9.7 \times 10^{-5}\,\frac{dR_p}{R_p}$$

so that a 10 per cent change in the plate resistances of the amplifier tubes in the three-phase oscillator would result in a frequency change of 9.7×10^{-4} per cent, or 0.0097 cps.

2-4 Application of Radius of Curvature

EXAMPLE. In a disk recording system, the stylus of the cutting head traces a sine wave on the recording blank which can be represented as

$$y = D \sin \frac{2\pi x}{\lambda}$$

where λ is the wavelength. In this case,

$$\frac{dy}{dx} = \frac{2\pi D}{\lambda} \cos \frac{2\pi x}{\lambda} \qquad \text{and} \qquad \frac{d^2y}{dx^2} = \frac{-4\pi^2 D}{\lambda^2} \sin \frac{2\pi x}{\lambda}$$

and, according to the standard expression for radius of curvature,

$$\begin{aligned}
\rho &= \frac{[1 + (4\pi^2D^2/\lambda^2) \cos^2 (2\pi x/\lambda)]^{3/2}}{(-4\pi^2D/\lambda^2) \sin (2\pi x/\lambda)} \\
&= \frac{(4\pi^2D^2/\lambda^2)^{3/2}[\lambda^2/4\pi^2D^2 + \cos^2 (2\pi x/\lambda)]^{3/2}}{(-4\pi^2D/\lambda^2) \sin (2\pi x/\lambda)} \\
&= \frac{(8\pi^3D^3/\lambda^3)[\lambda^2/4\pi^2D^2 + \cos^2 (2\pi x/\lambda)]^{3/2}}{(-4\pi^2D/\lambda^2) \sin (2\pi x/\lambda)} \\
&= \frac{(-2\pi D^2/\lambda)[\lambda^2/4\pi^2D^2 + \cos^2 (2\pi x/\lambda)]^{3/2}}{\sin (2\pi x/\lambda)}
\end{aligned}$$

2-5 Applications of Integration

EXAMPLE 1. In the derivation of the luminous emittance of a source L in lumens per square meter, it is found that L, according to Lambert's law for a point source, is

$$L = \int_0^{2\pi} \int_0^{\pi/2} \frac{I_0(\cos \alpha) R^2 \sin \alpha \, d\alpha \, d\varphi}{A_0R^2}$$

where R is the distance from the source; I_0 is the luminous intensity of source along a normal to the surface of the source A_0; and α is the angle between the direction under consideration and the normal to the source, where $R^2 \sin \alpha \, d\alpha \, d\varphi$ is the element of area on the hemisphere which surrounds the point source. If $B = I_0/A_0$, which is luminance in candles per square meter, show that $L = \pi B$.

$$\begin{aligned}
L &= \int_0^{2\pi} \int_0^{\pi/2} \frac{I_0(\cos \alpha) R^2 \sin \alpha \, d\alpha \, d\varphi}{A_0R^2} \\
&= \frac{I_0}{2A_0} \int_0^{2\pi} \int_0^{\pi/2} 2 \cos \alpha \sin \alpha \, d\alpha \, d\varphi \\
&= \frac{I_0}{2A_0} \int_0^{2\pi} \int_0^{\pi/2} \sin 2\alpha \, d\alpha \, d\varphi \\
&= \frac{I_0}{2A_0} \int_0^{2\pi} \left[-\frac{\cos 2\alpha}{2} \right]_0^{\pi/2} d\varphi = -\frac{I_0}{4A_0} \int_0^{2\pi} (-1 - 1) \, d\varphi \\
&= \frac{I_0}{2A_0} \int_0^{2\pi} d\varphi = \frac{I_0}{2A_0} [2\pi - 0] = \frac{\pi I_0}{A_0}
\end{aligned}$$

But $B = I_0/A_0$; hence it follows that

$$L = \pi B$$

Example 2. The square of the noise voltage produced by thermal agitation of the molecules in an input grid impedance considered as a two-terminal network is

$$E^2 = 4KT \int_0^\infty R(f)\, df$$

where K is Boltzmann's constant and is 1.38×10^{-23} joule/deg abs, T is the absolute temperature, and $R(f)$ is the equivalent series resistance of the network at some frequency f.

If it is desired to calculate the voltage across the terminals of a two-terminal network consisting of a resistance and capacitance in parallel, $R(f)$ must first be determined. $R(f)$ in this case is the equivalent series resistance of R in parallel with a reactance $1/j\omega C$, so that

$$R(f) = \frac{R}{1 + \omega^2 C^2 R^2}$$

The square of the noise voltage is

$$E^2 = 4KT \int_0^\infty \frac{R\, df}{1 + 4\pi^2 C^2 R^2 f^2} = \frac{4KTR}{2\pi CR} \int_0^\infty \frac{2\pi CR\, df}{1 + 4\pi^2 C^2 R^2 f^2}$$
$$= \frac{2KT}{\pi C} \Big[\tan^{-1} 2\pi CRf \Big]_0^\infty = \frac{KT}{C}$$

Thus the noise voltage in this case is independent of the shunt resistance.

2-6 Mean or Average Values

In the measurement of periodic wave forms, many meters indicate average rather than peak values. The average value of a wave form can

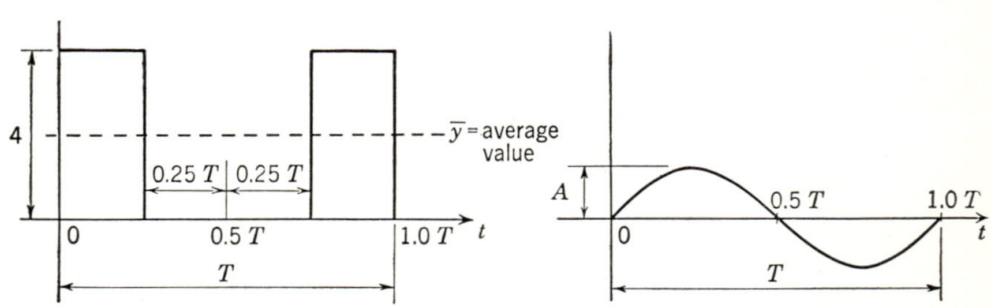

Fig. 2-5 Recurrent wave forms. (a) Rectangular wave. (b) Sine wave.

be related to its peak value in the following manner. Consider Fig. 2-5a, in which is shown one cycle of a recurrent square wave.

The peak value of the wave is 4. The total area of each square wave is $4 \cdot 0.25T$, or $1.00T$. The total area in one cycle is $2.00T$. The average ordinate value of the wave form $\bar{y}$ is equal to 2.00. In general, $\bar{y}$, the average value of any wave form which occurs in an interval from 0 to T, is proportional to the quotient of the area bounded by the wave form and the t axis divided by the interval T. From this, it can be seen that the average value of the sine wave $y = A \sin \omega t$, where $\omega = 2\pi f$ or $2\pi/t$, shown in Fig. 2-5*b*, is zero, because the area under the wave for $0 \leq t \leq T/2$ is equal in magnitude but opposite in algebraic sign to the area under the wave for $T/2 \leq t \leq T$.

Mathematically this can be verified by finding the area under the curve.

$$\text{Area} = \int_0^T A \sin\left(\frac{2\pi}{T}\right) dt = A\,\frac{T}{2\pi}\int_0^T \frac{\sin 2\pi t}{T}\left(\frac{2\pi}{T}\right) dt$$

$$= \frac{-AT}{2\pi}\left[\cos\frac{2\pi t}{T}\right]_0^T = \frac{-AT}{2\pi}(\cos 2\pi - \cos 0) = 0$$

EXAMPLE. Consider the wave form produced by a half-wave rectifier which consists of the positive half of a sine wave, as shown in Fig. 2-6, which is $y = A \sin (2\pi t/T)$ for $0 < t < T/2$ and $y = 0$ for $T/2 < t < T$.

The area under the wave is given by

$$\text{Area} = \int_0^{T/2} A \sin\left(\frac{2\pi t}{T}\right) dt$$

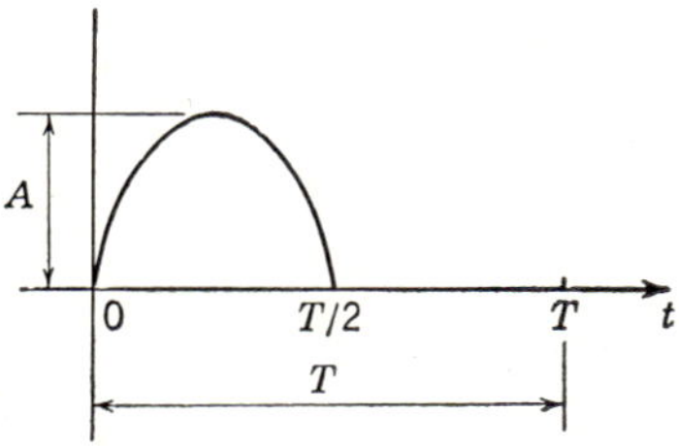

FIG. 2-6 Half-sinusoid wave form.

since there is no area contributed in the interval from $t = T/2$ to $t = T$.

$$\text{Area} = \frac{-AT}{2}\left[\cos\frac{2\pi t}{T}\right]_0^{T/2} = \frac{-AT}{2\pi}[\cos \pi - \cos 0]$$

$$= \frac{-AT}{2\pi}[-1 - 1] = \frac{AT}{\pi}$$

The average value is then

$$\bar{y}T = \frac{AT}{\pi}$$

$$\bar{y} = \frac{1}{T}\frac{AT}{\pi} = \frac{A}{\pi} = 0.318A$$

The *effective* or *root-mean-square* (rms) value of voltage (or current) is equal to the direct voltage (or current) that produces the same average power dissipation in a fixed resistance. The average power P in a resistance is directly proportional to the square of the voltage, or

$$P_{av} = \frac{(E^2)_{av}}{R}$$

where $(E^2)_{av}$ is the average of the square of the applied voltage. If this voltage is to be replaced by an equivalent voltage E_{rms} which is to develop the same power, then

$$P_{av} = \frac{(E_{rms})^2}{R} = \frac{(E^2)_{av}}{R} \qquad \text{and} \qquad (E_{rms})^2 = (E^2)_{av}$$

Hence
$$E_{rms} = +\sqrt{(E^2)_{av}}$$

In other words, the rms value of a voltage (or current) is the square root of the average value of the squared instantaneous voltage (or current).

The rms value of a sine wave is obtained as follows: Let the sine wave be represented by

$$e = E \sin \omega t$$

where E is the peak value of the wave and $\omega = 2\pi f$, or $\omega = 2\pi/T$ where T is the period. The value of e^2 is therefore

$$e^2 = E^2 \sin^2 \omega t$$

The average value of e^2 over the period T is therefore

$$(e^2)_{av} = \frac{E^2 \int_0^T \sin^2 (2\pi t/T)\, dt}{T} = \frac{E^2}{T} \int_0^T \frac{[1 - \cos (4\pi t/T)]\, dt}{2}$$
$$= \frac{E^2}{2T}\left[\int_0^T dt - \int_0^T \cos \frac{4\pi t}{T}\, dt\right] = \frac{E^2}{2T}\left[t - \frac{T}{4} \sin \frac{4\pi t}{T}\right]_0^T = \frac{E^2}{2}$$

Therefore
$$E_{rms} = +\sqrt{(e^2)_{av}} = \frac{E}{\sqrt{2}} = 0.707E$$

Problems

Prob. 2-1* An expression for the transit time t_0 of an electron in a velocity-modulated tube is given as

$$t_0 \approx \frac{L}{5.97 \times 10^7 \sqrt{V_d}} \frac{1}{1 + (V_m/2V_d) \sin \omega t_1}$$

where t_1 is the entrance time of a second electron into the electron stream, L is the length of the electron drift space, and $V_d + V_m \sin \omega t_1$ is the entering kinetic energy (in volts). Find dt_0/dt_1.

Prob. 2-2† In determining the "squint" of a linear antenna array, the function

$$\sin \theta = L[1 - L_c^2]^{1/2} - [1 - L_c^2 L^2]^{1/2}$$

* W. C. Hahn and G. F. Metcalf, Velocity-modulated Tubes, *Proc. I.R.E.*, **27** (2): 115, February, 1939.

† A. Roberts, "Radar Beacons," Radiation Laboratory Series, vol. 3, p. 370, McGraw-Hill Book Company, Inc., New York, 1947.

occurs. Find $d\theta/dL$, where L_c is constant, for the case where $L = 1$, $L_c = 0.7$, and $\theta \to 0$.

Prob. 2-3 The resistance R of materials used in the thermistor is generally expressed over limited temperature range by the relationship

$$R = R_0 \epsilon^{\beta(1/T - 1/T_0)}$$

where β is a characteristic of the material and R_0 is the value of the resistance R when the temperature $T = T_0$. If the temperature coefficient of resistance α is defined as the quotient or the rate of change of resistance with respect to temperature, find an expression which mathematically relates α, β, and T.

Prob. 2-4 Show that if the resistance R as a function of θ, the angle of potentiometer rotation, is

$$R = \frac{a + b \cos \theta}{2 \cos \theta} - \sqrt{\left(\frac{a - b \cos \theta}{2 \cos \theta}\right)^2 + ab}$$

then the potentiometer card shape must conform to

$$\frac{dR}{d\theta} = \frac{a \sin \theta}{2 \cos^2 \theta} \left\{1 - \frac{(a - b \cos \theta)/2 \cos \theta}{\sqrt{[(a - b \cos \theta)/2 \cos \theta]^2 + ab}}\right\}$$

Prob. 2-5* In the study of antennas used in loran systems it is found that for electrical antenna lengths not exceeding 90°, the reactive component of the antenna input impedance is given by the approximate formula

$$X_A = -Z_0 \cot \theta$$

where θ is the electrical length of antenna in degrees, X_A is antenna input reactance in ohms, and Z_0 is the average characteristic impedance in ohms. Show that the rate of change of the input reactance of the antenna with frequency is

$$\frac{dX_A}{df} = \frac{Z_0}{\sin^2 \theta} \frac{d\theta \text{ (in radians)}}{df}$$

Prob. 2-6† The plate current in a pentode amplifier tube is given by

$$I_b = \frac{2.33 \times 10^{-6} MA(E_{C_1} + E_{C_2}/\mu_{12})^{3/2}}{a^2 \left(1 + \dfrac{1}{\mu_{12}} \dfrac{a + b}{a}\right)^{3/2}}$$

* J. A. Pierce, A. A. McKenzie, and R. H. Woodward, "Loran," Radiation Laboratory Series, vol. 4, pp. 316–318, McGraw-Hill Book Company, Inc., New York, 1948.

† G. T. Ford, Characteristics of Vacuum Tubes for Radar Intermediate Frequency Amplifiers, *Bell System Tech. J.*, **25** (3): 391, July, 1946.

where M is the ratio of the plate current to the cathode current, A is the active area of the cathode, a is the distance from the cathode to the plane of the grid, b is the distance from the plane of the grid to the plate, E_b is the plate voltage, E_{C_1} is the effective control grid voltage, μ_{12} is the mu of the control grid with respect to the screen grid, and E_{C_2} is the screen voltage. Find the transconductance of the tube.

Prob. 2-7* In a receiver equipped with an avc system which contains a d-c amplifier, the r-f output voltage e_2 of the i-f amplifier is given as

$$e_2 = \mu(E + \beta e_2)$$

where β is the transmission factor of the avc detector which feeds the d-c amplifier, E is a voltage obtained from the audio-frequency amplifier, and μ is the transmission factor of the d-c amplifier and the i-f amplifier combined. Show that the degree of suppression of disturbances in the μ circuit, or de_2/e_2, is

$$\frac{de_2}{e_2} = \frac{1}{1 - \mu\beta}\frac{d\mu}{\mu}$$

Prob. 2-8† In a circular-sweep circuit developed for a cathode-ray tube, the voltage E is expressed by

$$E = \frac{V}{1 - \omega^2 LC + j\omega RC}$$

Find dE/E in terms of dC/C, where L is the circuit inductance, C the circuit capacitance, and R the circuit resistance. Assume L and R constant.

Prob. 2-9‡ (*a*) The attenuation constant for a circular wave guide which is used as an attenuator is given by the relation

$$\alpha = \frac{2\pi}{\lambda_c}\sqrt{1 - \left(\frac{\lambda_c}{\lambda}\right)^2}$$

where α is expressed in nepers per unit length, λ_c is the cutoff wavelength, and λ is the operating wavelength. Express $d\alpha/\alpha$ as a function of dr/r, if λ_c varies directly with the radius of the guide r. Explain why the tubing for such an attenuator must be manufactured carefully if $\lambda_c \ll \lambda$.

(*b*) Find $d\alpha/\alpha$ in terms of $d\lambda/\lambda$.

Prob. 2-10 The initial horizontal velocity of an electronically guided missile is 200 miles/min; the vertical rate of rise is 400 ft/min. The

* B. M. Oliver, Automatic Volume Control as a Feedback Problem, *Proc. I.R.E.*, **36** (4): 466–473, April, 1948.

† B. Chance, V. Hughes, E. MacNichol, D. Sayre, and F. Williams, "Waveforms," Radiation Laboratories Series, vol. 19, McGraw-Hill Book Company, Inc., New York, 1949.

‡ C. G. Montgomery, *op. cit.*, pp. 687, 834.

horizontal velocity increases 10 miles/min for each rise of 1,000 ft, and the vertical rate of rise increases 100 ft/min for each 100 miles traversed in the horizontal direction. Locate the rocket after 10 sec of flight.

Prob. 2-11* The attenuation α in a rectangular wave guide is

$$\alpha = \sqrt{\frac{E\omega}{2\sigma}}\frac{1}{b}\left[\frac{1 + (2b/a)(\lambda/2a)^2}{\sqrt{1 - (\lambda/2a)^2}}\right]$$

Find the minimum value of α as λ is varied.

Prob. 2-12† The playing time in minutes of a long-playing (LP) record is

$$T = \frac{\pi n(Dd - d^2)}{120V_{\text{min}}}$$

where n is the number of grooves per inch, V_{min} is the minimum permissible linear groove velocity in inches per second, D is the outside groove diameter in inches, and d is the inside groove diameter in inches.

(*a*) Show that the maximum playing time for a given inner-groove linear velocity is obtained when the inner diameter of the record equals one-half the outer diameter.

(*b*) If $T = 20$ min, $V_{\text{min}} = 10$ in./sec, and $D = 11.5$ in., show that the number of grooves per inch under these circumstances will be 230.

Prob. 2-13‡ In a particular circuit arrangement, the potentiometer error P is found to be

$$P = \frac{EARX(K - X + X^2)}{(A + RK)^2}$$

Show that the value of maximum error is

$$\frac{EAR}{27(A + RK)^2}\,[9K - 2 \mp 2(1 - 3K)^{3/2}]$$

Prob. 2-14§ A function which was developed in connection with an analysis of the operation of the traveling-wave tube is

$$Y_1 = \frac{-d^4}{4}\left[\frac{1}{4}\left(\beta^2 + \frac{1}{\beta^2}\right)^2 - 1\right]$$

* C. G. Montgomery, R. Dicke, and E. Purcell, "Principles of Microwave Circuits," Radiation Laboratory Series, vol. 8, p. 47, McGraw-Hill Book Company, Inc., New York, 1948.

† P. C. Goldmark, R. Snepvangers, and W. S. Bachman, The Columbia Long-playing Microgroove Recording System, *Proc. I.R.E.*, **37** (8): 923–924, August, 1949.

‡ L. A. Nettleton and F. E. Dole, Reducing Potentiometer Loading Error, *Rev. Sci. Instr.*, **18** (5): 340, May, 1947.

§ L. S. Nergaard, Analysis of a Simple Model of a Two-beam Growing-wave Tube, *RCA Rev.*, **9** (4): 600, December, 1948.

Find the minimum value of the function Y_1. Assume α a constant and β the independent variable.

Prob. 2-15* The Q of an electronic inductance is found to be

$$Q = \frac{X_c RG}{R^2 + X_c^2(1 + G)}$$

where G is the voltage gain of reactance tube, R is the resistance of phase-shifting network, and X_c is the capacitive reactance of phase-shifting network.

Show (a) by differentiation that the condition for a maximum Q is

$$\frac{R}{X_c} = \sqrt{1 + G}$$

(b) that the maximum Q is

$$Q_{\max} = \frac{G}{2\sqrt{1 + G}}$$

Prob. 2-16 In the analysis of standard-broadcast-antenna radiation characteristics, the function $f(\theta)$ is found to be

$$f(\theta) = \frac{\cos (G \sin \theta) - \cos G}{(1 - \cos G) \cos \theta}$$

Show that $f(\theta)$ has a minimum value with variations in θ when G, the antenna height in degrees, is held constant.

Prob. 2-17† Frequency stability in the cathode-coupled oscillator can be studied with the aid of the correction factor

$$F = 1 - \frac{L}{16C}\left(\alpha - \frac{1}{R_c}\right)^2$$

Show that the minimum value of F with respect to α is expressed by

$$\alpha = \frac{1}{R_c}$$

Prob. 2-18‡ The nominal loss in a shunt-type voltage regulator is given as

$$L = E_0 \left\{\frac{(\frac{1}{2}I_p)^2 + \frac{1}{2}kI_0I_p + kI_0^2}{\frac{1}{2}I_p - \frac{1}{2}kI_p - kI_0}\right\}$$

* J. E. Hayes, A New Type of Practical Distortion Meter, *Proc. I.R.E.*, **31** (3): 112–117, March, 1943.

† P. G. Sulzer, Frequency and Amplitude Stability of the Cathode-coupled Oscillator, *Proc. I.R.E.*, **38** (5): 541, May, 1950.

‡ W. G. Hoyle, Voltage Regulators of Shunt Type, *Rev. Sci. Instr.*, **19** (4): 245–246, April, 1948.

Show that a point of minimum nominal loss occurs for

$$I_p = \frac{2I_0}{k^{-1/2} - 1}$$

Prob. 2-19* If the power p received by an antenna is directly proportional to the square of the antenna power gain in the direction of a target, for the crossover angle, where $\theta = c/2$, then

$$p \propto V^2 \sin^4 1.392 \frac{c}{V}$$

Show that the received power is a maximum when $V = c/0.8373$.

Prob. 2-20† In the analysis of bridged reactance-resistance networks, the function P is obtained, where $P = 4Q + 2/Q + 6$. Show that the minimum value of P is 11.656.

Prob. 2-21‡ The expression for the characteristic impedance of a high-impedance coaxial cable is given as

$$Z = \frac{133Nd}{\sqrt{k}} \sqrt{\left[1 - \left(\frac{d}{a}\right)^2\right]\left[\ln\left(\frac{a}{d}\right)\right]}$$

Show that the impedance is a maximum when

$$2 \ln\left(\frac{a}{d}\right) = \left[\frac{(a/d)^2 - 1}{(a/d)^2 - 2}\right]$$

or approximately when $a = 2.06d$.

Prob. 2-22§ In a cascaded distributed amplifier, the total number of electron tubes is $N = mn$, where n is the number of tubes per stage and m is the number of cascaded stages. If the gain per stage $G^{1/m}$ is

$$G^{1/m} = n\left(\frac{f_0}{2f_c}\right)$$

where f_c is the high-frequency cutoff of the amplifier and f_0 is Wheeler's bandwidth-index frequency $G_m/\pi\sqrt{C_gC_p}$, show that (*a*) the smallest N is obtained when $m = \ln G$, (*b*) the corresponding value for n is

$$2\frac{f_c}{f_0}\epsilon$$

and (*c*) the gain per stage is equal to ϵ.

* D. Levine, Volume Scanning with Conical Beams, *Proc. I.R.E.*, **38** (3): 291, March, 1950.

† G. R. Harris, Bridged Reactance-resistance Networks, *Proc. I.R.E.*, **37** (8): 486, August, 1949.

‡ M. R. Winkler, Discussion of 'High Impedance Cable," *Proc. I.R.E.*, **35** (10): 1097, October, 1947.

§ E. L. Ginzton, W. R. Hewlett, J. H. Jasberg, and J. D. Noe, Distributed Amplification, *Proc. I.R.E.*, **36** (8): 966–967, August, 1948.

Prob. 2-23* In a certain oscillator circuit, the frequency shift measured in cycles per second due to small changes in the interelectrode capacitances is given by the expression

$$\frac{f(\omega M_g)^2 C_g}{2L} + \frac{f(\omega M_p)^2 C_p}{2L} \tag{1}$$

in which f is the oscillator frequency in cycles per second, ω is $2\pi f$, the inductance L and the mutual inductances M_g and M_p are measured in henrys, and the capacitances are in farads. Let

$$x = \frac{C_p}{C_g} \qquad \text{and} \qquad y = \frac{M_p}{M_g}$$

Show (*a*) that expression (1) may be written in the form

$$\frac{f}{2}\,\frac{\omega^2 M_g M_p}{L}\,\sqrt{C_g C_p}\left(\frac{1}{xy} + xy\right)$$

(*b*) that the minimum value for $1/xy + xy$ is 2 with x fixed, and (*c*) that this minimum value occurs when $y = 1/x$, this is, when $M_p/M_g = C_g/C_p$.

Prob. 2-24† In the frequency stabilization of a microwave oscillator, the rate of change of discriminator voltage with frequency is

$$\frac{dV}{df} = DP_0 \frac{Q_0}{V_0} \frac{2\alpha}{(1+\alpha)^2}$$

Show that this derivative is maximum for $\alpha = 1$.

Prob. 2-25‡ In the analysis of the operation of a certain r-f high-voltage supply, the expression for $1/G_{2e}$ is

$$\frac{1}{G_{2e}} = \frac{1}{k^2}\left(1 + \frac{k^2 f_{02}}{2\alpha} - \frac{k^2}{2}\right)\frac{L_1}{L_2}\left(R_2 + \frac{16\pi^2 L_2^2 \alpha^2}{R_2}\right)$$

Show that

$$\alpha = \left(\frac{k^2 R_2^2 f_{02}}{16\pi^2 L_2^2 (2 - k^2)}\right)^{1/3}$$

gives a maximum value.

Prob. 2-26§ (*a*) The absolute square of the coupled impedance in a coupled circuit is given as

* W. Van B. Roberts, The Limits of Inherent Frequency Stability, *RCA Rev.*, **4** (4): 478–484, April, 1940.

† R. V. Pound, Frequency Stabilization of Microwave Oscillator, *Proc. I.R.E.*, **35** (12): 1410, December, 1947.

‡ G. W. Mathers, Some Additions to the Theory of Radio Frequency High Voltage Supplies, *Proc. I.R.E.*, **37** (2): 203, February, 1949.

§ J. B. Sherman, Some Aspects of Coupled and Resonant Circuits, *Proc. I.R.E.*, **30** (11): 505–510, November, 1942.

$$|Z_c|^2 = \frac{M^4}{L_2{}^3C_2}\frac{\gamma^6}{\gamma^4 + \alpha\gamma^2 + 1}$$

where $\gamma = \omega/\omega_0$ and $\alpha = 1/(Q_0{}^2 - 2)$. Show that the maximum value of $|Z_c|^2$ is obtained when $\gamma^4 + 2\alpha\gamma^2 + 3 = 0$, and solve for γ.

(*b*) The coupled-resistance component of the circuit given in (*a*) is expressed as

$$R_c = \frac{M^2R_2}{L_2{}^2}\frac{\gamma^4}{\gamma^4 + \alpha\gamma^2 + 1}$$

Show that the coupled resistance is a maximum when $\gamma^2 + 2/\gamma = 0$ and determine the value of γ.

(*c*) The coupled inductance of the circuit is given as

$$L_c = \frac{M^2}{L_2}\frac{\gamma^2 - \gamma^4}{\gamma^4 + \alpha\gamma^2 + 1}$$

Show that L_c is a maximum when

$$(\alpha + 1)\gamma^4 + 2\gamma^2 - 1 = 0$$

(*d*) The absolute square of the parallel impedance is given as

$$|Z_p|^2 = R^2\frac{1 + Q_0{}^2\gamma^2}{\gamma^4 + \alpha\gamma^2 + 1}$$

Show that the maximum occurs when

$$\gamma = \sqrt{\left(1 + \frac{2}{Q_0{}^2}\right)^{1/2} - \frac{1}{Q_0{}^2}}$$

Prob. 2-27 Show that in a parallel resonant circuit the reactive component of the impedance is

$$X = \frac{\omega[L(1 - \omega^2LC) - CR^2]}{(1 - \omega^2LC)^2 + (\omega CR)^2}$$

that X has a maximum at frequencies which are $1/2Q$ below and above resonance, and that the reactance maximum has an amplitude exactly 50 per cent of the impedance at resonance.

Prob. 2-28* In the study of the energy storage in cavity resonators used in traveling-wave tubes, the following expression for the energy-storage ratio W_1/W_2 is obtained:

$$\left[\frac{W_1}{W_2}\right]^{1/3} = \left[\frac{2\sin^2(\theta/2)}{\theta_t(\theta/2)}\right]^{1/3}$$

* J. R. Pierce, Circuits for Traveling-wave Tubes, *Proc. I.R.E.*, **37** (5): 513, May, 1949.

Show that this has a maximum at $\theta = 2.33$ radians and is

$$\left(\frac{W_1}{W_2}\right)^{1/3} = \left(\frac{1.450}{\theta_t}\right)^{1/3}$$

Prob. 2-29* The voltage E_R measured by a high-input-impedance receiving system is given as

$$E_R = K \frac{R_0^{2/3}}{R_0 + R_t}$$

where R_0 is the output impedance of the transformer, R_t is the input impedance of the transducer, and K denotes a constant. Determine the value of R_0 for which E_R is a maximum.

Prob. 2-30 Show that the radius of curvature obtained in text example has a minimum value $\rho_{\min}$ where

$$\rho_{\min} = \frac{\lambda^2}{4\pi^2 D}$$

Prob. 2-31† In the analysis of a vacuum tube with parallel elements, Poisson's equation in rectangular coordinates is

$$\frac{d^2E}{dx^2} = -4\pi\rho$$

Also

$$I = \rho v$$

and

$$v = \left(\frac{2eE}{m}\right)^{1/2}$$

where ρ is the space-charge density, E the potential at any point a distance x from the cathode, v the velocity of the electrons at x, I_1 the current per unit area, and e and m the charge and mass of the electron.

(*a*) Show that

$$\frac{d^2E}{dx^2} = -4\pi \frac{I}{(2eE/m)^{1/2}}$$

(*b*) Solve for $\frac{1}{2}(dE/dx)^2$ by integration.

Prob. 2-32‡ In a klystron, the electric field strength in the space d between the two buncher grids is

$$E = \frac{V_1 \sin \omega t}{d}$$

* E. L. Carstensen, Self-reciprocity of Electroacoustic Transducers, *J. Acoust. Soc. Amer.*, **19** (6): 963, November, 1947.

† B. J. Thompson, Space-current Flow in Vacuum Tube Structures, *Proc. I.R.E.*, **31** (9): 486, September, 1943.

‡ H. B. Phillips and L. A. Ware, Transit-time Effect in Klystron Gaps, *Proc. I.R.E.*, **35** (10): 1077, October, 1947.

where V_1 is the voltage between the grids. The acceleration of an electron between the bunchers at any time t is

$$A = \frac{dv}{dt} = \frac{eV_1 \sin \omega t}{md}$$

(*a*) Find the velocity between some time t and the initial time t_1 at which the electron crosses the first buncher grid.

(*b*) Find the distance traversed by the electron in the interval $t - t_1$.

Prob. 2-33* In the analysis of the operation of a rectifier tube with a resistance load the following integral is obtained:

$$W_p = \frac{1}{\pi} \frac{I_m^{5/3}}{k^{2/3}} \int_0^{\pi/2} \sin^{5/3} \theta \, d\theta$$

Show that an equivalent expression for W_p is given by

$$W_p = \frac{0.27 I_m^{5/3}}{k^{2/3}}$$

Prob. 2-34† In the theory of an electronic fluxmeter, the integral

$$\int_{t_1}^{t_2} e \, dt = RC \int_{t_1}^{t_2} \frac{i}{C} \, dt$$

occurs. Show that this integral is equal to $RC[(q_2 - q_1)/C] = RC \, \Delta v$.

Prob. 2-35‡ The root-mean-square (rms) voltage of a rectified frequency-modulated carrier envelope is given by

$$\text{rms voltage} = e_1 \sqrt{\frac{1}{\pi} \int_0^{\pi} (1 + x^2 + 2x \cos \theta) \, d\theta}$$

Show that this voltage is equal to $e_1 \sqrt{1 + x^2}$.

Prob. 2-36§ The d-c resistance of the metallic area of a short-circuiting disk placed at the end of a coaxial-cable assembly is given as

$$R = \frac{r}{2\pi} \int_{d/2}^{D/2} \frac{dx}{x}$$

where r is the resistivity of the disk in ohms per square centimeter. Evaluate R.

* T. H. Rogers, A New Rectifier Tube for Extremely High Power and Voltage Levels, *Proc. Natl. Electronics Conf.*, **5**: 486.

† R. H. Dick, An Electronic Fluxmeter, *Rev. Sci. Instr.*, **19** (8): 534, August, 1948.

‡ M. S. Corrington, Frequency Modulation Distortion Caused by Common- and Adjacent-channel Interference, *RCA Rev.*, **7** (4): 547, December, 1946.

§ R. A. Kempf, Coaxial Impedance Standards, *Bell System Tech. J.*, **30** (3): 698, July, 1951.

Prob. 2-37* The potential V_F at any point r between the anode and cathode cylinders of an electrostatically focused radial-beam tube is given by the equation

$$V_F = \int_r^{R_2} \frac{V_c\, dr}{r \log (R_2/R_1)}$$

Find V_F, if V_c is the potential difference between anode and cathode, if R_1 and R_2 are the radii of the cathode and anode, respectively.

Prob. 2-38† If the element of power per unit solid angle reflected by an element of area dX from the surface of the moon toward the earth is

$$dP_E = 2\rho P_0 \frac{(R - X)^2}{R}\, dX$$

show that the power returned from the moon from the front to the depth X is

$$P_E = 2\rho P_0 \left[RX - X^2 + \frac{X^3}{3R} \right]$$

and determine P_E when $X = R$, that is, the whole visible surface.

Prob. 2-39‡ In an analysis of a barretter used in a wave guide, it is found that

$$\frac{R - R_a}{R_a} = \frac{1}{a} \int_0^a \frac{T - T_a}{T_a}\, dR$$

If
$$T - T_a = -PJ \left(\frac{h^2 - a^2}{4\pi a^2 lk} \right)$$

show that
$$R = R_a \left(1 + \frac{0.013P}{lkT_a} \right)$$

where R is the d-c resistance of the wire and R_a is the high-frequency resistance of the wire; l and a are the length and the radius, respectively, of the wire; T_a is the absolute temperature that is required to dissipate the generated heat; J is 0.239 cal/joule; P is the total dissipation; and k is the thermal conductivity of the wire.

Prob. 2-40§ Maxfield and Albersheim give the power gain factor for

* A. M. Skellett, Electrostatically Focused Radial-beam Tube, *Proc. I.R.E.*, **36** (11): 1354–1357, November, 1948.

† D. D. Grieg, S. Metzger, and R. Waer, Considerations of Moon-relay Communication, *Proc. I.R.E.*, **36** (5): 662, May, 1948.

‡ C. G. Montgomery, *op. cit.*, p. 167.

§ J. P. Maxfield and W. J. Albersheim, An Acoustical Constant of Enclosed Spaces Correlatable With Their Apparent Liveness, *J. Acoust. Soc. Amer.*, **20**, July, 1948.

the velocity or pressure-gradient microphone as

$$G_p = \frac{4\pi \cos^2 \varphi}{\int_0^\pi (\cos^2 \varphi) 2\pi \sin \varphi \, d\varphi}$$

where φ is the angle between the polar axis of sensitivity and the line that joins the sound source and the speaker. Show that $G_p = 3 \cos^2 \varphi$.

Prob. 2-41* Show that in a series RC circuit which has a d-c generator voltage of E, the energy dissipated in the resistance during charging is $E^2C/2$ (by integrating the expression for the instantaneous power dissipated in the resistance).

Prob. 2-42† In a straight-wire inductor used in a radio interference filter,

$$L = \int_{R_1}^{R_2} \frac{2I}{10r} \, dr \, \mu_\Delta \times 10^{-8}$$

where R_1 is the inner radius, R_2 is the outer radius, I is the current, and μ_Δ is the incremental permeability. Find L.

Prob. 2-43 (*a*) Find the average value of

1. The output of a full-wave rectifier which consists of two positive sinusoids in a cycle
2. A "sawtooth" or triangular equilateral wave form which occurs once per cycle
3. A parabolic wave form which occurs once per cycle

(*b*) Show that the rms value of a triangular wave which has a peak amplitude of 2 is $\sqrt{4/3}$.

Prob. 2-44‡ The electron-beam current I_s delivered to the phosphor screen through the shadow mask of a three-gun color kinescope is given as

$$I_s = I_0 \left[1 - \epsilon^{-M^2/4b^2}\right] \frac{\pi}{18\sqrt{3}} \left(\sqrt{3} - \frac{M}{S}\right)^2$$

where I_0 is the total beam current before the beam reaches the mask, M is the beam cross-section diameter, S is the separation of the axis of the electron beam from the tube axis, and B is the beam radius at which the beam current density is ρ_0/ϵ. Show that the condition for maximum

* V. Wouk, Energy Wasted in Charging a Condenser, *Communications*, **24** (4): 48, April, 1944.

† L. J. Knudson, The Design of Reactors for Radio Interference Filters, *Proc. A.I.E.E.*, **69**, 1950.

‡ H. B. Law, A Three-gun Shadow-mask Color Kinescope, *Proc. I.R.E.*, **39** (10): 1186–1194, October, 1951.

efficiency of operation (maximum I_s/I_0) occurs when

$$\frac{S}{M} = \frac{1}{\sqrt{3}}\left[1 + \frac{4b^2}{M^2}(\epsilon^{M^2/4b^2} - 1)\right]$$

By expanding the exponential term, show also that $S/M \to 2/\sqrt{3}$ as $b \to \infty$.

Prob. 2-45* In the design of parallel-T networks, the function

$$\Delta = \frac{M - 1 - n}{M} \cdot \frac{(1 - n)x + M + 1 - n}{\sqrt{-nx^2 + (M - 1 - n)x - 1}}$$

is obtained.

(a) Determine the value of x for which Δ is minimum.

(b) What is the minimum value of Δ for x obtained in (a)?

(c) When Δ is a minimum, show that the minimum value of M for a prescribed $\Delta(>2)$ is

$$M = 1 + n + 2\sqrt{n} \cdot \frac{\Delta}{\sqrt{\Delta^2 - 4}}$$

Prob. 2-46† The available power gain A_w of a grounded-grid triode is

$$A_w = \frac{g_m{}^2 G_s \gamma_p}{(g_m + G_s + G_t + G_1)(G_s + G_t + G_1)}$$

Find the optimum value of G_s.

Prob. 2-47‡ It is desired to calculate the indication of an average reading meter on Gaussian noise. The rectified d-c voltage V_0 from noise will be

$$V_0 = 2\int_0^\infty Vp(V)\,dV$$

If the noise is Gaussian, the probability $p(V)\,dV$ that the instantaneous voltage lies between V and $V + dV$ is

$$p(V)\,dV = \frac{1}{\sigma\sqrt{2\pi}}\,\epsilon^{-\frac{V^2}{2\sigma^2}}\,dV$$

where σ is the rms noise voltage.

(a) Determine V_0 in terms of σ.

(b) If the meter is calibrated to read the rms value of a sine wave, the

* Y. Oono, Design of Parallel-T Resistance-Capacitance Networks, *Proc. I.R.E.*, **43** (5): 617–619, May, 1955.

† R. S. Brown, UHF Tuner Design for the 6BA4, *Trans. I.R.E.* Professional Group on Broadcasting and TV Receivers, **7**: 6, July, 1954.

‡ B. M. Oliver, Some Effects of Waveform on VTVM Readings, *Hewlett-Packard J.*, **6** (10): 4, June, 1955.

indicated voltage V_i is

$$V_i = 1.11V_0 = \frac{\pi}{2\sqrt{2}}V_0$$

Determine V_i in terms of σ and show that the average reading meter will read 1.05 db low on Gaussian noise provided that no overload occurs on the peaks.

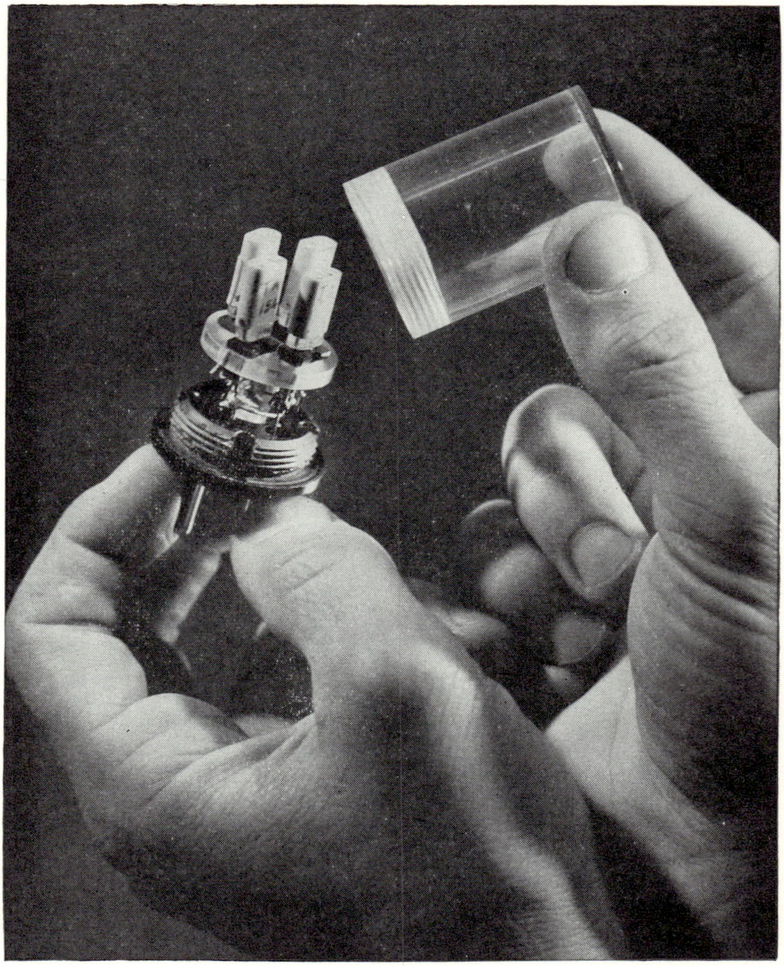

Plug-in all-transistor audio amplifier, with gain sufficient to operate a loudspeaker, developed at the David Sarnoff Research Center. (*Radio Corp. of America.*)

PART 2

Equation Testing Prior to Mathematical Operations

Chapter 3

DIMENSIONAL FORMULAS AND SYSTEMS

3-1 Introduction

In the realm of pure mathematics, the mathematician's symbols are purposely made general in order that the results obtained will be as abstract as possible. In fact, as Bertrand Russell has commented, "a mathematician is never so happy as when he doesn't know what he is talking about." On the other hand, every term in an equation or formula which an engineer writes must have a physical significance in terms of some measurable quantity. For an equation to be valid, the terms which appear on opposite sides of the equality sign must not only be equal in magnitude, but must also represent the same *kind* of quantity. To put it another way, the qualities or properties of all terms in an equation must be identical, and this similarity of kind is a necessary (but not sufficient) condition for a formula or equation to be correct.

Basic units change from age to age and from trade to trade. The cubit and league of olden days have yielded to the foot and the mile of today. The horse trader has his hand, the surveyor his chain, the mariner his fathom, and the physicist his angstrom unit. All these particular units of measure have one property or quality in common. Each represents the fundamental *dimension* of length. By dimension is meant the general property of the unit and not its magnitude.

3-2 Dimensional Analysis

The analysis of equations to determine the dimensionality of the terms contained is known as dimensional analysis. In the solution of engineering problems, a check on the dimensions of the various mathematical expressions used will oftentimes reveal hidden errors. On numerous occasions, a problem is so complex that a sufficiently comprehensive formula cannot be worked out. In certain of these cases, dimensional analysis can be of assistance in the indication of what is the probable relationship of the variables involved.

The dimensional approach to problems has proved profitable in many diverse fields of engineering. For example, the physicist through the use of dimensional analysis is able to predict to a certain extent the behavior of metal cut by a cutting tool,[1] and the aeronautical engineer predicts the behavior of a model plane in a wind tunnel. Other applications pertinent to the fields of television engineering and electronics will be described later.

3-3 Dimensional Formulas and Single-system Dimensions

A dimensional formula is obtained by replacing all the quantities in an equation by their basic dimensions. Consider the equations which have been previously encountered in the study of elementary mechanics. By custom, the fundamental dimensions in the mechanical system have been selected as mass, length, and time.

The dimensions of mass, length, and time are represented as M, L, and T, respectively. The dimensional formula for velocity becomes L/T, for area L^2, for volume L^3, etc. These dimensional formulas can be written symbolically as

$$\begin{aligned} [\text{Length}] &= [L] \\ [\text{Area}] &= [L^2] \\ [\text{Volume}] &= [L^3] \\ [\text{Velocity}] &= [LT^{-1}] \end{aligned}$$

where the brackets denote dimensionality.

Consider the formula for force in a mechanical system:

$$F = ma \tag{3-1}$$

which states that unit force is equal to unit mass times unit acceleration. Dimensionally this becomes

$$[F] = [M^1][M^0LT^{-2}]$$

so that upon combining the bracketed quantities the dimensional formula for force is

$$[F] = [MLT^{-2}] \tag{3-2}$$

As long as the system considered is a purely mechanical one, these three dimensions, M, L, and T, will be sufficient to define completely all quantities encountered. However, where a mechanical system is to be related to an electric, acoustic, hydraulic, or magnetic system, addi-

[1] D. C. Drucker and H. Ekstein, A Dimensional Analysis of Metal Cutting, *J. Appl. Phys.*, **21** (2), February, 1950.

tional provisions must be made to permit a complete dimensional analysis, as will be described later in Sec. 3-4.

In a like manner, all the quantities which appear in any one of the other systems mentioned above can also be defined dimensionally in terms of three basic dimensions. For instance, all equations which are developed in connection with the analysis of electric networks can be analyzed dimensionally using as basic dimensions voltage V, charge Q, and time T.

Current represents the rate of flow of charge, or charge per unit time. It may therefore be expressed dimensionally as

$$[I] = [QT^{-1}] \tag{3-3}$$

The formula for resistance according to Ohm's law is

$$R = \frac{V}{I} \tag{3-4}$$

Written dimensionally, resistance becomes

$$[R] = [VQ^{-1}T] \tag{3-5}$$

The charge stored in a capacitor is

$$Q = CV \tag{3-6}$$

where C is the capacitance and V is the voltage across the capacitor. From this equation, it follows that capacitance is

$$C = \frac{Q}{V} \tag{3-7}$$

which dimensionally is

$$[C] = [V^{-1}Q] \tag{3-8}$$

The instantaneous voltage V developed across an inductance L which conducts a current I is

$$V = L\frac{di}{dt} \tag{3-9}$$

so that, dimensionally, inductance is

$$L = \frac{[V][T]}{[QT^{-1}]} = [VQ^{-1}T^2] \tag{3-10}$$

The reciprocal of frequency is a period in seconds, so that the dimensions of frequency are

$$[f] = [T^{-1}] \tag{3-11}$$

It should be noted that although quantities like impedance and reactance differ somewhat electrically from resistance, the dimensions of all three will be the same since each is a quotient of a voltage and a current. Likewise, since the formula for admittance is

$$Y = \frac{I}{V} \tag{3-12}$$

the dimensions of admittance in the system chosen must be

$$[Y] = [QT^{-1}][V^{-1}] = [V^{-1}QT^{-1}] \tag{3-13}$$

as are the dimensions of conductance and susceptance.

The dimensions of power are of interest. The formula for electric power is

$$P = VI \tag{3-14}$$

The dimensions of electric power are therefore

$$[P] = [V][QT^{-1}] = [VQT^{-1}] \tag{3-15}$$

3-4 Dimensional Formulas and Universal Systems

As the sciences develop, new dimensional systems evolve. In the realm of mechanics, for example, no fewer than six different three-dimensional systems have appeared over the years. The reader may recall that three of these are absolute systems in which the fundamental dimensions are *mass, length,* and *time;* and three are gravitational systems in which the fundamental dimensions are *force, length,* and *time.* The basic units of the three absolute systems and their gravitational counterparts are the centimeter-gram-second (cgs), the meter-kilogram-second (mks), and the foot-pound-second (fps). However, in the absolute system, the gram, kilogram, and pound are measures of *mass,* while in the gravitational systems they are measures of *force* or *weight.*

In the same way, a multitude of dimensional systems have been designed to provide a common set of dimensions for the mechanical, electrical, and magnetic systems. Some of these are the cgs electromagnetic, the cgs electrostatic, the Gaussian, the Heaviside-Lorentz, the unrationalized and rationalized mks, the "international," and the "practical" systems.

All dimensional systems which are designed to define mechanical and electrical quantities must fulfill certain basic criteria. If the system is to be absolute, then by definition the first three basic dimensions in it must be mass, length, and time. In addition, a fourth basic dimension may be arbitrarily selected for maximum convenience providing that, whatever choice is made, the dimensions of μ_0 (permeability) and ϵ_0 (permittivity) in the system are such that the quantity $(\mu_0\epsilon_0)^{-1/2}$ has the dimensions of velocity.[1]

Choices for this fourth dimension have included current, resistance, charge, etc. In certain dimensional systems, no particular fourth dimension is specified. However, in all these, the limitation regarding μ_0 and ϵ_0 cited above must always be respected.

Of the many universal systems itemized earlier, only the cgs and the rationalized mks are of current interest. The cgs electromagnetic and electrostatic systems, or a mixture of the two called the Gaussian, are ordinarily employed by the engineer in the study of classical physics, while the rationalized mks system is used in the majority of recent publications on electromagnetic-wave propagation and has been accepted as the standard system of dimensions by the Institute of Radio Engineers.

The fourth dimension in each of these systems and the corresponding dimensional formulas[2] for μ_0 and ϵ_0 are shown in Table 3-1.

TABLE 3-1

Fourth dimension	Dimensional formulas for:	
	μ_0	ϵ_0
cgs electromagnetic:		
Permeability μ_0	$[\mu_0]$	$[L^{-2}T^2\mu_0^{-1}]$
cgs electrostatic:		
Permittivity ϵ_0	$[L^{-2}T^2\epsilon_0^{-1}]$	$[\epsilon_0]$
Rationalized mks:		
Charge Q	$[MLQ^{-2}]$	$[M^{-1}L^{-3}T^2Q^2]$

The reader will recall that the cgs systems are rather awkward to use because quantities like voltage and current are expressed in abvolts and statvolts, abamperes and statamperes, respectively, rather than in practical volts and amperes. Consequently the rationalized mks, or Giorgi, system has become the more popular, because in this system all the units important to the engineer have practical magnitudes. As J. A. Stratton has commented, "There is still the feeling among many physicists that this system is being forced upon them by a subversive group

[1] $(\mu_0\epsilon_0)^{-1/2} = c$, where c is the velocity of light. This relationship must hold true in any dimensional system, from physical considerations.

[2] For tables of units and conversion factors, see "Electric Circuits," M.I.T. Series, appendix C, John Wiley & Sons, Inc., New York, 1943; "Reference Data for Radio Engineers," 3d ed., pp. 26–27, Federal Telephone & Radio Co., New York, 1949; "American Standard Acoustical Terminology," p. 41, American Standards Association, New York, 1951.

of engineers. Perhaps it is, although it was Maxwell himself who first had the idea."[1]

Nevertheless, no better recommendation for the standardization of this system can be presented than that offered by S. A. Schelkunoff:[2]

> Although the rationalized m.k.s. system of units was first suggested in the middle of the last century, it remained almost unknown until about 15 years ago. In the last 15 years, however, it has made astonishingly rapid conquests. The reasons for this are many. In the m.k.s. system the electrical units are those already in common use in laboratory measurements: the volt, the ampere, the ohm, etc. If the system is of the rationalized variety, Maxwell's equations assume a form which is merely a generalization of the one-dimensional transmission-line equations. In recent years, the gap between circuit and transmission-line theories on the one hand and field theory on the other hand has been closed by waveguides and microwave circuits in general. This made it essential that there be no clash between the ideas, terminology, and units employed in these theories. Since the rationalized m.k.s. system fulfills this requirement, it is only natural that it should enjoy rapidly increasing popularity. An added factor in this popularity is that the rationalized m.k.s. system is equally well adapted to electromechanical theories. It has begun to appeal to many physicists as well as to engineers—which is particularly fortunate, since the engineer of today must be somewhat of a physicist, and the physicist somewhat of an engineer.

It should be noted that the cgs systems described earlier are "unrationalized," whereas the mks system is of the "rationalized" variety. By rationalization is meant the absorption of the numerical factors 4π and c (the velocity of light) into the constants μ_0 (permeability) and ϵ_0 (permittivity).[3]

Table 3-2

	cgs electromagnetic unrationalized	cgs electrostatic unrationalized	mks rationalized
μ_0	Unity	$\frac{1}{c^2}$ sec²/cm²	$4\pi \times 10^{-7}$ henry/meter
ϵ_0	$\frac{1}{c^2}$ sec²/cm²	Unity	$\frac{1}{36\pi} \times 10^{-9}$ farad/meter

Table 3-2 shows the magnitudes of μ_0 and ϵ_0 in the three systems.

[1] J. A. Stratton, "Electromagnetic Theory," preface, McGraw-Hill Book Company, Inc., New York, 1941.

[2] S. A. Schelkunoff, The End Is in Sight, *Proc. I.R.E.*, **36** (7): 827, July, 1948.

[3] See also G. W. O. Howe, The Rationalized M.K.S. System, *Wireless Engr.*, November, 1948, and January, 1950.

TABLE 3-3

Quantity	mks rationalized dimensions					Equivalent units
	M	L	T	Q	θ	
1. Acceleration	0	1	−2	0	0	meters/second^2
2. Acceleration, angular	0	0	−2	0	0	radians/second^2
3. Admittance	−1	−2	1	2	0	mhos
4. Angle	0	0	0	0	0	degrees, radians
5. Area	0	2	0	0	0	meters^2
6. Capacitance	−1	−2	2	2	0	farads
7. Capacitance, distributed	−1	−3	2	2	0	farads/meter
8. Capacity, thermal	0	2	−2	0	−1	joules/kilogram °C
9. Charge	0	0	0	1	0	coulombs
10. Charge density, line	0	−1	0	1	0	coulombs/meter
11. Charge density, surface	0	−2	0	1	0	coulombs/meter^2
12. Charge density, volume	0	−3	0	1	0	coulombs/meter^3
13. Conductance	−1	−2	1	2	0	mhos
14. Conductance, distributed	−1	−3	1	2	0	mhos/meter
15. Conductivity	−1	−3	1	2	0	mhos/meter
16. Conductivity, thermal	1	1	−3	0	−1	joules/meter-second °C
17. Current	0	0	−1	1	0	amperes
18. Current density, line	0	−1	−1	1	0	amperes/meter
19. Current density, surface	0	−2	−1	1	0	amperes/meter^2
20. Density	1	−3	0	0	0	kilograms/meter^3
21. Density, sound-energy	1	−1	−2	0	0	newtons/meter^2
22. Diffusivity, thermal	0	2	−1	0	0	meters^2/second
23. Discharge, hydraulic	0	3	0	0	0	meters^3
24. Displacement, dielectric	0	−2	0	1	0	coulombs/meter^2
25. Elastance	1	2	−2	−2	0	darafs
26. Elasticity, modulus of Young's, shear, bulk	1	−1	−2	0	0	newtons/meter^2
27. Emissivity	1	0	−3	0	−1	joules/meter-second °C
28. Energy	1	2	−2	0	0	joules (watt-second)
29. Energy density, volume	1	−1	−2	0	0	joules/meter^3
30. Enthalpy	1	2	−2	0	0	joules
31. Entropy	1	2	−2	0	−1	joules/degree
32. Expansion, coefficient of thermal	0	0	0	0	−1	per °C
33. Field intensity, electric	1	1	−2	−1	0	volts/meter
34. Field intensity, magnetic	0	−1	−1	1	0	amperes/meter
35. Flux, dielectric	0	0	0	1	0	coulombs
36. Flux, luminous	1	2	−3	0	0	lumens
37. Flux, magnetic	1	2	−1	−1	0	webers (volt-second)
38. Flux, sound-energy	1	2	−3	0	0	watts
39. Flux, specific sound-energy	1	0	−3	0	0	watts/meter^2
40. Flux density, dielectric	0	−2	0	1	0	coulombs/meter^2
41. Flux density, magnetic	1	0	−1	−1	0	webers/meter^2

TABLE 3-3 (*Continued*)

Quantity	mks rationalized dimensions					Equivalent units
	M	L	T	Q	θ	
42. Force	1	1	−2	0	0	newtons (joules/meter)
43. Force (emf)	1	2	−2	−1	0	volts
44. Force, magnetizing	0	−1	−1	1	0	amperes/meter
45. Force, magnetomotive	0	0	−1	1	0	amperes
46. Frequency	0	0	−1	0	0	cycles/second
47. Head	1	−1	−2	0	0	newtons/meter^2
48. Heat	1	2	−2	0	0	joules (calories)
49. Heat of fusion, vaporization, or combustion	0	2	−2	0	0	joules/kilogram
50. Heat transfer, coefficient of	1	0	−3	0	−1	joules/meter
51. Illumination, incident	1	0	−3	0	0	foot-candles
52. Impedance	1	2	−1	−2	0	ohms
53. Impedance, acoustic	1	−4	−1	0	0	acoustical ohms
54. Impedance, mechanical	1	0	−1	0	0	mechanical ohms
55. Impedance, specific acoustic	1	−2	−1	0	0	acoustical ohm-meters^2
56. Inductance	1	2	0	−2	0	henrys (volt-second/ampere)
57. Inductance, distributed	1	1	0	−2	0	henrys/meter
58. Intensity, luminous	1	2	−3	0	0	candles (lumens/steradian)
59. Intensity, sound	1	0	−3	0	0	watts/meter^2
60. Length	0	1	0	0	0	meters
61. Luminance	1	0	−3	0	0	candles/foot^2
62. Magnetization	1	0	−1	−1	0	webers/meter^2
63. Mass	1	0	0	0	0	kilograms
64. Moment, electric	0	1	0	1	0	coulomb-meters
65. Moment of inertia	1	2	0	0	0	kilogram-meters^2
66. Moment, magnetic	0	2	−1	1	0	ampere-meters^2
67. Momentum	1	1	−1	0	0	kilogram-meters/second
68. Momentum, angular	1	0	−1	0	0	kilograms/second
69. Permeability	1	1	0	−2	0	henrys/meter
70. Permittivity	−1	−3	2	2	0	farads/meter
71. Piezoelectric strain coefficient	−1	−1	2	1	0	coulombs/newton
72. Piezoelectric stress coefficient	0	−2	0	1	0	coulombs/meter^2
73. Polarization, electric	0	−2	0	1	0	coulombs/meter^2
74. Polarization, magnetic	0	−1	−1	1	0	amperes/meter
75. Potential, electric	1	2	−2	−1	0	volts
76. Potential, gradient	1	1	−2	−1	0	volts/meter
77. Power	1	2	−3	0	0	watts (joules/second)
78. Power density, surface	1	0	−3	0	0	watts/meter^2
79. Power density, volume	1	−1	−3	0	0	watts/meter^3
80. Pressure	1	−1	−2	0	0	newtons/meter^2
81. Reactance	1	2	−1	−2	0	ohms
82. Reactance, acoustic	1	−4	−1	0	0	acoustical ohms
83. Reactance, mechanical	1	0	−1	0	0	mechanical ohms

TABLE 3-3 (*Continued*)

Quantity	mks rationalized dimensions					Equivalent units
	M	L	T	Q	θ	
84. Reluctance, magnetic	−1	−2	0	2	0	amperes/weber
85. Resistance	1	2	−1	−2	0	ohms
86. Resistance, acoustic	1	−4	−1	0	0	acoustical ohms
87. Resistance, distributed	1	1	−1	−2	0	ohms/meter
88. Resistance, mechanical	1	0	−1	0	0	mechanical ohms
89. Resistivity	1	3	1	2	0	ohm-meters
90. Strain	0	0	0	0	0	
91. Stress, pressure	1	−1	−2	0	0	newtons/meter²
92. Susceptance	−1	−2	1	2	0	mhos
93. Temperature	0	0	0	0	1	°C
94. Tension, surface	1	0	−2	0	0	newtons/meter
95. Thalpance	0	2	−1	0	0	meters²/second
96. Thermoelectricity, coefficient of	1	2	−2	−1	−1	volts/°C
97. Time	0	0	1	0	0	seconds
98. Torque	1	2	−2	0	0	joules
99. Velocity, angular	0	0	−1	0	0	cycles/second
100. Velocity, linear	0	1	−1	0	0	meters/second
101. Velocity, volume	0	3	−1	0	0	meters³/second
102. Viscosity	1	−1	−1	0	0	poises
103. Viscosity, kinematic	0	2	−1	0	0	meters²/second
104. Voltage	1	2	−2	−1	0	volts
105. Volume	0	3	0	0	0	meters³
106. Weight	1	1	−2	0	0	newtons
107. Work	1	2	−2	0	0	joules

It can be seen that only in the rationalized system do both the numerical constants 4π and c appear in both μ_0 and ϵ_0.

In the rationalized mks system, certain of the units will at first appear a bit strange to a reader accustomed to the cgs system. In this system, the unit of force is the newton, which equals 10^5 dynes and is sometimes called the "dyne-five" for that reason. Magnetic field intensity H is expressed in ampere turns per meter, which has a good physical correlation, while magnetic flux density B is measured in units of webers per square meter, each one of which is 10^4 gauss. In general, however, the mks units are extremely practical and most convenient to use.

In order that the dimensional system employed be sufficiently general to accommodate any quantity normally encountered in physical problems, one additional dimension, temperature θ, must be added. The five-dimensional system, therefore, should be adequate to define com-

TABLE 3-4

Dimensions					Quantity number
M	L	T	Q	θ	
1	3	−1	−2	0	89
1	2	0	0	0	65
1	2	0	−2	0	56
1	2	−1	−1	0	37
1	2	−1	−2	0	52, 81, 85
1	2	−2	0	0	28, 30, 48, 98, 107
1	2	−2	0	−1	31
1	2	−2	−1	0	43, 75, 104
1	2	−2	−1	−1	96
1	2	−2	−2	0	25
1	2	−3	0	0	36, 38, 58, 77
1	1	0	−2	0	57, 69
1	1	−1	0	0	67
1	1	−1	−2	0	87
1	1	−2	0	0	42, 106
1	1	−2	−1	0	33, 76
1	1	−3	0	−1	16
1	0	0	0	0	63
1	0	−1	0	0	54, 68, 83, 88
1	0	−1	−1	0	41, 62
1	0	−2	0	0	94
1	0	−3	0	0	39, 51, 59, 61, 78
1	0	−3	0	−1	27, 50
1	−1	−1	0	0	102
1	−1	−2	0	0	21, 26, 29, 47, 80, 91
1	−1	−3	0	0	79
1	−2	−1	0	0	55
1	−3	0	0	0	20
1	−4	−1	0	0	53, 82, 86
0	3	0	0	0	23, 105
0	3	−1	0	0	101
0	2	0	0	0	5
0	2	−1	0	0	22, 95, 103
0	2	−1	1	0	66
0	2	−2	0	0	49
0	2	−2	0	1	8
0	1	0	1	0	64
0	1	0	0	0	60
0	1	−1	0	0	100
0	1	−2	0	0	1
0	0	1	0	0	97
0	0	0	1	0	9, 35
0	0	0	0	1	93
0	0	0	0	0	4, 90

TABLE 3-4 (*Continued*)

Dimensions					Quantity number
M	L	T	Q	θ	
0	0	0	0	−1	32
0	0	−1	1	0	17, 45
0	0	−1	0	0	46, 99
0	0	−2	0	0	2
0	−1	0	1	0	10
0	−1	−1	1	0	18, 34, 44, 74
0	−2	0	1	0	11, 24, 40, 72, 73
0	−2	−1	1	0	19
0	−3	0	1	0	12
−1	−1	2	1	0	71
−1	−2	2	2	0	6
−1	−2	1	2	0	3, 13, 92
−1	−2	0	2	0	84
−1	−3	2	2	0	70, 7
−1	−3	1	2	0	14, 15

pletely all quantities which belong to the mechanical, electrical, magnetic, hydraulic, acoustical, and thermal systems.

Table 3-3 summarizes the dimensional relationships which exist in the rationalized mks system. The numbers listed in the columns below the basic dimensions M, L, T, Q, and θ represent the exponents which each of the dimensions will have in the dimensional formulas for the quantities listed on the left. The column on the right gives the corresponding units for these quantities.

Table 3-4 provides a means for finding a quantity when only its dimensions are known. The numbers in the column on the right refer to the quantities listed in Table 3-3. (These tables are expanded versions of those originally prepared by Winzemer.[1]) Certain specialized quantities used in acoustical engineering have been omitted from Tables 3-3 and 3-4, such as acoustic mass, or inertance, $[M^1L^{-4}T^0Q^0\theta^0]$ and acoustic compliance, or capacitance, $[M^{-1}L^4T^2Q^0\theta^0]$. These quantities should not be confused with mechanical mass $[M^1L^0T^0Q^0\theta^0]$ and mechanical compliance $[M^{-1}L^0T^2Q^0\theta^0]$.

3-5 Dimensionless and Dimensional Constants

In the checking of equations dimensionally, there will frequently be encountered quantities which are pure ratios and which, by definition, must have no dimensions at all. A partial list of these might include:

[1] A. M. Winzemer, Dimensional Analysis of Electromagnetic Equations, *Proc. I.R.E.*, **35** (11): 1383–1384, November, 1947.

Amplification factor	Luminosity
Atomic number	Mach number
Atomic weight	Nusselt's number
Avogadro's number	Poisson's ratio
Coefficient of coupling	Power factor
Coefficient of friction	Prandtl's number
Coefficient of reflection	Q
Compression ratio	Refractive index
Dielectric constant	Relative humidity
Efficiency	Reynold's number
Froude's number	Specific gravity
Gain	Specific heat
Grashof's number	Weber's number

On the other hand, certain numerical constants found in physical equations which are apparently pure numbers actually contain dimensions implicitly. Some of these which are frequently encountered by the electronics engineer include:

*Constant**	*Dimensions (mks rationalized)*
Boltzmann's constant	$M^1L^2T^{-2}Q^0\theta^{-1}$
Electron charge/mass	$M^{-1}L^0T^0Q^1\theta^0$
Electron volt	$M^1L^2T^{-2}Q^0\theta^0$
Gas constant per molecule	$M^1L^2T^{-2}Q^0\theta^{-1}$
Gravitational acceleration	$M^0L^1T^{-2}Q^0\theta^0$
Planck's constant	$M^1L^2T^{-1}Q^0\theta^0$
Stefan-Boltzmann constant	$M^1L^0T^{-3}Q^0\theta^{-4}$
Velocity of light	$M^0L^1T^{-1}Q^0\theta^0$
Wien displacement-law constant	$M^0L^1T^0Q^0\theta^1$

* For the magnitude of these constants, see "Reference Data for Radio Engineers," 3d ed., p. 25, Federal Radio & Telephone Co., New York, 1949; J. W. M. Dumond and E. R. Cohen, Our Knowledge of the Atomic Constants, *F*, *N*, *m*, and *h* in 1947, and of Other Constants Derivable Therefrom, *Revs. Mod. Phys.*, **20**, January, 1948.

Consider the formula for the force per unit length exerted between the two parallel conductors which are carrying currents of I_1 and I_2 amperes, respectively, and which are separated by a distance D:

$$F = k\frac{I_1I_2}{D} \qquad \text{newtons/meter} \tag{3-16}$$

where k is a numerical proportionality factor.

Writing Eq. (3-16) dimensionally,

$$[M^1L^0T^{-2}Q^0\theta^0] = \frac{[k][M^0L^0T^{-1}Q^1\theta^0]^2}{[L^1]} \tag{3-17}$$

the dimensions of the factor k are seen to be

$$[k] = [M^1L^1T^0Q^{-2}\theta^0] \tag{3-18}$$

Reference to Table 3-4 indicates that k has the dimensions of quantity 69 in Table 3-3. Evidently, then, k is a dimensional constant and must represent permeability dimensionally even though it customarily appears as a purely numerical factor.

The general form of the Child-Langmuir law[1] for the current I in a space-charge-limited plane diode is usually written as

$$I = kE_b^{3/2} \tag{3-19}$$

where E_b is the diode plate voltage and k is a proportionality factor commonly referred to as the perveance of the tube.[2] No name has been assigned for the unit of perveance,[3] so that it is generally stated in the units amperes (or milliamperes) per volt to the three-halves power. It is interesting to analyze Eq. (3-19) to determine just what the dimensions of perveance must be. Thus

$$[k] = [E]^{-3/2}[I]^1 = [M^{-3/2}L^{-3}T^2Q^{5/2}\theta^0] \tag{3-20}$$

so that the basic units of perveance are

$$\frac{\text{coulombs}^{5/2} \cdot \text{seconds}^2}{\text{kilograms}^{3/2} \cdot \text{meters}^3}$$

A check into the derivation of Child's law shows how this peculiar set of dimensions is obtained. The actual value of the perveance is[4]

$$k = \frac{\epsilon}{9\pi}\left(\frac{2e}{m}\right)^{1/2}\frac{A}{d_{kp}^2} \tag{3-21}$$

where ϵ is the permittivity, e/m is the electron charge per unit mass, A is the electrode area, and d_{kp} is the cathode plate spacing. In Eq. (3-21), k has the dimensions cited in Eq. (3-20), and its numerical value is equal to $2.33 \times 10^{-6}A/d_{kp}^2$.

In general, then, each numerical factor which is present in an equation which is to be checked dimensionally must be held suspect unless it is definitely known that it contains no dimensions implicitly.

[1] C. D. Child, Discharge from Hot CaO, *Phys. Rev.*, **58,** May, 1911.

[2] See for example Y. Kusunose, Calculation of the Characteristics and the Design of Triodes, *Proc. I.R.E.*, **17** (10): 1706, October, 1929; G. R. Partridge, Factors Influencing the Perveance of Power-output Triodes, *Proc. I.R.E.*, **37** (1): 87, January, 1949.

[3] It has been proposed that the unit of perveance be designated the Langmuir. See C. D. O'Neill, The Unit of Perveance, *Proc. I.R.E.*, **37** (11): 1295, November, 1949.

[4] B. J. Thompson, Space-current Flow in Vacuum-tube Structures, *Proc. I.R.E.*, **31** (9): 485, September, 1943.

Problems

Prob. 3-1 Analyze the following combinations of resistance R, inductance L, and capacitance C, using $[VQT]$, then identify

(*a*) RC	(*g*) $L^{3/2}/R^2C^{3/2}$
(*b*) L/R	(*h*) L^2/R^2C
(*c*) $1/\sqrt{LC}$	(*i*) L/RC
(*d*) R^2C	(*j*) L^2/R^4C
(*e*) L/R^2	(*k*) $L^{1/2}RC^{1/2}$
(*f*) L/R^3C^2	

Prob. 3-2 The tangent of the phase angle φ of the alternating current in an RLC circuit is given by

$$\tan \varphi = \frac{\omega^2 LC - 1}{\omega RC}$$

where $\omega = 2\pi f$. Show that $\tan \varphi$ is dimensionless.

Prob. 3-3 Experiments with the circuit shown in Fig. 3-1 indicate that if the value of R is selected properly, the circuit will act like a pure

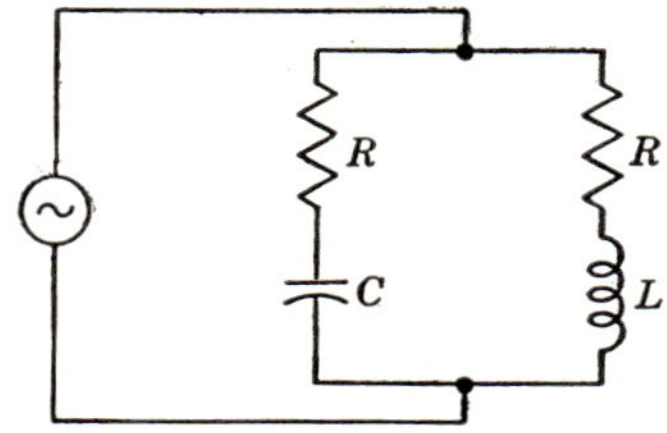

FIG. 3-1

resistance to an applied voltage of any frequency and will exhibit no transient response when excited with a square wave. How must R be related to L and C?

Prob. 3-4 In the cgs electromagnetic system, the force in dynes exerted between two magnetic poles m_1 and m_2 separated by a distance r in centimeters in a medium of permeability μ_0 is

$$f = \frac{m_1 m_2}{\mu_0 r^2}$$

Also, the magnetic field intensity H in oersteds or dynes per unit pole developed by a solenoid of N turns and length l which conducts a current I is

$$H = \frac{f}{m} = \frac{4\pi NI}{l}$$

If, in the cgs electrostatic system, the force in dynes developed between charges q_1 and q_2 separated by a distance r in centimeters in a medium of permittivity ϵ_0 is

$$f = \frac{q_1 q_2}{\epsilon_0 r^2} \qquad \text{dynes}$$

show that the dimensions of $[\mu_0\epsilon_0]^{-1/2}$ are always those of velocity.

Prob. 3-5 Show that in the cgs electromagnetic system, if μ_0 is a basic dimension, then because of the dimensional relation proved in Prob. 3-4 the dimensions of ϵ_0 must be $M^0L^{-2}T^2Q^0T^0$.

Prob. 3-6 From the equation for the force between charges (see Prob. 3-4), find the dimensions of charge in the cgs electromagnetic system.

Prob. 3-7 In the cgs system, resistivity is often stated in terms of ohms per centimeter cube. Another popular unit is the ohm per circular mil–foot. Are these units correct dimensionally? Check the definition of resistivity and explain.

Prob. 3-8 If power is defined as work per unit time, why does the product of voltage times current yield power?

Prob. 3-9 Confirm dimensionally Einstein's relation between mass, energy, and the velocity of light. What power would be generated by the transformation of 1 gram of matter into energy in a time interval of 1 sec?

Prob. 3-10 The intrinsic impedance of space is known to depend on the permeability of space μ and its permittivity ϵ. Determine the correct formula for this impedance dimensionally and its exact value in ohms.

Chapter 4

CHECKING EQUATIONS AND PREDICTING SOLUTIONS

4-1 Checking Equations in Single Systems

When a complicated equation has been evolved in the analysis of a problem, it is quite useful to have a means for checking it independently of the conditions of the problem. Dimensional analysis provides this means, for the validity of an equation can be checked to a certain extent by establishing that it is dimensionally consistent, in accord with the principle originally established by Fourier. The fact that an equation does check dimensionally, of course, does not mean that it is correct, but, if the dimensional check fails, the equation must be in error.

Before equations can be broken down into their dimensional equivalents, it is important to note that

(*a*) Pure numbers are dimensionless.

(*b*) All transcendental functions are dimensionless, for example, $\sin \theta$, $\cos x$, $\tan \phi$, $\sinh A$, $\cosh B$, $\log y$, etc.

(*c*) All quantities which are the arguments of transcendental functions must be dimensionless, that is, θ, x, ϕ, A, B above.

(*d*) All exponents must be dimensionless.

With these facts in mind, relationships may now be analyzed dimensionally provided the variables involved can be defined by a single system of three basic dimensions, such as voltage, time, and charge.

Suppose it is necessary to check an expression for the power factor of a series RC circuit which is given as

$$\text{Power factor} = \frac{\omega CR}{\sqrt{1 - \omega^2 C^2 R^2}} \tag{4-1}$$

where $\omega = 2\pi f$.

The reader will recall that the power factor of any network represents the cosine of the phase angle of the network. The power factor, therefore, must be dimensionless, according to (*b*) above. Substituting dimensions in (4-1), there is obtained

$$[\mathrm{PF}] = \frac{[T^{-1}][V^{-1}Q][VQ^{-1}T]}{\sqrt{1 + [T^{-1}]^2[V^{-1}Q]^2[VQ^{-1}T]^2}} \tag{4-2}$$

It can easily be seen that both the numerator and the quantities under the radical in the denominator are dimensionless, so that Eq. (4-1) is dimensionally correct.

A student has derived the following expression for the rate of change of current with time in a series RL circuit to which is applied a constant voltage V:

$$\frac{di}{dt} = VL\epsilon^{-Rt/L} \tag{4-3}$$

From (*d*) above the exponent of the quantity $\epsilon^{-Rt/L}$ should be and is dimensionless:

$$\frac{[R][T]}{[L]} = \frac{[VQ^{-1}T][T]}{[VQ^{-1}T^2]} \tag{4-4}$$

The dimensions of the derivative di/dt are the same as those of $\Delta I/\Delta t$, or

$$\frac{di}{dt} = \frac{[QT^{-1}]}{[T]} = [QT^{-2}] \tag{4-5}$$

However, the dimensions of the product VL are

$$[VL] = [V][VQ^{-1}T^2] = [V^2Q^{-1}T^2] \tag{4-6}$$

and therefore

$$\left[\frac{di}{dt}\right] = [QT^{-2}] \neq [VL] = [V^2Q^{-1}T^2] \tag{4-7}$$

so that the derivation of Eq. (4-3) is in error and should be corrected. If the factor VL is changed to V/L, then

$$\left[\frac{di}{dt}\right] = \left[\frac{V}{L}\right] = \frac{[V]}{[VQ^{-1}T^2]} = [QT^{-2}] \tag{4-8}$$

which checks dimensionally. The proper form of Eq. (4-3) evidently is

$$\frac{di}{dt} = \frac{V}{L}\epsilon^{-Rt/L} \tag{4-9}$$

The method of dimensional checking may also be applied to integrals and differential equations.[1] The dimensions of a derivative have just

[1] Dimensional analysis can also be used to solve certain differential equations. See G. Birkhoff, Dimensional Analysis of Partial Differential Equations, *Elec. Eng.*, **67** (12), December, 1948.

been determined in Eq. (4-5). In general, the dimensions of any derivative can be established as follows:

If the dependent variable is y, the dimensions of y in a derivative of any order will be $[y]^1$. If the independent variable is x, the dimensions of x, in a derivative of order n, will be $[x]^{-n}$. For example, if x represents displacement, and t time, the first derivative of x with respect to time will be dx/dt, and, according to the rule just stated,

$$\left[\frac{dx}{dt}\right] = [LT^{-1}] \tag{4-10}$$

It will be recalled from Sec. 3-3 that the dimensions of velocity in the mechanical system are $[LT^{-1}]$, to which those of Eq. (4-10) correctly correspond.

In the same way, the dimensions of the second derivative of x with respect to t are

$$\left[\frac{d^2x}{dt^2}\right] = [LT^{-2}] \tag{4-11}$$

which are also the dimensions of acceleration, $[LT^{-2}]$. The rate of change of acceleration d^3x/dt^3 would have the corresponding dimensions $[LT^{-3}]$, and so forth.

Consider the derivative of Eq. (4-9):

$$\frac{d^2i}{dt^2} = -\frac{VR}{L^2}\epsilon^{-Rt/L} \tag{4-12}$$

As before, the exponential term is dimensionless, and

$$\left[\frac{-VR}{L^2}\right] = \frac{[V][VQ^{-1}T]}{[V^2Q^{-2}T^4]} = [QT^{-3}] \tag{4-13}$$

and

$$\left[\frac{d^2i}{dt^2}\right] = [QT^{-1}][T^{-2}] = [QT^{-3}] \tag{4-14}$$

so that Eq. (4-12) is dimensionally correct.

In determining the dimensions of an integral, an inverse procedure is followed. If $f(x)$ is the dependent variable which makes up the integrand and if x is the independent variable in the integral

$$\phi = \int f(x)\,dx$$

the dimensions of the dependent variable will be $[f(x)]^1$, and the dimensions of the independent variable are $[x]^n$, where n is equal to the indicated number of integrations. For example, if v is velocity, then the displacement x over a time from 0 to t will be

$$x = \int_0^t v\,dt \tag{4-15}$$

The dimensions of this equation in the simple mechanical system used previously are, according to the rule just stated,

$$[x] = [LT^{-1}][T] = [L] \tag{4-16}$$

Similarly, the integral

$$x = \int_0^t \int_0^t a\, dt\, dt = \int_0^t \int_0^t a\, (dt)^2 \tag{4-17}$$

where a represents acceleration, will be

$$[x] = [LT^{-2}][T^2] = [L] \tag{4-18}$$

The expression for the voltage e_C developed across a capacitor C which is charged by a current i over a time interval from 0 to t is

$$e_C = \frac{1}{C}\int_0^t i\, dt \tag{4-19}$$

The dimensions of this expression according to the rule for integrals are

$$[V] = [Q^{-1}V][QT^{-1}][T^1] = [V] \tag{4-20}$$

The differential equation for a series RLC network in terms of current i is

$$L\frac{d^2i}{dt^2} + R\frac{di}{dt} + \frac{i}{C} = 0 \tag{4-21}$$

Term by term, according to the rules cited above for derivatives, this equation becomes

$$\left[L\frac{d^2i}{dt^2}\right] = [VQ^{-1}T^2]\frac{[QT^{-1}]}{[T^2]} = [VT^{-1}] \tag{4-22}$$

$$\left[R\frac{di}{dt}\right] = [VQ^{-1}T]\frac{[QT^{-1}]}{[T]} = [VT^{-1}] \tag{4-23}$$

$$\left[\frac{i}{C}\right] = \left[\frac{QT^{-1}}{V^{-1}Q}\right] = [VT^{-1}] \tag{4-24}$$

It can be seen that every term of Eq. (4-21) has the dimensions $[VT^{-1}]$, which is reasonable to expect since the equation was obtained by differentiating with respect to time the sum of a number of voltage drops around a network mesh or loop which is equal to zero. This indicates that quantities equated to zero are not necessarily dimensionless. It is readily realized that zero has either no dimensionality at all or can assume any dimension desired.

In addition to the peculiar dimensional behavior of zero, there are a number of quantities of interest to the television and electronics engineer which are dimensionless, or pure ratios. These pure ratios include such quantities as power factor, gain, amplification factor, Q, and any quantity expressed in decibels. There are some other quantities which are commonly encountered in applications which at first appear to be dimensionless or *pure* numerics, but which actually contain dimensions implicitly which must be accounted for when dimensional checks are made. These were discussed in detail in Sec. 3-5.

4-2 Checking Equations in the Universal System

The use of the universal mks rationalized system described in Sec. 3-4 permits a dimensional check of any equation which represents physical magnitudes. In dimensional checking by means of a universal system of units, the principles evolved for the single system in Sec. 4-1 are observed. However, in the case of the universal system, five rather than three basic dimensions are involved, and frequent reference to Tables 3-3 and 3-4 of Sec. 3-4 will be helpful.

An example of the dimensional analysis of combined electrical and mechanical units can be found in a check on the expression for the power loss incurred in an iron-cored choke or saturable reactor which is conducting alternating current. This power consumption (which neglects I^2R loss) is given as[1]

$$P = \pi BHVf \times 10^{-8} \qquad \text{watt} \tag{4-25}$$

When this formula was transferred to an engineer's notes, the variables involved were not identified. However, it is recollected that B is the flux density in the core, H is the magnetizing force, and f is the frequency of the current. The problem is to determine what V must represent. Obtaining the dimensions of P, B, H, and f from Table 3-2, it is found that dimensionally Eq. (4-25) is

$$[M^1L^2T^{-3}Q^0\theta^0] = [M^1L^0T^{-1}Q^{-1}\theta^0][M^0L^{-1}T^{-1}Q^1\theta^0][T^{-1}][V] \tag{4-26}$$

and therefore the dimensions of V are

$$[V] = [L^3] \tag{4-27}$$

so that V represents a volume, actually the volume of the core.

The expression for the transit time τ of a magnetron has been derived as

$$\tau = \frac{L}{\sqrt{2V_0e/m}} \tag{4-28}$$

where L is the axial length of the magnetron, V_0 is the beam voltage, and

[1] C. Helber, Designing Saturable-core Reactors for Specific Uses, *Electronic Inds.*, **6** (12), December, 1947.

e/m is the charge-to-mass ratio of the electron. Dimensionally, this becomes

$$[T]^1 = \frac{[L]^1}{([M^1L^2T^{-2}Q^{-1}][Q^1M^{-1}])^{1/2}} = [L^1][L^2T^{-2}]^{-1/2} = [T^1] \quad (4\text{-}29)$$

This establishes the dimensional correctness of Eq. (4-28).

It has been shown[1] that in a diode the plate current I_b is related to the temperature-limited current I_s by the expression

$$I_b = I_s\epsilon^{eE_m/kt} \quad (4\text{-}30)$$

where E_m is the potential at the virtual cathode, e is the charge of the electron, k is Boltzmann's constant (1.380×10^{-23} erg/deg), T is absolute temperature of the cathode in degrees Kelvin, and ϵ is the Naperian logarithmic base.

By definition, I_b and I_s are identical dimensionally, so that it remains only to check the exponent of ϵ, which must be dimensionless according to Sec. 4-1. Thus

$$[kT] = [eE_m] \quad (4\text{-}31)$$

From Table 3-1 the dimensions of Eq. (4-31) are

$$[M^1L^2T^{-2}Q^0\theta^{-1}][\theta^1] = [Q^1][M^1L^2T^{-2}Q^{-1}\theta^0] \quad (4\text{-}32)$$

so that Eq. (4-20) is correct dimensionally.

The heat generated on the inside surfaces of a water-cooled television transmitting tube such as the RCA type 8D21 is transferred to and carried away by the cooling liquid which is constantly circulated through the tube. From thermal engineering, it is known that, assuming turbulent flow, all factors involved in the mechanism of heat transfer from the internal surfaces of the tube to the liquid are connected by the following relation:[2]

$$\left(\frac{hD}{k}\right) = 0.024\left[\left(\frac{D\rho V}{\mu}\right)^{0.8}\right]\left[\left(\frac{C_p\mu}{k}\right)^{0.4}\right] \quad (4\text{-}33)$$

where h is the coefficient of heat transfer, k is the thermal conductivity of the fluid, ρ is the fluid density, μ is fluid viscosity, C_p is the thermal capacity of the fluid, V is the average fluid velocity, and D is the equivalent diameter of the cross section of the fluid channel.

It is desired to show that each of the parenthesized expressions in

[1] Saul Dushman, Thermionic Emission, *Revs. Mod. Phys.*, **2**: 381, 1930.

[2] I. E. Mouromtseff, Water and Forced-air Cooling of Vacuum Tubes, *Proc. I.R.E.*, **30** (4), April, 1942.

Eq. (4-33) is dimensionless. From Table 3-3,

$$\left[\frac{hD}{k}\right] = \frac{[M^1L^0T^{-3}Q^0\theta^{-1}][L^1]}{[M^1L^1T^{-3}Q^0\theta^{-1}]} \tag{4-34}$$

$$\left[\frac{D\rho V}{\mu}\right] = \frac{[L^1][M^1L^{-3}][L^1T^{-1}]}{[M^1L^{-1}T^{-1}]} \tag{4-35}$$

$$\left[\frac{C_p\mu}{k}\right] = \frac{[M^0L^2T^{-2}Q^0\theta^{-1}][M^1L^{-1}T^{-1}]}{[M^1L^1T^{-3}Q^0\theta^{-1}]} \tag{4-36}$$

and it can be seen that this is the case.

4-3 Dimensionless Groupings

In conducting an experiment on a new and untried system, an engineer is generally confronted with a number of dependent and independent variables. His usual goal is to determine the relationship which exists between them. Often the total number of these variables is so large as to render impracticable individual variation of each of them over a wide range of values to determine the part each plays in the experimental result. A number of methods exist whereby the significant variables may be sorted out from the insignificant under these circumstances. Of prime importance in this connection is the method of Graeco-Latin squares,[1] which enables an engineer to cover a range of many variables in a systematic fashion with a minimum number of tests.

When the significant variables in an experiment are known, or can be surmised, dimensional analysis can in many instances be used to reduce radically the range over which the quantities involved must be varied in order to obtain adequate experimental data. In order to accomplish this best, the quantities or variables involved in the experiment must be grouped into dimensionless combinations, or πs.[2]

As an example of how this technique has been successfully applied in the past, consider the dimensionless quantity $(D\rho V/\mu)$, which appears in Eq. (4-33).

Some 70 years ago, Osborne Reynolds established the fact that this π, or dimensionless quantity, now known as Reynolds' number, governs the flow behavior of fluids and that, when its value is numerically less than 2,100, the flow of a liquid is always viscous and, when greater than 4,000, the flow is turbulent. For the intermediate values, the flow exhibits instability between these two conditions. This law holds for

[1] H. Poritsky, Method of Graeco-Latin Squares, *Elec. Eng.*, **67** (11), November, 1948; F. Caplan, Statistical Methods in the Design of Experiments to Serve the Electronics Industry, *Proc. Natl. Electronics Conf.*, **9**: 841–848, 1954.

[2] π is a familiar dimensionless constant (and transcendental number), being the ratio of the circumference to the diameter of a circle. π is used here, however, solely as a symbol of nondimensionality rather than as a number.

any fluid (gas or liquid) and has been repeatedly verified by many experimenters. It will be seen that, when this particular dimensionless grouping is formed, full information on all possible types of fluid flow can be determined by selecting appropriate values of D, ρ, V, and μ which make the value of the ratio vary between perhaps 1,000 and 5,000, with some additional tests to determine the limiting conditions at extremely large or small Reynolds' numbers. Further simplification can be achieved by using a single fluid which has a fixed density, a fixed viscosity, and a fluid channel of constant cross section. Then, full experimental data can be secured by varying only one of the four variables, the average fluid velocity V, over the necessary range of values.

A similar example can be cited from the field of naval architecture. It has been found that all the variables which determine the drag of a ship in motion can be combined into a dimensionless quantity, or π, which is Froude's number and is V^2/gl, where V is the normal operating speed of the ship, l is its length, and g is the gravitational acceleration. Experiments have shown that if a ship is designed so that Froude's number is $1/n\pi$, where n is an integer, then the drag of the ship will be a *minimum* when n is even and a *maximum* when n is odd. This information provides the ship designer with a laborsaving method for rapidly determining the proper hull length for a fast passenger or slow cargo vessel.

In the field of electronics, the behavior of triode oscillators at ultrahigh frequencies has been successfully analyzed from the standpoint of dimensional analysis.[1] It has been determined that all the parameters which dictate the maximum efficiency which can be achieved in the operation of a triode oscillator at these frequencies may be combined in a dimensionless quantity, or π, which is

$$\phi = \frac{fd_{kp}\sqrt{m/e}}{\sqrt{E_b}} \tag{4-37}$$

where f is the operating frequency of the oscillator in megacycles per second, d_{kp} is the cathode to plate spacing in centimeters, E_b is the applied plate voltage, and e/m is the electron charge per unit mass. Investigation shows that the maximum operating efficiency possible for any oscillator will be that attained when ϕ is as small numerically as is possible. Omitting the factor $\sqrt{m/e}$, efficiency will remain substantially constant for values of ϕ up to 2.5, and operation can be achieved, with very low efficiencies, for values of ϕ exceeding 8.

[1] G. J. Lehmann and A. R. Vallarino, Study of Ultra-high-frequency Tubes by Dimensional Analysis, *Proc. I.R.E.*, **33** (10), October, 1945. See also G. Lehmann, Dimensional Analysis Applied to Very-high-frequency Triodes, *Elec. Commun.*, **24** (3): 391–405, September, 1947, and Very-high-frequency Triode Oscillator and Amplifier Circuits, *Elec. Commun.*, **25** (1): 50–61, March, 1948.

If the operating frequency is arbitrarily fixed, then the operating efficiency can be raised only by increasing the plate voltage on the tube and by decreasing the cathode-to-plate spacing. Tubes which have similar internal geometry may be compared directly with regard to possible efficiencies as oscillators at a given frequency by the use of ϕ as a criterion.

If the Child-Langmuir-law equation stated in the previous chapter (which is also valid for triodes) is rewritten to include the evaluation of the perveance k obtained in that chapter, a second dimensionless combination or π may be formed, which is

$$\psi = \epsilon \sqrt{\frac{e}{m}} \frac{A}{d_{kp}^2} \frac{E_b^{3/2}}{I} \tag{4-38}$$

where ϵ is the permittivity of the internal space.

If the maximum cathode current density J of which the tube is capable is substituted for I/A, then Eq. (4-38) becomes

$$\psi = \epsilon \sqrt{\frac{e}{m}} \frac{E_b^{3/2}}{J d_{kp}^2} \tag{4-39}$$

The common factor d_{kp} may now be eliminated between Eqs. (4-37) and (4-39). This yields

$$d_{kp}^2 = \frac{\epsilon \sqrt{e/m}\, E_b^{3/2}}{J\psi} = \frac{\phi^2 E_b(e/m)}{f^2} \tag{4-40}$$

and

$$\frac{E_b f^4}{J^2} = \frac{\phi^4 \psi^2 (e/m)}{\epsilon^2} \tag{4-41}$$

It will be observed that the right-hand member of (4-41) is a constant. Equation (4-41) can therefore be interpreted to mean that, if the maximum efficiency of the tube as an oscillator is held constant by fixing the value of ϕ, then the plate voltage E_b which is required for this efficiency is proportional to J^2/f^4. This demonstrates clearly that the useful plate voltage falls off rapidly with frequency, and the only remedy is to increase the maximum cathode emission, or J.

In the same way, the factor E_b may be eliminated between Eqs. (4-37) and (4-39). It will be found that the cathode-to-plate distance d_{kp} is proportional to J/f^3. Such a relation leads rapidly to impossibly small dimensions for the tube if the efficiency (or ϕ) is to be held constant as f increases, and it indicates why a given type of electronic structure can be used only within relatively narrow frequency limits. Note that in every case an increase in J has a beneficial effect on the efficiency.

By similar means, R. R. Law[1] has deduced an empirical relationship for anode efficiency η as a function of frequency, interelectrode spacing,

[1] R. R. Law, Electronics of Ultra-high-frequency Triodes, *Proc. I.R.E.*, **37** (3), March, 1949.

and operating voltage, which is given as

$$\eta = \eta_0 \left(1 - k \frac{fd}{E_b^{1/2}}\right) \tag{4-42}$$

where f is in megacycles, η is the anode efficiency, η_0 is the maximum anode efficiency at low frequencies where transit-time effects are negligible (typical value 70 per cent), and k is a constant which has values for oxide-coated-cathode tubes of 1.75 for oscillator service and 1.20 for amplifier service. It should be noted that the factor d in this case is measured in *inches* and represents the *effective cathode-to-grid* spacing. For closely spaced triodes the grid-to-virtual-cathode spacing may be appreciably less than the geometric spacing, and it is the former value which is to be used. The importance of keeping the cathode-to-grid spacing small in achieving relatively high efficiencies at the ultrahigh frequencies can be appreciated from the fact that the practical upper limit on E_b in Eq. (4-42) for c-w operation of oxide-coated-cathode tubes appears to be near 1,000 volts.

Dimensionless combinations of variables are also useful in the engineering procedure called scaling. Many engineering experiments are too expensive and too difficult to perform on a full-scale basis. Through the proper use of scaling, experimental results obtained from measurements on scaled-down models may be used to predict full-scale performance. As Osborne Reynolds, who was responsible for Reynolds' number, observed: "Experimenting with models seems to afford a ready means of investigating and determining beforehand the effects of any proposed estuary or harbor works; a means, after what I have seen, I should feel it madness to neglect before entering upon any costly undertaking."

Scaling is most accurately accomplished when the physical laws under consideration are expressed as pure numbers or dimensionless groupings. Then, to compare the performance of a model to a full-scale system, it is only necessary to hold constant the proper dimensionless number. For example, the results obtained previously for the behavior of amplifiers and oscillators at uhf may also be applied to establish scaling factors for such tubes, and these have been tabulated elsewhere[1] for reference purposes.

Scaling is also used in ripple tanks[2] and in rubber-sheet models used to simulate electron trajectories in tube structures.[3] Scale factors are

[1] "Reference Data for Radio Engineers," 3d ed., pp. 221–222, Federal Telephone and Radio Co., New York, 1949.

[2] A. H. Schooley, Electronically Driven Ripple-tank, *Radio and Electronic Eng. Suppl., Radio and Television News,* **47** (7), June, 1950.

[3] J. W. Clark and R. E. Neuber, A Dynamic Electron Trajectory Tracer, *Proc. I.R.E.,* **38** (5), May, 1950.

such that a few cycles per minute in the model can correspond to many hundreds of megacycles in the corresponding tube. Scaling has been used to establish model theory for antenna designs,[1] the principle of similitude for cavity resonators,[2] and designs for microwave magnetrons,[3] loud speakers, horns, and studios.

4-4 Prediction of Relationships and the Buckingham π Theorem

If all the physical quantities which influence an experimental result are known, it will often be possible to formulate directly an equation which relates cause to effect in correct form (except for a proportionality factor or constant which must be determined by experiment). The fact that dimensional considerations limit the number of ways that variables can be combined to produce a desired dimensional result is the principle upon which the method of dimensional formulation is based. As an example, consider the following applications which involve the electrical-system dimensions $[VTQ]$.

The energy stored in the magnetic field of a filter choke or a magnetic deflection yoke is known from experimentation to be a function of the current in the inductor and its inductance, so that

$$\text{Stored energy} = kF(I,L) \tag{4-43}$$

where k is a proportionality constant. Equation (4-43) may be written dimensionally as

$$[\text{Energy}] = [I]^x[L]^y \tag{4-44}$$

where x and y are the exponents of I and L which are to be determined. Now fundamentally, energy represents the product of power and time, so that the dimensions of energy in the electrical system are $[VQ]$.

Substituting dimensions for I and L, Eq. (4-44) becomes

$$[VQ] = [QT^{-1}]^x[VQ^{-1}T^2]^y \tag{4-45}$$

By writing the factors in reduced form, Eq. (4-45) is

$$[VQ] = [V^yQ^{x-y}T^{-x+2y}] \tag{4-46}$$

[1] G. Sinclair, Theory of Models of Electromagnetic Systems, *Proc. I.R.E.*, **36** (11): 1364–1370, November, 1948.

[2] I. G. Wilson, C. W. Schramm, and J. P. Kinzer, High Q Resonant Cavities for Microwave Testing, *Bell System Tech. J.*, **25** (3): 421, July, 1946; H. Konig, The Laws of Similitude of the Electromagnetic Field and Their Application to Cavity Resonators, *Hochtech u. Elek:akus*, **58**: 174–180, 1941; also *Wireless Engr.*, **19** (1304): 216–217, 1942.

[3] G. B. Collins, "Microwave Magnetrons," Radiation Laboratory Series, vol. 6, p. 414, McGraw-Hill Book Company, Inc., New York, 1948; "Industrial Electronics Reference Book," sec. 5.24, John Wiley & Sons, Inc., New York, 1948.

If this expression is to check dimensionally, the exponents of V, Q, and T on each side of Eq. (4-46) must be equal; hence there are three simultaneous equations in x and y:

$$1 = y$$
$$1 = x - y$$
$$0 = -x + 2y$$

From these, $y = 1$ and $x = 2$, so that the correct form for Eq. (4-43) is

$$\text{Energy stored} = kLI^2 \tag{4-47}$$

The constant k can be determined by experiment and in this case is equal to ½.

In studio television systems, synchronizing pulses which are transmitted by coaxial cable from point to point are delayed to an extent

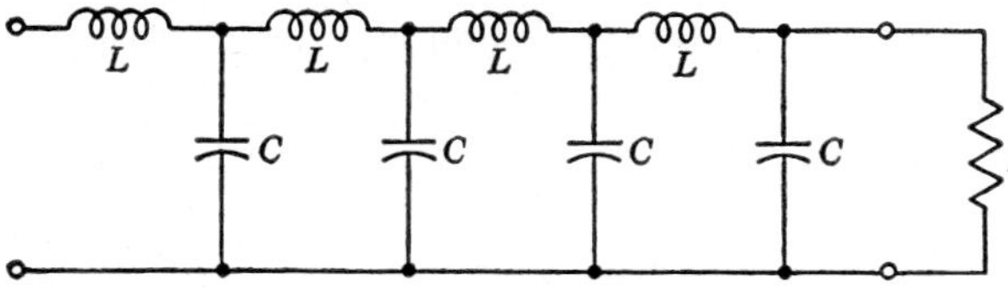

FIG. 4-1 Delay network.

depending on the length of cable path they traverse. A delay of 0.077 μsec, for example, is encountered over a 50-foot path length of RG-11/U 75-ohm cable. To equalize these delays, LC delay networks are inserted in the system. The time delay τ introduced by each section of such a delay network (Fig. 4-1) can be obtained dimensionally. Assume that

$$\tau = kF(L,C) \tag{4-48}$$

It is evident that

$$\tau = [L]^x[C]^y \tag{4-49}$$

where x and y are to be determined. Substituting dimensions for $[L]$ and $[C]$, there is obtained

$$[T^1] = [VQ^{-1}T^2]^x[V^{-1}Q]^y \tag{4-50}$$
$$[T^1] = [V^{x-y}Q^{-x+y}T^{2x}] \tag{4-51}$$

from which it follows that

$$0 = x - y$$
$$0 = -x + y$$
$$1 = 2x$$

Also $x = \frac{1}{2}$ and $y = \frac{1}{2}$, so that

$$\tau = kL^{\frac{1}{2}}C^{\frac{1}{2}} = k\sqrt{LC} \tag{4-52}$$

where k in practice is unity.

The reader will appreciate that these examples of dimensional formulation have been chosen for their simplicity. It is easy to see that more complicated problems can contain several derived quantities all of which can be expressed in terms of the three basic dimensions which define the single system, that is, $[M,L,T]$ or $[V,Q,T]$. For each of these derived quantities (velocity, acceleration, kinetic energy, momentum, etc., in the mechanical system; or current, inductance, capacitance, resistance, transconductance, power, etc., in the electrical system) an unknown exponent must be assigned and then determined from the dimensional equation. If the number of these derived quantities exceeds three, there will be more unknown exponents than there are equations relating them and solutions for the exponents will be ambiguous, since they will not be unique. If the universal system is involved, there will be, correspondingly, five equations with which to deal. It is obvious that if the number of unknown exponents exceeds the number of possible equations (or the number of dimensions) involved, then the usual unique algebraic solutions for the exponents cannot be expected.

In order to deal with such a situation, Buckingham[1] devised a theorem which is known as the Buckingham π theorem, because it is based on the nondimensional or π grouping. The sense of this theorem can be expressed as follows:

If an equation is dimensionally homogeneous, it can be reduced to an expression which consists of one or more factors, each of which is a π *(that is, dimensionless product).*

It can also be shown that *if the equation contains n variables in a system of r dimensions, the number of* π*'s necessary will be* $n - r$. For a single system, $r = 3$. For the universal system $r = 5$, unless one of the dimensions such as Q or θ does not appear in the analysis of any of the variables. One way of verifying the number of π's required is to use the matrix notation described in Chap. 7. The number of π's required is then equal to the number of variables minus the rank of the matrix formed by tabulating the dimensions of the variables involved.

The application of these ideas can best be illustrated by undertaking a problem.

It is desired to relate the quantities voltage V, inductance L, capacitance C, power P, and resistance R, all of which have to do with the power consumed by a high-Q tank circuit to which V is applied.

All these variables can be expressed dimensionally by the single system of dimensions $[VQT]$. The number of variables involved is 5 and the number of dimensions 3. $n - r$ is therefore 2, and two π's can be formed.

[1] E. Buckingham, On Physically Similar Systems; Illustrations of the Use of Dimensional Equations, *Phys. Rev.*, **4** (4): 345, 1914; P. W. Bridgman, "Dimensional Analysis," Yale University Press, New Haven, Conn., 1931.

To confirm this, establish the dimensional matrix for these variables and find its rank from

$$\begin{array}{c|ccccc} & V & L & C & P & R \\ \hline V & 1 & 1 & -1 & 1 & 1 \\ Q & 0 & -1 & 1 & 1 & -1 \\ T & 0 & 2 & 0 & -1 & 1 \end{array} \tag{4-53}$$

As explained in Chap. 7, if a matrix contains a nonzero determinant of order r, and if all determinants of order greater than r are zero, the rank of the matrix is r.

For a first trial, the determinant formed by the last three columns of Eq. (4-53) can be used. In this case,

$$\begin{vmatrix} -1 & 1 & 1 \\ 1 & 1 & -1 \\ 0 & -1 & 1 \end{vmatrix} = -2 \tag{4-54}$$

Since this is a third-order determinant different from zero, the rank is 3, which confirms the value of r based on the number of dimensions.

Since two π's can be formed, the following equation can be written:

$$k\pi_1\pi_2 = V^v L^w C^x P^y R^z \tag{4-55}$$

where π_1 and π_2 are the unknown dimensionless groupings, k is a proportionality factor, and v, w, x, y, and z are exponents to be determined. The dimensions of Eq. (4-55) are

$$[\pi_1\pi_2] = [V^0Q^0T^0] = [V]^v[VQ^{-1}T^2]^w[V^{-1}Q]^x[VQT^{-1}]^y[VQ^{-1}T]^z \tag{4-56}$$

or, by simplifying,

$$[\pi_1\pi_2] = [V^0Q^0T^0] = [V^{v+w-x+y+z}Q^{-w+x+y-z}T^{2w-y+z}] \tag{4-57}$$

Since the exponents of V, Q, and T on the left are zero, the exponents of V, Q, and T on the right must equal zero, so that

$$\begin{aligned} v + w - x + y + z &= 0 \\ - w + x + y - z &= 0 \\ 2w - y + z &= 0 \end{aligned} \tag{4-58}$$

Note that the coefficients of each equation in (4-58) are the same as the rows in the dimensional matrix (4-53), so that these equations can be written directly from the matrix, bypassing Eqs. (4-55) to (4-57).

Equation (4-58) can be solved for three of the variables in terms of the other two, and

$$\begin{aligned} y &= -\frac{v}{2} \\ w &= -x \\ z &= 2x - \frac{v}{2} \end{aligned} \tag{4-59}$$

Equation (4-59) possesses an infinite number of solutions dependent on the values which are arbitrarily assigned to v and x. In general, the number of variables in any problem which must be arbitrarily assigned will be $n - r$. The technique used in forming the two π's in this case is to assign a nonzero value for v and a zero value for x to form the first π and, in reverse, a zero value for v and a nonzero value for x to form the second.

In selecting an appropriate value for v, it is necessary to resort to the physical conditions of the problem. Voltage is usually squared when associated with power. Therefore, as a first choice, assign the values $v = 2$, $x = 0$. Then $y = -1$, $w = 0$, $z = -1$. From Eq. (4-55), the first π consists of

$$\pi_1 = \frac{V^2}{PR} \tag{4-60}$$

For the second π, $v = 0$, and let $x = 1$. Then $y = 0$, $w = -1$, and $z = 2$. π_2 is

$$\pi_2 = \frac{CR^2}{L} \tag{4-61}$$

The complete equation is the product of the π's, or

$$k\pi_1\pi_2 = \frac{V^2}{PR}\frac{CR^2}{L} = \frac{V^2CR}{PL} \tag{4-62}$$

which is the complete dimensionless grouping of all the original variables. If the formula for the power is desired, then, from Eq. (4-62),

$$P = \frac{V^2RC}{L} \tag{4-63}$$

if $k = 1$. This equation checks with the physical facts since, in general, $P = V^2/R_{eff}$ and R_{eff} for a high-Q tank circuit has the form L/CR.

Consider a problem which involves the universal system of dimensions. It is necessary to determine the relationship between the power density P_d and the peak magnetizing force H required in the electronic induction heating of a solid metallic cylinder. Besides the magnetizing force developed in the cylinder by the field produced by the coupling coils of the induction heater, additional factors which must be taken into account are the permeability μ, the mean resistivity ρ, the diameter d of the cylinder, and the operating frequency f of the induction heater.

The dimensional matrix for these variables is formed from

	P_d	H	μ	ρ	f	d
M	1	0	1	1	0	0
L	−1	−1	1	3	0	1
T	−3	−1	0	−1	−1	0
Q	0	1	−2	−2	0	0

(4-64)

The dimension θ has not been included because it does not appear in any of the variables. In this case, $n = 6$ and $r = 4$, so that $n - r = 2$ and there will be two π's. To check this, establish the rank of Eq. (4-64). Considering the last elements of the four columns,

$$\begin{vmatrix} 1 & 1 & 0 & 0 \\ 1 & 3 & 0 & 1 \\ 0 & -1 & -1 & 0 \\ -2 & -2 & 0 & 0 \end{vmatrix} = 4 \tag{4-65}$$

Since there is a nonzero fourth-order determinant, the rank is definitely 4. Writing the equations directly from the matrix,

$$\begin{aligned} \mu \quad + w + x \quad &= 0 \\ -\mu - v + w + 3x \quad + z &= 0 \\ -3\mu - v \quad - x - y \quad &= 0 \\ v - 2w - 2x \quad &= 0 \end{aligned} \tag{4-66}$$

from which

$$\begin{aligned} \mu &= -\frac{v}{2} \\ y &= w \\ x &= \frac{v}{2} - w \\ z &= -v + 2w \end{aligned} \tag{4-67}$$

Once again we have $n - r = 2$ variables to assign arbitrarily (v and w). For the first π, let $v = 1$ and $w = 0$. Then $\mu = -\frac{1}{2}$, $y = 0$, $x = \frac{1}{2}$, and $z = -1$, so that

$$\pi_1 = \frac{H\rho^{1/2}}{P_d^{1/2}d} \tag{4-68}$$

For the second π, set $v = 0$, and $w = 1$; then $\mu = 0$, $y = 1$, $x = -1$, and $z = 2$, so that

$$\pi_2 = \frac{\mu f d^2}{\rho} \tag{4-69}$$

π_1 and π_2 now represent a set of dimensionally correct relationships between the given variables. This does not mean, however, that the product $\pi_1\pi_2$ will give directly an expression which is in accord with the physical facts. In other words, the choices for v and w were arbitrary and not related to a working knowledge of induction heating. It happens that, actually, P_d varies as the square of H. This relationship can be achieved by squaring π_1 in Eq. (4-68). (Note that πs can be raised to a power *without changing their nondimensionality*.) The new value for π_1 is

$$\pi_1^2 = \frac{H^2\rho}{P_d d^2} \tag{4-70}$$

In addition, by experiment it is known that frequency must appear as

$f^{1/2}$ in the final expression. This can be arranged by taking the square root of π_2, which also does not alter its nondimensionality. The new value for π_2 is

$$\sqrt{\pi_2} = \frac{\mu^{1/2} f^{1/2} d}{\rho^{1/2}} \tag{4-71}$$

The final expression is the product of the revised πs and Eqs. (4-70) and (4-71), or

$$k\pi_1{}^2 \sqrt{\pi_2} = \frac{H^2 \rho}{P_d d^2} \frac{\mu^{1/2} f^{1/2} d}{\rho^{1/2}} \tag{4-72}$$

$$= \frac{H^2 \sqrt{\rho\mu f}}{P_d d}$$

Solving for P_d, an expression is obtained for the power density which is in accord with practical experience:[1]

$$P_d = \frac{kH^2 \sqrt{\rho\mu f}}{d}$$

The process of manipulating the πs used above is convenient to employ when the original assignments made for the first $n-r$ variables do not jibe with experimental facts. Manipulations performed with each π, such as squaring, etc., are perfectly acceptable provided that the nondimensionality of the π remains unaffected and the relationship of the variables within each π is unaltered.

In closing, it is worthwhile to bring to the attention of the reader some additional references which relate to the general subject of dimensional analysis by Hessler and Robb,[2] Henvald and McCann,[3] Moon and Spencer,[4] Nicholson,[5] Morton and Ryder[6] and Langhaar.[7]

Problems

Prob. 4-1 Analyze dimensionally the equation for the vertical displacement s of a falling object as a function of time t,

$$s = v_0 t + \tfrac{1}{2} g t^2$$

[1] For example J. P. Jordan, Design of Electronic Heaters for Induction Heating, *Proc. I.R.E.*, **32** (8), August, 1944.

[2] V. P. Hessler and D. D. Robb, Generalized Electrical Formulas, *Elec. Eng.*, **70** (4): 332–336, April, 1951.

[3] S. W. Henvald and G. D. McCann, Dimensionless Analysis of Servomechanisms by Electrical Analogy, *Elec. Eng.*, **65** (10), October, 1946.

[4] P. Moon and D. E. Spencer, A Modern Approach to Dimensions, *J. Franklin Inst.*, **251** (12), December, 1949.

[5] G. F. Nicholson, An Introduction to the Rationalized MKS System of Units, *Brit. J. Appl. Phys.*, **2**: 177–182, July, 1951.

[6] J. A. Morton and R. M. Ryder, Design Factors of the BTL Triode, *Bell System Tech. J.*, **29** (3): 496–530, October, 1950.

[7] H. L. Langhaar, "Dimensional Analysis and Theory of Models," John Wiley & Sons, Inc., New York, 1951.

where v_0 is the initial velocity of the object in a vertical direction, and g the acceleration due to gravity.

Prob. 4-2 Check dimensionally the following differential equation for the response of a series-parallel RC network to an applied pulse:

$$E = R_1R_2C_1\frac{d^2q}{dt^2} + \left(R_1 + R_2 + \frac{C_1}{C_2}R_2\right)\frac{dq}{dt} + \frac{q}{C_2}$$

Prob. 4-3 Two equations which were developed in the transient analysis of a network are

$$e + \frac{1}{RC}\int e\,dt = 2Ri_1 + \frac{q_C}{C} + L\frac{di_2}{dt}$$

$$e + LR\frac{de}{dt} = 2Ri_2 + \frac{q_C}{C} + L\frac{di_2}{dt}$$

Check each dimensionally and make any corrections necessary.

Prob. 4-4 The resonant frequency of an RLC circuit depends on all three variables in the circuit when $Q \ll 10$. The resonant frequency f has been derived as

$$f = \frac{1}{2\pi}\sqrt{\frac{1}{L^2C} - \frac{R^2}{4L^2}}$$

Check this formula dimensionally and make any necessary corrections.

Prob. 4-5* It has been shown that a combination of resistance and capacitance may be used to compensate for the internal inductance of an ammeter as shown in the circuit in Fig. 4-2. The expression for the net inductance of the network has been derived as

$$L_T = L - \frac{R^2C}{R^2\omega^2C + 1}$$

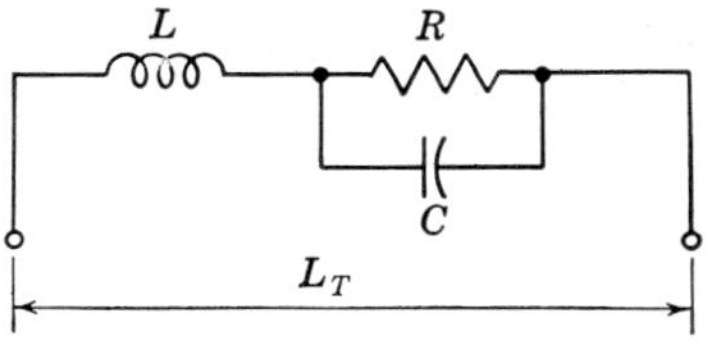

FIG. 4-2 Ammeter inductance compensator.

Check the expression dimensionally and make any corrections necessary.

Prob. 4-6† The absorption cross section of an antenna in square meters has been derived as

$$A_0 = \frac{h_e}{R_a}\sqrt{\frac{\mu}{\epsilon}}$$

where h_e is the effective height of the antenna for the particular angle of wave incidence, R_a is the input resistance of the antenna, and μ and ϵ

* J. M. Whittenton and C. A. Wilkinson, Frequency Compensated A. C. Ammeters and Voltmeters, *Elec. Eng.*, **65** (11), November, 1946.

† D. D. King, The Measurement and Interpretation of Antenna Scattering, *Proc. I.R.E.*, **37** (7), July, 1949.

are the permeability and permittivity of space, respectively. Check the expression and make any corrections necessary.

Prob. 4-7* The cathode-to-grid transit angle θ for a lighthouse triode tube designed to operate at ultrahigh frequencies is given as

$$\theta = \frac{126}{\lambda}\left(\frac{x}{I_0}\right)^{1/3}$$

where θ is in radians, λ is the wavelength in centimeters, I_0 is the cathode current density in amperes per square centimeter, and x is the cathode-to-grid distance in centimeters. What are the implicit dimensions of the factor 126 in this expression?

Prob. 4-8† For a parallel resonant circuit, there is an equivalent definition of Q given as

$$Q = \tfrac{1}{2} R\omega_0 \left(\frac{dB}{d\omega}\right)_{\omega=\omega_0}$$

where R is the resonant resistance of the circuit, $\omega_0 = 2\pi f_0$, f_0 is the resonant frequency of the circuit, and B is the circuit susceptance. Check the expression dimensionally.

Prob. 4-9‡ In the investigation of cosmic static an expression for the energy liberated when an electron encounters an ion under the proper circumstances is given as

$$\rho = \frac{32\pi^2 z^2 e^6}{3\sqrt{3}\, c^3 m^2 U}\, ns\, \Delta v$$

where ρ is expressed in ergs per cubic centimeter second; n and s are numbers of electrons and ions per cubic centimeter, respectively; and e, m, and U are the charge, mass, and the velocity of the electrons, respectively. Also, z is the atomic number, c is the velocity of light, and Δv is the band width. Check the expression dimensionally.

Prob. 4-10§ A broad-band method for the measurement of the power transmitted through a wave guide makes use of the forces exerted on the sides of the wave guide by the electromagnetic fields in the guide. The force per unit area (dynes per square meter) at the center of the bottom of a wave guide is given as

$$F_m = \frac{\beta S \times 10^5}{ba}$$

* N. T. Lavoo, Transadmittance and Input Conductance of a Lighthouse Triode at 3000 Mc., *Proc. I.R.E.*, **35** (11), November, 1947.

† W. W. Harman, R-Q Factor, *Proc. I.R.E.*, **37** (11): 1295, November, 1949.

‡ G. Reber, Cosmic Static, *Proc. I.R.E.*, **28** (2), February, 1940.

§ L. E. Norton, Broad-band Power-measuring Methods at Microwave Frequencies, *Proc. I.R.E.*, **37** (7), July, 1949.

where $\beta = \left[\left(\frac{\omega}{c}\right)^2 - \left(\frac{\pi}{b}\right)^2\right]^{1/2}$

$\omega = 2\pi f$

f = frequency

a = small wave-guide dimension

b = large wave-guide dimension

S = power transmitted through the guide

c = velocity of light

Check the expression dimensionally.

Prob. 4-11 Planck's equation for E, the energy per cubic centimeter associated with the wavelength interval $d(\lambda)$ is given as

$$E\,d(\lambda) = \frac{8\pi ch[d(\lambda)]}{\lambda^5 \epsilon^{ch/k\lambda T - 1}}$$

where c is the velocity of light, h is Planck's constant, k is the gas constant per molecule, λ is the wavelength, and T is the absolute temperature. Check the expression dimensionally.

Prob. 4-12 Develop an expression for the holding force exerted by an electromagnet in terms of the pole-piece area, flux density, and permeability.

Prob. 4-13 The energy stored in a capacitor is known to depend on the voltage V applied to the capacitor and on the capacitance C. Find the relationships between these quantities.

Prob. 4-14 The deflection yoke of a remote monitor kinescope in a studio television system is to be energized by coaxial cable. The inductance L and the resistance R of the yoke are known. It is planned to make the yoke appear purely resistive in order to match the cable by shunting it with a series RC combination, the R of which is to equal that of the yoke. Derive dimensionally the proper value of C in terms of the yoke inductance and resistance.

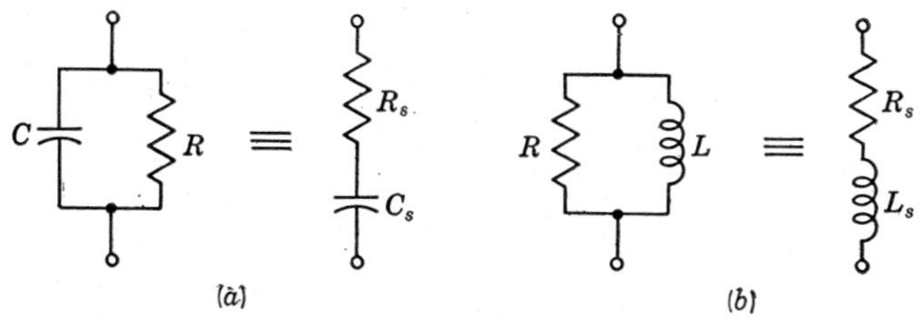

FIG. 4-3 Series-parallel equivalents.

Prob. 4-15 Every parallel RC or RL two-terminal network has a corresponding series RC or RL equivalent at a given frequency. Derive dimensionally the expressions for R_s, L_s, and C_s if these are the equivalent series elements for Fig. 4-3*a* and *b*, respectively. (Note: For $Q < 1$, $R = R_s$.)

Prob. 4-16 In the process of steel hardening by induction heating, it is necessary to determine the power per unit area required to produce a fixed rise in the surface temperature of a slab of metal. If the power required to raise the temperature of an infinite slab of metal $\Delta\theta°$ depends on the thermal conductivity of the metal, its density, its thermal capacity, and the time interval over which the power is applied, find the relationship between these quantities.

Prob. 4-17 The so-called "skin effect" is the diminishing of current density from the surface to the interior of a conductor carrying alternating current. The depth of penetration δ is defined as that depth at which the current density in a conductor has fallen to $1/\epsilon$, or approximately 37 per cent, of the surface current density. Find the relationship between δ, the resistivity and the permeability of the conductor, and the frequency of the alternating current.

Prob. 4-18* The fundamental frequency of the filament vibration encountered in the subminiature tubes which are used in the electronic control units of proximity-fused artillery shells and guided missiles is known to depend on the filament tension, the filament length, the filament density, and the flexural stiffness of the filament, which for the nickel or tungsten strands of 1 mil or less in diameter that are used for the purpose can be neglected. Derive the equation for the resonant frequency of these filaments dimensionally. From your knowledge of elementary physics, what is the probable value of the constant proportionality factor?

Prob. 4-19 An absolute voltmeter, originally designed by Lord Kelvin, consists of a balance, one pan of which is replaced by a circular metal electrode or plate equipped with an electrostatic guard ring. A second fixed plate is placed immediately below the balance electrode, and the force developed by a potential applied between the plates is then counterbalanced and measured by weights placed in the opposite pan of the balance. Determine the expression for the electric potential difference between the plates if the known factors which influence the experimental result are the distance between the plates, the mass added in the opposite balance pan, the permittivity of the medium in which the voltmeter is located, the area of the movable plate, and the gravitational constant at the balance location.

Prob. 4-20 A cylindrical cavity is drilled parallel to the axis of a long, cylindrical conductor of permeability μ but is displaced from its

* V. W. Cohen and A. Bloom, Microphonism in a Subminiature Triode, *Proc. I.R.E.*, **36** (8): 1039, August, 1948.

center at a distance a. A current which has a constant current density J flows axially through the remaining conductor. Find an expression for the magnetic flux density at the center of the cavity. (It is to be noted that, by experimentation, the proportionality factor for this problem is found to be equal to 0.5.)

Prob. 4-21 Under certain circumstances, the impedance looking into the grid circuit of a triode amplifier which has an inductive plate load will have a negative resistance component. If this negative resistance depends on the tube transconductance, the plate load inductance, the operating frequency, and the plate-to-grid interelectrode capacitance, determine the relationship between these variables.

NOTE: Frequency normally appears as angular frequency $2\pi f$ in mathematical relationships.

Prob. 4-22* Vacuum tubes operated at ultrahigh frequencies tend to have lower grid input impedances due in part to the lead inductance of the cathode structure, which is not negligible at these frequencies. Find the relationship between grid input admittance, tube transconductance, cathode lead inductance, and grid-to-cathode capacitance.

Prob. 4-23† An additional cause of the reduced input impedances of vacuum tubes at ultrahigh frequencies is a loading effect due to transit-time losses. Evolve a formula for the resistive loading due to the transit time of the electrons from cathode to anode, the tube transconductance, and the operating frequency.

Prob. 4-24 Form the dimensionless groupings which interrelate the air resistance of a wing F, the wing size L, the wing velocity V, the air density ρ, and the air viscosity μ, if the dimensional array is

	F	L	V	ρ	μ
M	1	0	0	1	1
L	1	1	1	−3	−1
T	−2	0	−1	0	−1

Prob. 4-25‡ Develop an expression for the peak current delivered by the output stage of a television transmitter in terms of the power output,

* M. J. O. Strutt and A. Van der Zeil, The Causes for the Increase of the Admittances of Modern High-frequency Amplifier Tubes on Short-waves, *Proc. I.R.E.*, **26** (8): 1011, August, 1938.

† W. R. Ferris, Input Resistance of Vacuum Tubes as Ultra-high-frequency Amplifiers, *Proc. I.R.E.*, **24** (1): 82, January, 1936.

‡ See E. Labin, Design of the Output Stage of a High Power Television Transmitter, *Elec. Commun.*, vol. 20, no. 3, 1942.

the total shunt capacitance of the stage, and the modulation band width. The proportionality factor in this case is 0.4.

Prob. 4-26 Find the relationship between the velocity of sound in a gas, the gas pressure, and the gas density, if the dimensional array is

	V	p	ρ
M	0	1	1
L	1	−1	−3
T	−1	−2	0

Prob. 4-27* In the analysis of mechanoelectronic transducers in which a voltage is developed by the motion of one or more of the control elements in an electron tube, it is found that the effective mass of a conical bar with its base as a fulcrum is

$$m = \int_0^L \pi r^2 \rho \left(\frac{x}{L}\right)^2 \frac{(L - x)^2}{2}\, dx$$

where ρ is the density of the bar, L is its length, r is the radius, and x also has the dimensions of length. Check this equation for dimensional correctness.

Prob. 4-28† The equivalent network of a reactance tube is given as

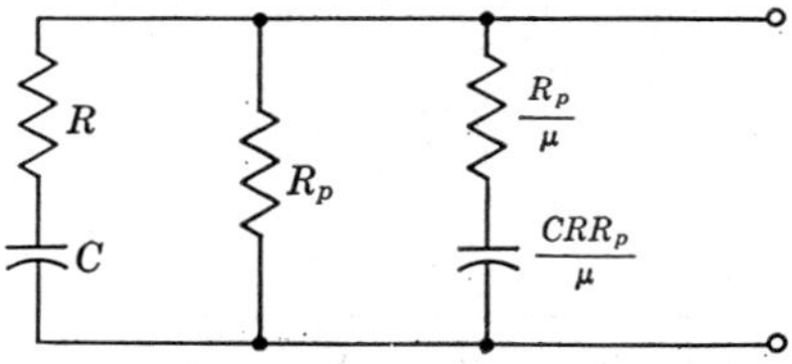

FIG. 4-4 Equivalent network of reactance tube.

shown in Fig. 4-4. Make a dimensional analysis of the values shown for the circuit elements and correct the diagram.

Prob. 4-29‡ Every cavity resonator, regardless of shape, has a series of resonant frequencies, infinite in number and more closely spaced as

* H. F. Olson, Mechano-electronic Transducers, *J. Acoust. Soc. Amer.*, **19** (2): 312, March, 1947.

† See F. W. Smith, A Note on Network Representation of Input and Output Admittances of Amplifiers, *Proc. I.R.E.*, **39** (4): 439, April, 1951. (An exponent of 2 has been omitted by the printer from one of the quantities given in this note.)

‡ I. G. Wilson, C. W. Schramm, and J. P. Kinzer, High *Q* Resonant Cavities for Microwave Testing, *Bell System Tech. J.*, **25** (3): 409, July, 1946.

the frequency increases. Relate N, the total number of these having a resonant frequency less than f, to V, the volume of the cavity in cubic meters, to c, the velocity of electromagnetic waves in the cavity dielectric in meters per second, and to f, the frequency in cycles per second. (The proportionality factor is $8\pi/3$.)

Analog computer used to simulate the flight of an airplane. (*Boeing Airplane Co.*)

PART 3

The Algebra of Circuit Analysis

Chapter 5

THEORY OF DETERMINANTS

5-1 Introduction

Network analysis is one of two great fields of mathematical application in the science of electronics, the other being electric- and magnetic-field theory. Most of the practical problems which confront the electronics engineer can be solved with the aid of equivalent networks. The analysis of the behavior of the electric currents in these networks inevitably requires the solution of linear simultaneous equations if the networks are at all complicated. These equations are linear because the network elements are presumed, for the sake of convenience, to behave linearly; they are differential equations if transient conditions are important and algebraic if steady-state conditions are to be calculated.

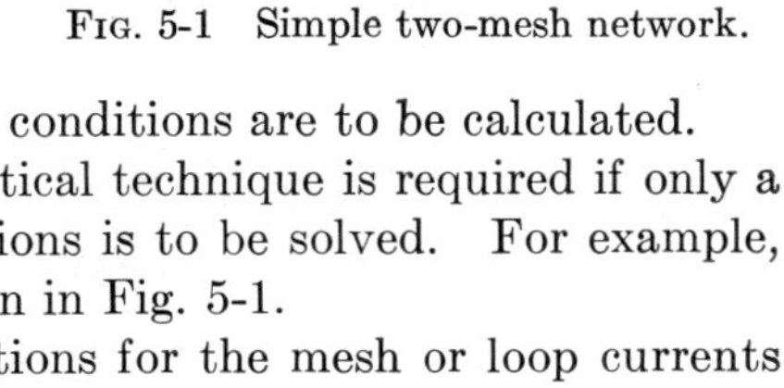

Fig. 5-1 Simple two-mesh network.

Very little in the way of mathematical technique is required if only a pair of simultaneous algebraic equations is to be solved. For example, consider the two-mesh network shown in Fig. 5-1.

By Kirchhoff's first law, the equations for the mesh or loop currents I_1 and I_2 are

$$16I_1 - 8I_2 = 4 \tag{5-1}$$
$$8I_1 - 20I_2 = -10 \tag{5-2}$$

These equations can be solved by substitution or by the elimination of a variable by addition and subtraction. By the use of the latter method,

$$\begin{aligned} 16I_1 - \ \ 8I_2 &= \ \ 4 \\ -16I_1 + 40I_2 &= 20 \\ \hline 32I_2 &= 24 \\ I_2 &= 3/4 \end{aligned} \tag{5-3}$$

from which it follows that $I_1 = 5/8$.

Solution by substitution in this simple two-mesh network is relatively uncomplicated. However, where a multimesh circuit, such as the one in Fig. 5-2, is to be analyzed by means of Kirchhoff's first law and Maxwell's cyclic current as shown, there will be three equations relating I_1, I_2, and I_3. In an n-mesh network, there will correspondingly be n equations. Some experience with solution of a multimesh network by the method of substitution or elimination of variables by addition and

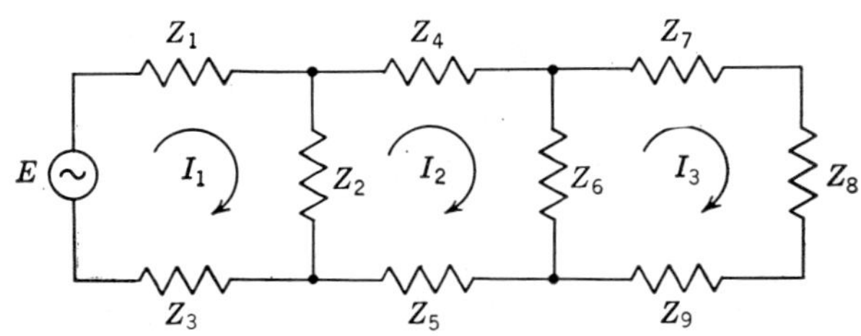

FIG. 5-2 Three-mesh network.

subtraction will indicate the inordinate amount of time required in such a process.

There exists, however, a more concise and less tedious mathematical method which can be applied to the solutions of simultaneous equations. Moreover, the mathematics introduced by this method has far-reaching applications, not only in network analysis, but in many other branches of engineering and applied mathematics not directly concerned with linear simultaneous algebraic equations.

5-2 Basic Determinant Concepts and Linear Algebraic Equations

Consider the simultaneous equations

$$a_{11}x + a_{12}y = c_1 \tag{5-4}$$
$$a_{21}x + a_{22}y = c_2 \tag{5-5}$$

where x and y are the quantities for which solutions are to be obtained in terms of the constants a_{11}, a_{12}, c_1, a_{21}, a_{22}, and c_2.

If Eq. (5-4) is multiplied by a_{22} and Eq. (5-5) by a_{12}, respectively, then

$$a_{22}a_{11}x + a_{22}a_{12}y = a_{22}c_1 \tag{5-6}$$
$$a_{12}a_{21}x + a_{12}a_{22}y = a_{12}c_2 \tag{5-7}$$

and if Eq. (5-7) is subtracted from (5-6), there results the equation

$$(a_{22}a_{11} - a_{12}a_{21})x = a_{22}c_1 - a_{12}c_2 \tag{5-8}$$

from which there is obtained

$$x = \frac{a_{22}c_1 - a_{12}c_2}{a_{11}a_{22} - a_{12}a_{21}} \tag{5-9}$$

In like manner, y may be obtained as

$$y = \frac{a_{11}c_2 - a_{21}c_1}{a_{11}a_{22} - a_{12}a_{21}} \tag{5-10}$$

The values of x and y as stated in relations (5-9) and (5-10), respectively, are unique [i.e., these are the only values which are capable of satisfying Eqs. (5-4) and (5-5)], provided $a_{11}a_{22} - a_{12}a_{21}$ is different from zero. The case where the number of variables exceeds the number of linear equations will be discussed in detail in Chap. 7, where matrix theory will be considered.

The solution of the simultaneous equations of (5-4) and (5-5) may also be written as

$$x = \frac{\begin{vmatrix} c_1 & a_{12} \\ c_2 & a_{22} \end{vmatrix}}{\begin{vmatrix} a_{11} & a_{12} \\ a_{21} & a_{22} \end{vmatrix}} \tag{5-11}$$

and

$$y = \frac{\begin{vmatrix} a_{11} & c_1 \\ a_{21} & c_2 \end{vmatrix}}{\begin{vmatrix} a_{11} & a_{12} \\ a_{21} & a_{22} \end{vmatrix}} \tag{5-12}$$

if

$$\begin{vmatrix} a_{11} & a_{12} \\ a_{21} & a_{22} \end{vmatrix} \neq 0$$

where the following relations hold:

$$\begin{vmatrix} c_1 & a_{12} \\ c_2 & a_{22} \end{vmatrix} = c_1a_{22} - a_{12}c_2 \qquad \begin{vmatrix} a_{11} & c_1 \\ a_{21} & c_2 \end{vmatrix} = a_{11}c_2 - c_1a_{21}$$

and

$$\begin{vmatrix} a_{11} & a_{12} \\ a_{21} & a_{22} \end{vmatrix} = a_{11}a_{22} - a_{12}a_{21}$$

Consider

$$3x + 2y = 4 \tag{5-13}$$
$$8x - 5y = 21 \tag{5-14}$$

By the method above,

$$x = \frac{\begin{vmatrix} 4 & 2 \\ 21 & -5 \end{vmatrix}}{\begin{vmatrix} 3 & 2 \\ 8 & -5 \end{vmatrix}} = \frac{(4)(-5) - (2)(21)}{(3)(-5) - (2)(8)} = \frac{-62}{-31} = 2$$

$$y = \frac{\begin{vmatrix} 3 & 4 \\ 8 & 21 \end{vmatrix}}{\begin{vmatrix} 3 & 2 \\ 8 & -5 \end{vmatrix}} = \frac{(3)(21) - (4)(8)}{(3)(-5) - (2)(8)} = \frac{31}{-31} = -1$$

Hence the solutions to (5-13) and (5-14) are $x = 2$ and $y = -1$.

Similarly, the linear simultaneous equations in x, y, and z

$$a_{11}x + a_{12}y + a_{13}z = c_1 \tag{5-15}$$

$$a_{21}x + a_{22}y + a_{23}z = c_2 \tag{5-16}$$

$$a_{31}x + a_{32}y + a_{33}z = c_3 \tag{5-17}$$

have a unique solution expressible as

$$x = \frac{\begin{vmatrix} c_1 & a_{12} & a_{13} \\ c_2 & a_{22} & a_{23} \\ c_3 & a_{32} & a_{33} \end{vmatrix}}{\begin{vmatrix} a_{11} & a_{12} & a_{13} \\ a_{21} & a_{22} & a_{23} \\ a_{31} & a_{32} & a_{33} \end{vmatrix}} \tag{5-18}$$

$$y = \frac{\begin{vmatrix} a_{11} & c_1 & a_{13} \\ a_{21} & c_2 & a_{23} \\ a_{31} & c_3 & a_{33} \end{vmatrix}}{\begin{vmatrix} a_{11} & a_{12} & a_{13} \\ a_{21} & a_{22} & a_{23} \\ a_{31} & a_{32} & a_{33} \end{vmatrix}} \tag{5-19}$$

$$z = \frac{\begin{vmatrix} a_{11} & a_{12} & c_1 \\ a_{21} & a_{22} & c_2 \\ a_{31} & a_{32} & c_3 \end{vmatrix}}{\begin{vmatrix} a_{11} & a_{12} & a_{13} \\ a_{21} & a_{22} & a_{23} \\ a_{31} & a_{32} & a_{33} \end{vmatrix}} \tag{5-20}$$

provided that the denominator

$$\begin{vmatrix} a_{11} & a_{12} & a_{13} \\ a_{21} & a_{22} & a_{23} \\ a_{31} & a_{32} & a_{33} \end{vmatrix} \tag{5-21}$$

is not equal to zero.

It is important at this point to define for the reader the word *determinant* and to discuss the properties of the determinant symbol. The symbol

$$\begin{vmatrix} a_{11} & a_{12} \\ a_{21} & a_{22} \end{vmatrix} \tag{5-22}$$

is called a *symbol of the determinant* (of second order). When it is expanded, it becomes

$$a_{11}a_{22} - a_{12}a_{21} \tag{5-23}$$

which is known as the *determinant.* When Eq. (5-23) is evaluated, the value so obtained is known as the *value of the determinant.*

5-3 Determinants of Second and Third Orders

The symbol of a second-order determinant from Sec. 5-2 may be illustrated by

$$\begin{vmatrix} 5 & 9 \\ -3 & 7 \end{vmatrix} \tag{5-24}$$

The *determinant* is then $(5)(7) - (9)(-3)$, from the previous section, and its *value* is 62. The method of evaluation is to multiply the elements in the diagonal starting with 5, and subtract the product of the elements starting with 9. Using arrows to illustrate,

$$\begin{vmatrix} 5 & 9 \\ -3 & 7 \end{vmatrix} = (5)(7) - (9)(-3) = 35 + 27 = 62 \tag{5-25}$$

Every determinant symbol consists of n rows and n columns, or n^2 elements, arranged in a *square array.* The positive integral value of n is known as the *order* of the determinant. It is now obvious why (5-22) represented a second-order determinant symbol.

Before consideration is given to general theorems for the simplification and the evaluation of determinants of any order, it will be the immediate purpose of this section to show how to evaluate third-order determinants.

Consider the symbol of the third-order determinant,

$$\begin{vmatrix} a_{11} & a_{12} & a_{13} \\ a_{21} & a_{22} & a_{23} \\ a_{31} & a_{32} & a_{33} \end{vmatrix} \tag{5-26}$$

which can be expressed in determinant form by the sums and differences of certain terms made up of products of three element factors—(see dotted lines).

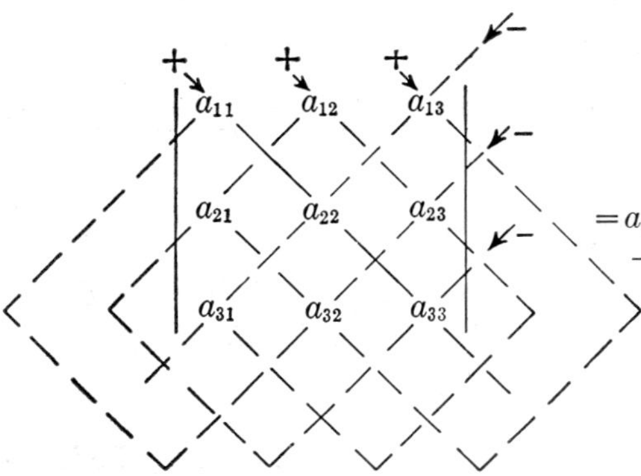

$$= a_{11}a_{22}a_{33} + a_{12}a_{23}a_{31} + a_{13}a_{32}a_{21} - a_{13}a_{22}a_{31} - a_{23}a_{32}a_{11} - a_{33}a_{21}a_{12} \tag{5-27}$$

The following example of the evaluation of a third-order determinant makes this process clearer.

$$\begin{aligned} \begin{vmatrix} 3 & 5 & 1 \\ 2 & -2 & 2 \\ 1 & 9 & 5 \end{vmatrix} &= (3)(-2)(5) + (5)(2)(1) + (1)(9)(2) \\ &\quad - (1)(-2)(1) - (2)(9)(3) - (5)(2)(5) \\ &= -30 + 10 + 18 + 2 - 54 - 50 \\ &= -104 \end{aligned} \tag{5-28}$$

5-4 Number of Terms in a Determinant

It will be observed that the second-order determinant (i.e., a determinant symbol of 2^2 elements) has two terms; a third-order determinant (i.e., a determinant symbol of 3^2 elements) has six terms. It will be seen that a determinant symbol of nth order will have $n!$ terms in its corresponding determinant.

5-5 Simplification Theorems for Determinants

Theorem 5-5A: *If, in a determinant symbol, the rows are made into columns so that the first row will become the first column, the second row becomes the second column, etc., and correspondingly the first column becomes the first row, the second column becomes the second row, etc., then the value of the determinant remains unchanged.*

EXAMPLE. Consider as an example of this theorem,

$$\begin{vmatrix} 3 & 5 & 7 \\ 2 & 4 & 6 \\ 1 & 3 & 5 \end{vmatrix} \tag{5-29}$$

and

$$\begin{vmatrix} 3 & 2 & 1 \\ 5 & 4 & 3 \\ 7 & 6 & 5 \end{vmatrix} \tag{5-30}$$

The value of (5-29) is

$$(3)(4)(5) + (5)(6)(1) + (7)(3)(2) - (7)(4)(1) - (6)(3)(3) - (5)(2)(5) \tag{5-31}$$

The value of (5-30) becomes

$$(3)(4)(5) + (2)(3)(7) + (1)(6)(5) - (1)(4)(7) - (3)(6)(3) - (5)(5)(2) \tag{5-32}$$

A comparison of the terms in (5-31) and (5-32) will show that the values of these expressions must be identical.

Theorem 5-5B: *If two rows (or columns) of a determinant symbol are interchanged, then the resulting determinant symbol will when expanded have the same numerical magnitude, but will require a negative sign prefixed to it.*

EXAMPLE

$$D_1 = \begin{vmatrix} 3 & 1 & 0 \\ 2 & 2 & 1 \\ 1 & 3 & 5 \end{vmatrix} \tag{5-33}$$

$$D_2 = \begin{vmatrix} 1 & 3 & 0 \\ 2 & 2 & 1 \\ 3 & 1 & 5 \end{vmatrix} \tag{5-34}$$

$$D_1 = (3)(2)(5) + (1)(1)(1) + (0)(3)(2) - (0)(2)(1) - (1)(3)(3) - (5)(2)(1) = 30 + 1 - 9 - 10 = 12$$

$$D_2 = (1)(2)(5) + (3)(1)(3) + (0)(1)(2) - (0)(2)(3) - (1)(1)(1) - (5)(2)(3) = 10 + 9 - 1 - 30 = -12$$

Hence, $D_1 = -D_2$

Theorem 5-5C: *If two rows (or columns) are alike, the determinant equals zero.*

EXAMPLE. Consider a special case, i.e., a third-order determinant.

$$\begin{vmatrix} a_{11} & a_{12} & a_{12} \\ a_{21} & a_{22} & a_{22} \\ a_{31} & a_{32} & a_{32} \end{vmatrix} \tag{5-35}$$

Suppose column 2 and column 3 are interchanged; then by Theorem 5-5B it immediately follows that the new determinant is the negative of the original determinant, or $-D = D$, or $D = 0$. The same reasoning applies to adjacent identical rows. If the rows are not adjacent, then by Theorem 5-5B they may be made adjacent. The same process may be applied to a determinant of order n.

Theorem 5-5D: *If each element in a column (or row) of a determinant symbol is multiplied by the same factor, the result is to multiply the determinant by the same multiplying factor.*

EXAMPLE. Consider the determinant symbol

$$D_3 = \begin{vmatrix} 3 & 1 & 6 \\ 2 & 2 & -3 \\ 5 & 9 & 12 \end{vmatrix} \tag{5-36}$$

Inspection of (5-36) will reveal that the third column contains the factor 3. In accord with Theorem 5-5D, it follows that (5-36) can be written

$$D_4 = 3\begin{vmatrix} 3 & 1 & 2 \\ 2 & 2 & -1 \\ 5 & 9 & 4 \end{vmatrix} \tag{5-37}$$

Suppose (5-36) and (5-37) are then evaluated:

$$D_3 = (3)(2)(12) + (1)(-3)(5) + (6)(9)(2) - (6)(2)(5) - (-3)(9)(3) - (12)(2)(1)$$

$$= 162$$

$$D_4 = 3[(3)(2)(4) + (1)(-1)(5) + (2)(9)(2) - (2)(2)(5) - (-1)(9)(3) - (4)(2)(1)]$$

$$= 3[54] = 162$$

5-6 Minors of a Determinant

If there is removed from a determinant the row and column which contain a particular element, there remains a determinant which is designated as the *minor* of that specific element.

In the symbol, for example,

$$\begin{vmatrix} a_{11} & a_{12} & a_{13} \\ a_{21} & a_{22} & a_{23} \\ a_{31} & a_{32} & a_{33} \end{vmatrix} \tag{5-38}$$

the minor of a_{13} is

$$\begin{vmatrix} a_{21} & a_{22} \\ a_{31} & a_{32} \end{vmatrix} \tag{5-39}$$

Theorem 5-6A: *Any determinant symbol may be evaluated by taking the elements of its first column, forming the product of each and its respective minor, and, starting with the first element times its minor, prefixing alternate positive and negative algebraic signs.*

It can also be shown that a determinant symbol is expandable by the method of minors using any row or column for cofactors. A consideration of signs required for expansion by any row or column will yield the result that the signs are arranged in checkerboard pattern in the square array thus:

$$\begin{matrix} + & - & + & - & + & - \\ - & + & - & + & - & + \\ + & - & + & - & + & - \\ - & + & - & + & - & + \\ + & - & + & - & + & - \\ - & + & - & + & - & + \end{matrix}$$

Expansion by the third row of (5-38) would yield $a_{31}B_{31} - a_{32}B_{32} + a_{33}B_{33}$.

The reader will notice that, if the sum of the subscripts of the cofactors is odd, the sign is negative, and, if even, the sign is positive. It also is obvious that if all of the elements of any one of the rows or columns are zero, the determinant when evaluated will be equal to zero.

EXAMPLE. Consider as an example, once again,

$$D_3 = \begin{vmatrix} 3 & 1 & 6 \\ 2 & 2 & -3 \\ 5 & 9 & 12 \end{vmatrix} \tag{5-40}$$

By the method of minors, using the second column for cofactors,

$$\begin{aligned} D_3 &= -(1)\begin{vmatrix} 2 & -3 \\ 5 & 12 \end{vmatrix} + 2\begin{vmatrix} 3 & 6 \\ 5 & 12 \end{vmatrix} - 9\begin{vmatrix} 3 & 6 \\ 2 & -3 \end{vmatrix} \\ &= -[(2)(12) - (-3)(5)] + 2[(3)(12) - (6)(5)] \\ &\qquad - 9[(3)(-3) - (6)(2)] \\ &= -(24 + 15) + 2(36 - 30) - 9(-9 - 12) \\ &= 162 \end{aligned}$$

If (5-40) is again expanded by the method of minors but this time the cofactors are chosen to be the elements appearing in the first column, then

$$\begin{aligned} D_3 &= 3\begin{vmatrix} 2 & -3 \\ 9 & 12 \end{vmatrix} - 2\begin{vmatrix} 1 & 6 \\ 9 & 12 \end{vmatrix} + 5\begin{vmatrix} 1 & 6 \\ 2 & -3 \end{vmatrix} \\ &= 3[(2)(12) - (-3)(9)] - 2[(1)(12) - (6)(9)] + 5[(1)(-3) - (6)(2)] \\ &= 3[24 + 27] - 2[12 - 54] + 5[-3 - 12] \\ &= 153 + 84 - 75 \\ &= 162 \end{aligned}$$

It should be noted that the use of the minors of a determinant to simplify a high-order determinant is essential, because the diagonal method of solution is correct only for second- and third-order determinants. An alternate scheme to the method of minors which is preferred by some for the reduction of high-order determinants was developed by Gauss and is described as the pivotal-condensation or the pivotal-element method. For determinants of order greater than 6, the Gauss-Seidel iterative method is also used.[1]

5-7 Addition of Determinants: Other Theorems

The algebra of determinants involves the usual processes of addition, subtraction, multiplication, and division. In this section, the addition and subtraction of determinants are discussed. The mechanics of multiplication of determinants and the process of division by locating the inverse of a determinant and then the completion of the operation by multiplication will be shown in Chap. 7, when matrices and their determinants are presented.

Theorem 5-7A: *If any row (or column) has all its elements equal to zero, then it will follow that the evaluated determinant will have a value of zero.*

Example. Suppose the determinant symbol is

$$D = \begin{vmatrix} 3 & 5 & 0 \\ 2 & 4 & 0 \\ 9 & 2 & 0 \end{vmatrix} \tag{5-41}$$

Then by Theorems 5-6A and 5-7A it follows that

$$D = -0 \begin{vmatrix} 2 & 4 \\ 9 & 2 \end{vmatrix} + 0 \begin{vmatrix} 3 & 5 \\ 9 & 2 \end{vmatrix} - 0 \begin{vmatrix} 3 & 5 \\ 2 & 4 \end{vmatrix}$$

or

$$D = 0$$

Theorem 5-7B: *If D_1 and D_2 are two determinant symbols where all the rows and columns are identical with the exception of one row (or column), the sum $D_s = D_1 + D_2$, where D_s, D_1, and D_2 are symbolized as*

$$D_1 = \begin{vmatrix} a_{11} & a_{12} & \cdots & a_{1n} \\ a_{21} & a_{22} & \cdots & a_{2n} \\ a_{31} & a_{32} & \cdots & a_{3n} \\ \cdot & \cdot & & \cdot \\ \cdot & \cdot & & \cdot \\ \cdot & \cdot & & \cdot \\ a_{n1} & a_{n2} & & a_{nn} \end{vmatrix} \tag{5-42}$$

[1] See M. G. Salvadori and K. S. Miller, "The Mathematical Solution of Engineering Problems," chap. 4, Columbia University Press, New York, 1953.

$$D_2 = \begin{vmatrix} b_{11} & a_{12} & \cdots & a_{1n} \\ b_{21} & a_{22} & \cdots & a_{2n} \\ b_{31} & a_{32} & \cdots & a_{3n} \\ \cdot & \cdot & & \cdot \\ \cdot & \cdot & & \cdot \\ \cdot & \cdot & & \cdot \\ b_{n1} & a_{n2} & \cdots & a_{nn} \end{vmatrix} \tag{5-43}$$

$$D_s = \begin{vmatrix} (a_{11} + b_{11}) & a_{12} & \cdots & a_{1n} \\ (a_{21} + b_{22}) & a_{22} & \cdots & a_{2n} \\ (a_{31} + b_{31}) & a_{32} & \cdots & a_{3n} \\ \cdot & \cdot & & \cdot \\ \cdot & \cdot & & \cdot \\ \cdot & \cdot & & \cdot \\ (a_{n1} + b_{n1}) & a_{n2} & \cdots & a_{nn} \end{vmatrix} \tag{5-44}$$

EXAMPLE. Consider the sum of

$$\begin{vmatrix} 5 & 9 \\ 2 & 3 \end{vmatrix} \quad \text{and} \quad \begin{vmatrix} 8 & 9 \\ 9 & 3 \end{vmatrix}$$

$$D_s = \begin{vmatrix} (5 + 8) & 9 \\ (2 + 9) & 3 \end{vmatrix} = \begin{vmatrix} 13 & 9 \\ 11 & 3 \end{vmatrix}$$

Theorem 5-7C: If any row (or column) is multiplied by a constant and then added to the corresponding elements of any other row (or column), the value of the original determinant will remain unchanged.

$$\begin{vmatrix} a_{11} & a_{12} & \cdots & a_{1\mu} & \cdots & a_{1\nu} & \cdots & a_{1n} \\ a_{21} & a_{22} & \cdots & a_{2\mu} & \cdots & a_{2\nu} & \cdots & a_{2n} \\ a_{31} & a_{32} & \cdots & a_{3\mu} & \cdots & a_{3\nu} & \cdots & a_{3n} \\ \cdot & \cdot & & \cdot & & \cdot & & \cdot \\ \cdot & \cdot & & \cdot & & \cdot & & \cdot \\ \cdot & \cdot & & \cdot & & \cdot & & \cdot \\ a_{n1} & a_{n2} & \cdots & a_{n\mu} & \cdots & a_{n\nu} & \cdots & a_{nn} \end{vmatrix}$$

$$= \begin{vmatrix} a_{11} & a_{12} & \cdots & a_{1\mu} + \kappa a_{1\nu} & \cdots & a_{1n} \\ a_{21} & a_{22} & \cdots & a_{2\mu} + \kappa a_{2\nu} & \cdots & a_{2n} \\ a_{31} & a_{32} & \cdots & a_{3\mu} + \kappa a_{3\nu} & \cdots & a_{3n} \\ \cdot & \cdot & & \cdot & & \cdot \\ \cdot & \cdot & & \cdot & & \cdot \\ \cdot & \cdot & & \cdot & & \cdot \\ a_{n1} & a_{n2} & \cdots & a_{n\mu} + \kappa a_{n\nu} & \cdots & a_{nn} \end{vmatrix} \tag{5-45}$$

where $\mu, \nu = 1, 2, 3, 4, \cdots, n$ for $\mu \neq \nu$

EXAMPLE 2.

$$\begin{vmatrix} 3 & 5 & 1 \\ 2 & 3 & 6 \\ 7 & 8 & 9 \end{vmatrix} = 52$$

If there is added to the first column the corresponding elements of the third column multiplied by -3, there is obtained

$$\begin{vmatrix} 0 & 5 & 1 \\ -16 & 3 & 6 \\ -20 & 8 & 9 \end{vmatrix} = 52$$

5-8 Cramer's Rule

Determinants are normally employed in network analysis in connection with the method developed by Gabriel Cramer for the rapid solution of simultaneous equations.

Consider a set of n linear equations in n unknowns.

$$\begin{array}{ccccccccc} a_{11}x_1 & + & a_{12}x_2 & + & a_{13}x_3 & + \cdots + & a_{1n}x_n & = & k_1 \\ a_{21}x_1 & + & a_{22}x_2 & + & a_{23}x_3 & + \cdots + & a_{2n}x_n & = & k_2 \\ \cdot & & \cdot & & \cdot & & \cdot & & \cdot \\ \cdot & & \cdot & & \cdot & & \cdot & & \cdot \\ \cdot & & \cdot & & \cdot & & \cdot & & \cdot \\ a_{n1}x_1 & + & a_{n2}x_2 & + & a_{n3}x_3 & + \cdots + & a_{nn}x_n & = & k_n \end{array} \tag{5-46}$$

where $x_1, x_2, x_3, \ldots, x_n$ are the unknowns, and $a_{11}, a_{12}, \ldots, a_{nn}$, $k_1, k_2, \ldots, k_n$ are constants.

Suppose D represents the symbol (indicated below) formed by using as elements the coefficients of the unknowns as they appear in the above equations.

$$D = \begin{vmatrix} a_{11} & a_{12} & a_{13} & \cdots & a_{1n} \\ a_{21} & a_{22} & a_{23} & \cdots & a_{2n} \\ \cdot & \cdot & \cdot & & \cdot \\ \cdot & \cdot & \cdot & & \cdot \\ \cdot & \cdot & \cdot & & \cdot \\ a_{n1} & a_{n2} & a_{n3} & \cdots & a_{nn} \end{vmatrix} \tag{5-47}$$

Let D_1 be the following:

$$D_1 = \begin{vmatrix} k_1 & a_{12} & a_{13} & \cdots & a_{1n} \\ k_2 & a_{22} & a_{23} & \cdots & a_{2n} \\ \cdot & \cdot & \cdot & & \cdot \\ \cdot & \cdot & \cdot & & \cdot \\ \cdot & \cdot & \cdot & & \cdot \\ k_n & a_{n2} & a_{n3} & \cdots & a_{nn} \end{vmatrix} \tag{5-48}$$

which inspection will show is identical with D except for the elements in the first column, which have been replaced by the constants which appear on the right-hand side of the equations. Likewise

$$D_2 = \begin{vmatrix} a_{11} & k_1 & a_{13} & \cdots & a_{1n} \\ a_{21} & k_2 & a_{23} & \cdots & a_{2n} \\ \cdot & \cdot & \cdot & & \cdot \\ \cdot & \cdot & \cdot & & \cdot \\ \cdot & \cdot & \cdot & & \cdot \\ a_{n1} & k_n & a_{n3} & \cdots & a_{nn} \end{vmatrix} \tag{5-49}$$

where D_2 is identical with D except for the second column, where the elements are replaced by the constants which appear on the right-hand side of the equations. Similarly, $D_3, D_4, \ldots, D_n$ can be formed. Cramer's rule can then be stated as follows:

Theorem 5-8A: *In a set of n linear algebraic equations such as expressed by Eq.* (5-46), *if* $D \neq 0$, *then*

$$x_1 = \frac{D_1}{D} \qquad x_2 = \frac{D_2}{D} \qquad x_3 = \frac{D_3}{D} \qquad x_4 = \frac{D_4}{D} \qquad \cdots \qquad x_n = \frac{D_n}{D}$$

Example. Solve by Cramer's rule

$$\begin{aligned} 3x + 2y + 4z &= 15 \\ 2x + y - 2z &= 0 \\ 6x - 7y - z &= -10 \end{aligned}$$

$$x = \frac{\begin{vmatrix} 15 & 2 & 4 \\ 0 & 1 & -2 \\ -10 & -7 & -1 \end{vmatrix}}{\begin{vmatrix} 3 & 2 & 4 \\ 2 & 1 & -2 \\ 6 & -7 & -1 \end{vmatrix}} \qquad y = \frac{\begin{vmatrix} 3 & 15 & 4 \\ 2 & 0 & -2 \\ 6 & -10 & -1 \end{vmatrix}}{\begin{vmatrix} 3 & 2 & 4 \\ 2 & 1 & -2 \\ 6 & -7 & -1 \end{vmatrix}}$$

$$z = \frac{\begin{vmatrix} 3 & 2 & 15 \\ 2 & 1 & 0 \\ 6 & -7 & -10 \end{vmatrix}}{\begin{vmatrix} 3 & 2 & 4 \\ 2 & 1 & -2 \\ 6 & -7 & -1 \end{vmatrix}}$$

5-9 Non-network Applications of Determinants

It is often necessary to determine the interelectrode capacitances of an electron tube. Unfortunately, the calculation of these capacitances on a theoretical basis presents a problem which is too involved for an immediate solution.

Consider the equivalent circuit of the interelectrode capacitances of a pentode shown in Fig. 5-3*a* and *b*. The symbol p is used to designate the plate, k to designate the cathode, and g_1, g_2, and g_3, the control, screen, and suppressor grids, respectively. Because the suppressor is normally tied directly to the cathode, and the screen grid is heavily bypassed to the cathode, these three elements have been indicated as being directly connected in the equivalent network in Fig. 5-3*b*.

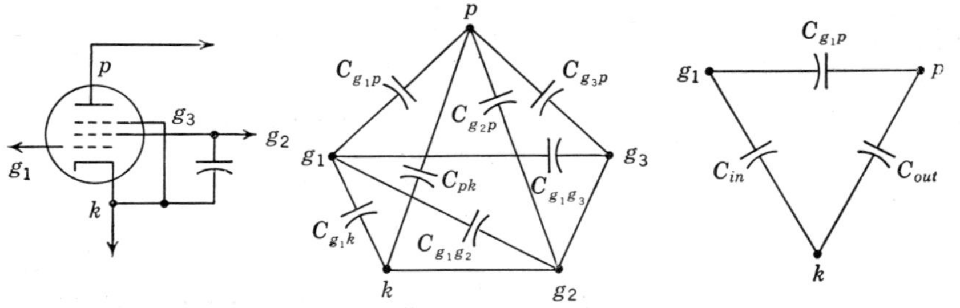

(a) (b) (c)

FIG. 5-3 Interelectrode capacitances of a pentode.

Since the capacitances from g_1 to g_3, g_2, and k and from p to g_3, g_2, and k are essentially in parallel, the equivalent network can be simplified somewhat to that shown in Fig. 5-3*c*, where

$$C_{\text{in}} = C_{g_1k} + C_{g_1g_2} + C_{g_1g_3}$$

and $C_{\text{out}} = C_{pk} + C_{g_2p} + C_{g_3p}$.

Assume that the following measurements are made with a capacitance bridge on a type 6AH6 pentode with no external shield:

(*a*) With g_1 and p connected together, the capacitance between this combination and the cathode, screen, and suppressor grids tied together is 12.00 $\mu\mu$f.

(*b*) With g_1, k, g_2, and g_3 connected together, the capacitance between this combination and the plate is 2.03 $\mu\mu$f.

(*c*) With p, k, g_2, and g_3 connected together, the capacitance between this combination and the control grid is 10.03 $\mu\mu$f.

The resulting simultaneous equations may be written as follows:

$$\begin{aligned} 12.00 &= \qquad\quad C_{\text{in}} + C_{\text{out}} \\ 2.03 &= C_{g_1p} \qquad\quad + C_{\text{out}} \\ 10.03 &= C_{g_1p} + C_{\text{in}} \end{aligned} \tag{5-50}$$

By Cramer's rule for the solution of simultaneous equations by determinant method,

$$C_{g_1p} = \frac{\begin{vmatrix} 12.00 & 1 & 1 \\ 2.03 & 0 & 1 \\ 10.03 & 1 & 0 \end{vmatrix}}{\begin{vmatrix} 0 & 1 & 1 \\ 1 & 0 & 1 \\ 1 & 1 & 0 \end{vmatrix}} \tag{5-51}$$

$$= \frac{12.06 - 12.00}{2} \tag{5-52}$$

$$= 0.03\ \mu\mu\text{f} \tag{5-53}$$

$$C_{\text{in}} = \frac{\begin{vmatrix} 0 & 12.00 & 1 \\ 1 & 2.03 & 1 \\ 1 & 10.03 & 0 \end{vmatrix}}{\begin{vmatrix} 0 & 1 & 1 \\ 1 & 0 & 1 \\ 1 & 1 & 0 \end{vmatrix}} \tag{5-54}$$

$$= \frac{22.03 - 2.03}{2} \tag{5-55}$$

$$= 10.00\ \mu\mu\text{f} \tag{5-56}$$

$$C_{\text{out}} = \frac{\begin{vmatrix} 0 & 1 & 12.00 \\ 1 & 0 & 2.03 \\ 1 & 1 & 10.03 \end{vmatrix}}{2} \tag{5-57}$$

$$= \frac{14.03 - 10.03}{2} \tag{5-58}$$

$$= 2.00\ \mu\mu\text{f} \tag{5-59}$$

Another newly developed field of determinant application is the analysis of the operation of color-television matrixing equipment and the transformation of color coordinates in color systems.[1]

Problems

Prob. 5-1 Evaluate

$$\begin{vmatrix} 3 & 2 & -1 \\ 7 & 8 & 3 \\ 2 & 1 & -1 \end{vmatrix}$$

Prob. 5-2 Evaluate

$$\begin{vmatrix} 5 & 2 & -3 \\ 7 & 4 & -3 \\ 9 & 6 & -3 \end{vmatrix}$$

[1] W. T. Wintringham, Color Television and Colorimetry, *Proc. I.R.E.*, **39** (10), October, 1951.

Prob. 5-3 Evaluate

$$\begin{vmatrix} 13 & 16 & 19 \\ 14 & 17 & 20 \\ 15 & 18 & 21 \end{vmatrix}$$

Prob. 5-4 Solve for x, y, and z:

$$\begin{aligned} x + y \phantom{{}+ z} &= 5 \\ y + z &= 6 \\ x \phantom{{}+y} - 5z &= -9 \end{aligned}$$

Prob. 5-5 Solve for x, y, and z:

$$\begin{aligned} 5x - 4y + 6z &= 11 \\ 4x - 2y + 8z &= 14 \\ 3x + 2y - 2z &= 17 \end{aligned}$$

Prob. 5-6 Solve for a, b, c, and d:

$$\begin{aligned} 4a + 6b + 12c - 3d &= 1 \\ 2a - 10b + 6c + 5d &= 5 \\ 4a + 6b + 12c - 3d &= 3 \\ 3a - 12b + 9c + 6d &= 4 \end{aligned}$$

Prob. 5-7 Solve for λ:

$$\begin{vmatrix} (15 - 2\lambda) & 11 & -20 \\ (11 - 3\lambda) & 17 & -32 \\ (7 - \lambda) & 14 & 39 \end{vmatrix} = 0$$

Prob. 5-8 Prove that the area A of a plane triangle with vertices at $P\ (x_1,y_1)$, $Q\ (y_2,y_2)$, and $R\ (x_3,y_3)$ is given by

$$A = \tfrac{1}{2} \begin{vmatrix} x_1 & y_1 & 1 \\ x_2 & y_2 & 1 \\ x_3 & y_3 & 1 \end{vmatrix}$$

Prob. 5-9 By the application of the relation of Prob. 5-8, show that the points $P\ (x_1,y_1)$, $Q\ (x_2,y_2)$, and $R\ (x_3,y_3)$ are collinear when

$$\begin{vmatrix} x_1 & y_1 & 1 \\ x_2 & y_2 & 1 \\ x_3 & y_3 & 1 \end{vmatrix} = 0$$

Prob. 5-10 If λ is one of the complex cube roots of unity, evaluate

$$\Delta = \begin{vmatrix} 1 & \lambda^3 & \lambda^2 \\ \lambda^3 & 1 & \lambda \\ \lambda^2 & \lambda & 1 \end{vmatrix}$$

Suggested solution:

$$\begin{aligned}\Delta &= 1 + (\lambda^2)(1) + \lambda^2(\lambda)(1) - \lambda^2(1)(\lambda^2) - \lambda(\lambda)(1) - (1)(1)(1)\\ &= 1 + \lambda^3 + \lambda^3 - \lambda^4 - \lambda^2 - 1\end{aligned}$$

But $\lambda^3 = 1$, since λ is a cube root of unity; since $\lambda^3 = 1$, it follows that

$$\lambda^3 - 1 = 0 \qquad \text{or} \qquad (\lambda - 1)(\lambda^2 + \lambda + 1) = 0$$

Hence either

$$\lambda = 0 \qquad \text{or} \qquad \lambda^2 + \lambda + 1 = 0$$

Since $\lambda \neq 0$ under the hypothesis, it is obvious that $\lambda^2 + \lambda + 1 = 0$. Since $\Delta = 1 + 1 + 1 - \lambda^4 - \lambda^2 - 1$,

$$\Delta = 2 - \lambda^2 - \lambda^3(\lambda)$$

But $\lambda^3 = 1$. Hence

$$\begin{aligned}\Delta &= 2 - \lambda^2 - \lambda = 2 - \lambda^2 - \lambda - 1 + 1\\ &= 3 - (\lambda^2 + \lambda + 1)\\ &= 3 \qquad \textit{Ans.}\end{aligned}$$

Prob. 5-11 If λ is one of the complex cubic roots of unity, evaluate

$$\begin{vmatrix} 1 & \lambda & \lambda^2 \\ \lambda & \lambda^2 & 1 \\ \lambda^2 & 1 & \lambda \end{vmatrix}$$

Prob. 5-12 Show that

$$\begin{vmatrix} (a^2 + \theta) & ab & ac & ad \\ ab & (b^2 + \theta) & bc & bd \\ ac & bc & (c^2 + \theta) & cd \\ ad & bd & cd & (d^2 + \theta) \end{vmatrix}$$

is divisible by θ^3. What is the other factor?

Prob. 5-13 Solve for x:

$$\begin{vmatrix} 1 & 1 & x \\ 3 & 3 & 3 \\ -2 & x & -2 \end{vmatrix} = 0$$

Suggested solution:

This can be simplified by using theorems of this chapter as follows:

$$\begin{vmatrix} 1 & 1 & x \\ 3 & 3 & 3 \\ -2 & x & -2 \end{vmatrix} = 3 \begin{vmatrix} 1 & 1 & x \\ 1 & 1 & 1 \\ -2 & x & -2 \end{vmatrix} = \begin{vmatrix} 1 - x & 1 & x \\ 0 & 1 & 1 \\ 0 & x & -2 \end{vmatrix} = 0$$

Hence $$(1 - x)\begin{vmatrix} 1 & 1 \\ x & -2 \end{vmatrix} = (1 - x)[-2 - x] = 0$$

Therefore $$x = 1 \text{ or } x = -2 \qquad \textit{Ans.}$$

Prob. 5-14 The following measurements are made on one section of a type 6J6 twin triode without an external shield:

(*a*) With grid and plate tied together, the capacitance between this combination and the cathode is 2.6.

(*b*) With grid and cathode tied together, the capacitance between this combination and the plate is 2.0.

(*c*) With plate and cathode tied together, the capacitance between the combination and the grid is 3.8.

Find C_{gk}, C_{pk}, and C_{gp}.

Chapter 6

NETWORK SOLUTIONS BY DETERMINANTS

6-1 Introduction

This chapter is devoted primarily to the applications of determinant theory in network analysis. It will be assumed that the reader is familiar with the solutions of simple one- and two-mesh series, parallel, and series-parallel networks. These cases require, from a mathematical point of view, only a good working knowledge of complex algebra. The multimesh network, however, demands the use of determinants for rapid solution.

Maxwell is responsible for the two basic methods of network analysis, which are known as the mesh method and the node method because they are based on the two fundamental laws of network theory established by Kirchhoff in 1847. These fundamental laws may be stated as follows:

Law 1. The total voltage drop around any mesh or loop of a network is zero.

Law 2. The total current at any junction or node of a network is zero.

A mesh is defined as a set of branches forming a closed path in a network, and a node is defined as a terminal common to two or more branches of a network. For example, in Fig. 6-1, there are three meshes and three nodes according to the definitions just given.

It should be understood that the laws just stated follow directly from the physical facts. A potential difference or voltage between two points is defined as the work done in moving a unit charge from the first point to the second. It follows, as given in the first law, that the work done in moving a unit charge over any closed path, regardless of the nature of the path, will be zero because the charge will eventually be returned to its starting point. In other words, the first law is a restatement of the law of conservation of energy. The second law merely reflects the fact that charge must be conserved. If all possible paths along which displacement of charge is possible are indicated in an equivalent network, and it is certain that actual electron emission or other unlooked for effects are not possible, then the net charge displaced toward a junction, or node, must equal that displaced away from it.

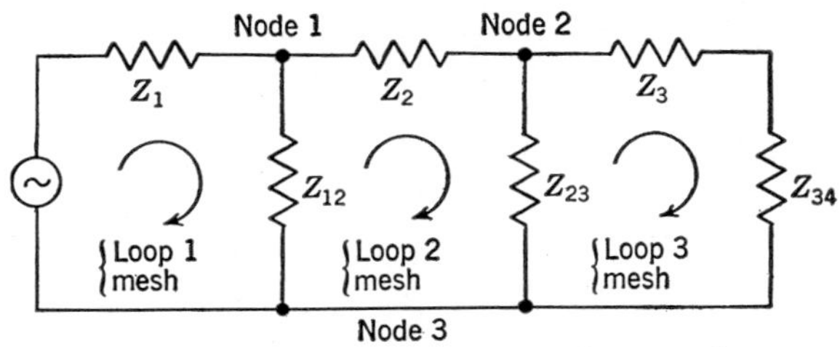

FIG. 6-1 Three-mesh–three-node network.

Accordingly, the total rate at which charge enters a node, or the incoming current, must equal the total rate at which charge leaves a node, or the outgoing current.

6-2 Mesh Analysis

In the *mesh* analysis, which is based on the first of the two laws, cyclic or mesh currents are assumed for each of the meshes in the network to be analyzed. Consider the bridge network shown in Fig. 6-2. There are three "windowpanes," or meshes, and three currents, namely, I_1, I_2, and I_3, have been assumed in the corresponding meshes. It is reasonable to inquire why separate currents have not been assumed in each of the branches instead of in each of the meshes. The answer, of course, lies in the fact that a separate current need not be assumed in Z_2, for example, because the current in this branch is already determined as $I_1 - I_2$, since the direction of I_1 in Z_2 is opposed to that of I_2, etc. Thus the assumption of "cyclic" currents for each mesh will ultimately provide the current in each branch. The equations for this network by Kirchhoff's first law are as follows:

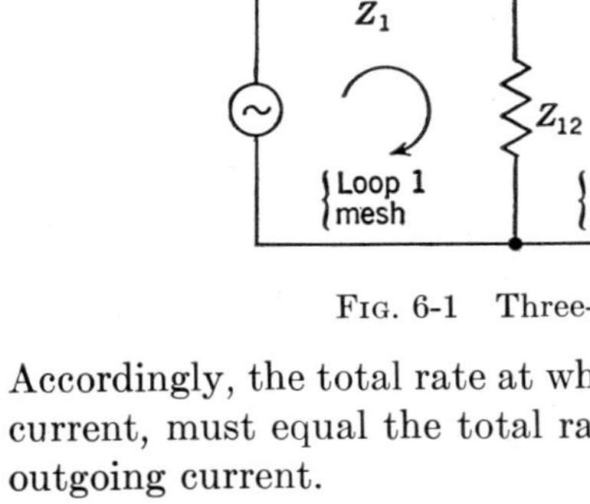

FIG. 6-2 Bridge network.

For mesh 1:

$$E = I_1(Z_1 + Z_2 + Z_5) - I_2Z_2 - I_3Z_5 \tag{6-1}$$

For mesh 2:

$$0 = -I_1(Z_2) + I_2(Z_2 + Z_3 + Z_6) - I_3(Z_6) \tag{6-2}$$

For mesh 3:

$$0 = -I_1(Z_5) - I_2(Z_6) + I_3(Z_4 + Z_5 + Z_6) \tag{6-3}$$

These three linear simultaneous equations must be solved to determine I_1, I_2, and I_3.

It can be seen that although additional cyclic currents could be assumed, such as I_4, these would be redundant and would not yield any additional information. It follows that sufficient cyclic currents will have been introduced when some cyclic current appears in every network element or branch. It will be observed that the direction of the cyclic current is always assumed as clockwise and that the voltage equations are always written in the direction of the reference mesh current of the mesh under consideration, i.e., the assumed polarity of any IZ voltage drop is from + to −, in the *positive* or clockwise direction of the cyclic current which is responsible for the drop.

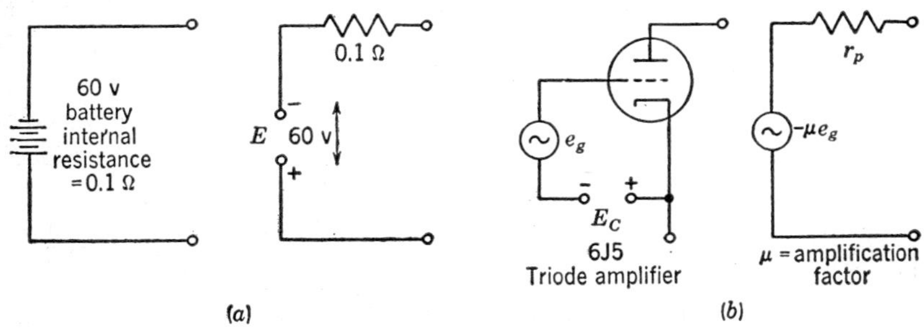

FIG. 6-3 Voltage-generator equivalents.

Also, the current directions and voltage polarities which are assumed are entirely *arbitrary*—but once a given convention is adopted, it must be used consistently. Since the impedances in Fig. 6-2 could be purely resistive, the current and voltage equations obtained are independent of whether the system emfs are alternating or unidirectional (d-c). In fact, the voltage sources may even be nonsinusoidal since such wave forms may be broken down into sinusoidal harmonic components by Fourier analysis and a solution achieved for each component. The reader should have no qualms about assumptions made concerning the cyclic-current directions and the voltage polarities. From the very nature of the linear equations obtained in this manner, the initial current directions and voltage polarities will correct themselves as the solution proceeds.

Regarding practical sources of voltage (generators) which must be represented in the mesh analysis, these are most conveniently shown as constant-voltage, zero-internal-impedance generators of voltage e, equal to the open-circuit voltage of the practical generator, placed in series with some impedance Z_i, which represents the internal impedance of the generator. For example, in Fig. 6-3a and b are shown some typical equivalents. These are based in both cases on Thévenin's theorem.

In the writing of mesh equations where magnetic coupling is involved, the polarity of the magnetically coupled elements such as transformer windings must first be determined. One convenient designation places a dot next to the terminals of corresponding polarity of the coils between which the mutual coupling exists, which is the customary marking of instrument transformers. A charge displacement, or current, entering either winding through the dot-marked terminal produces magnetic flux in the same direction. In Fig. 6-4, charge displacement I_1 through the dotted terminal on the primary will cause a flux in the same direction as displacement I_2 at the dotted terminal on the secondary, in accord with Lenz's law and the right-hand rule.

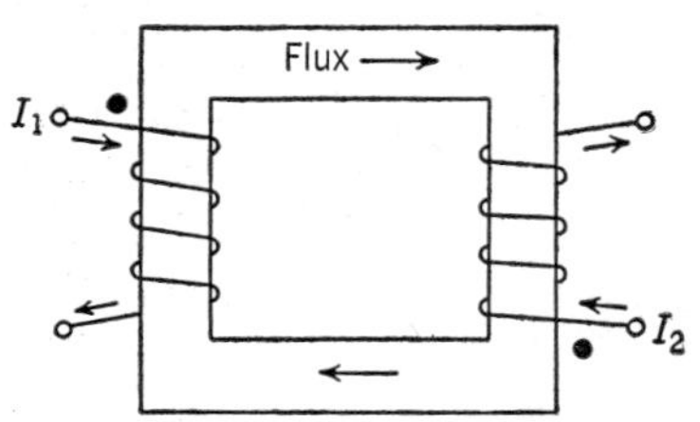

FIG. 6-4 Transformer-winding polarity designations.

Another way for establishing polarity is to determine which secondary terminal is positive and which is negative with a conventional voltmeter

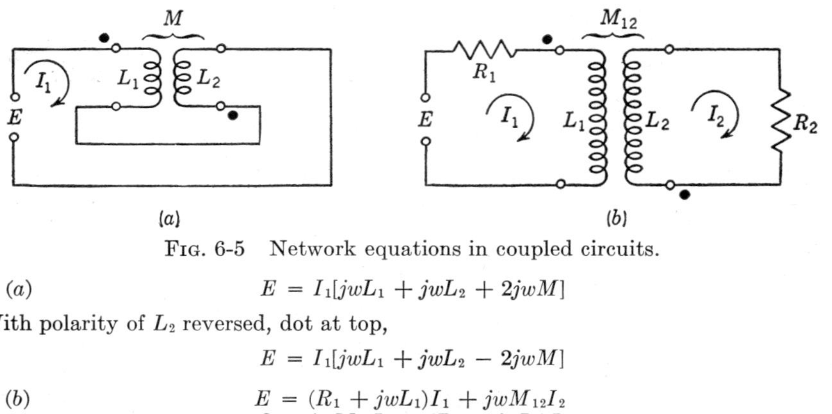

FIG. 6-5 Network equations in coupled circuits.

(a) $$E = I_1[jwL_1 + jwL_2 + 2jwM]$$

With polarity of L_2 reversed, dot at top,

$$E = I_1[jwL_1 + jwL_2 - 2jwM]$$

(b) $$E = (R_1 + jwL_1)I_1 + jwM_{12}I_2$$
$$O = jwM_{12}I_1 + (R_2 + jwL_2)I_2$$

With polarity of L_2 reversed, dot at top,

$$E = (R_1 + jwL_1)I_1 - jwM_{12}I_2$$
$$O = -jwM_{12}I_1 + (R_2 + jwL_2)I_2$$

when one primary terminal is made instantaneously positive with respect to the other by means of a battery, etc. The terminals then exhibiting like positive polarity would be dotted. Once the proper polarities of mutually coupled windings are established in a network either from tests on the practical device to be used or by arbitrary assignment as a design specification, the loop equations may be written as is shown in Fig. 6-5*a* and *b*. In this figure, double subscript notation might have been used in which the primary impedance $j\omega L_1$ would have been Z_{11}; the mutual impedance $j\omega M$, Z_{12}; and the secondary impedance $j\omega L_2$, Z_{22}.

In summary, the mesh analysis is performed by means of cyclic or mesh currents arbitrarily assigned in sufficient numbers so that full information on all the network elements is obtained. The passive network elements are impedances, resistances, and reactances. The active elements, or generators, are constant-voltage, zero-internal-impedance generators with suitable series impedances which represent the actual generators as obtained by means of Thévenin's theorem. The element connections in the mesh are in series, and the basic law is Kirchhoff's first law. The n-mesh network leads in general to n linear simultaneous equations, and each term in these equations has the dimensions of *voltage*.

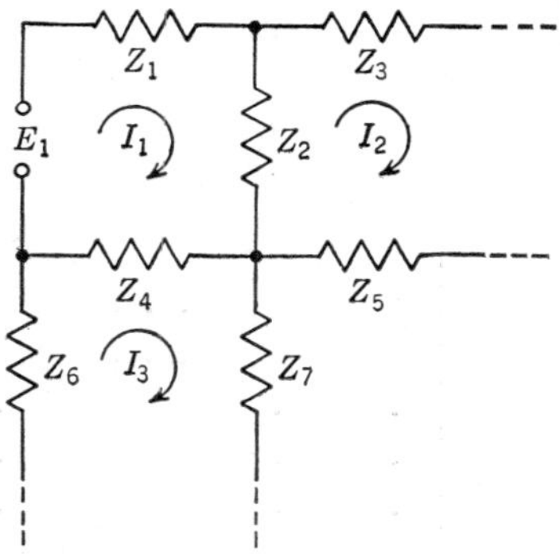

FIG. 6-6 Multimesh network.

6-3 The Multimesh Network and the Determinant

When a linear bilateral[1] n-mesh network is analyzed by the method of mesh analysis, the simultaneous linear voltage equations will appear as follows:

$$
\begin{aligned}
E_1 &= Z_{11}I_1 + Z_{12}I_2 + Z_{13}I_3 + \cdots + Z_{1n}I_n \\
E_2 &= Z_{21}I_1 + Z_{22}I_2 + Z_{23}I_3 + \cdots + Z_{2n}I_n \\
E_3 &= Z_{31}I_1 + Z_{32}I_2 + Z_{33}I_3 + \cdots + Z_{3n}I_n \\
\cdot \quad & \quad \cdot \qquad \cdot \qquad \cdot \qquad\qquad \cdot \\
E_j &= Z_{j1}I_1 + Z_{j2}I_2 + Z_{j3}I_3 + \cdots + Z_{jn}I_n \\
\cdot \quad & \quad \cdot \qquad \cdot \qquad \cdot \qquad\qquad \cdot \\
E_n &= Z_{n1}I_1 + Z_{n2}I_2 + Z_{n3}I_3 + \cdots + Z_{nn}I_n
\end{aligned}
\tag{6-4}
$$

Equations (6-4) are obtained from an analysis of the circuit in Fig. 6-6. In the first mesh, the generator voltage is E_1, the loop impedance which the cyclic or mesh current I must traverse is

$$Z_{11} = Z_1 + Z_2 + Z_4$$

and the voltage drops due to I_2 and I_3 which appear in mesh 1 are $-I_2Z_2$

[1] A *bilateral element* conducts to the same extent in both directions, whereas a *unilateral element* conducts differently in opposing directions. Familiar examples of unilateral network elements are electron tubes, transistors, copper-oxide rectifiers, germanium and silicon diodes, etc.

and $-I_3Z_4$, respectively, so that $Z_{12} = -Z_2$ and $Z_{13} = -Z_4$. Z_{12} is described as the mutual impedance or coupling between meshes 1 and 2, since this element conducts both of the mesh currents 1 and 2.

In general, then, Z_{11}, Z_{22}, Z_{33}, Z_{nn} represent the loop impedances which would exist around the individual meshes if all branches of the network other than those common to the mesh were open-circuited and Z_{12}, Z_{13}, Z_{14}; Z_{23}, Z_{24}, Z_{25}, etc., are all mutual impedances. It is important to note that reversing the order of the subscripts of mutual impedances does not alter their values, that is, $Z_{12} = Z_{21}$, $Z_{23} = Z_{32}$, etc., and that mutual coupling may result either from elements common to two meshes or from magnetic coupling between elements. It can be seen that if two mesh currents pass through a mutual impedance in opposite directions, then the value of the mutual impedance is the negative of the common impedance. As in the case cited previously, $Z_{12} = -Z_2$.

The solution by Cramer's rule for the set of equations given above for the mesh current I_k in the kth mesh is

$$I_k = \frac{\begin{vmatrix} Z_{11} & Z_{12} & \cdots & Z_{1,k-1} & E_1 & Z_{1,k+1} & \cdots & Z_{1n} \\ Z_{21} & Z_{22} & \cdots & Z_{2,k-1} & E_2 & Z_{2,k+1} & \cdots & Z_{2n} \\ Z_{31} & Z_{32} & \cdots & Z_{3,k-1} & E_3 & Z_{3,k+1} & \cdots & Z_{3n} \\ \cdot & \cdot & & \cdot & \cdot & \cdot & & \cdot \\ \cdot & \cdot & & \cdot & E_j & \cdot & & \cdot \\ \cdot & \cdot & & \cdot & \cdot & \cdot & & \cdot \\ Z_{n1} & Z_{n2} & & Z_{n,k-1} & E_n & Z_{n,k+1} & \cdots & Z_{nn} \end{vmatrix}}{\Delta} \tag{6-5}$$

$$\text{where} \qquad \Delta = \begin{vmatrix} Z_{11} & Z_{12} & \cdots & Z_{1n} \\ Z_{21} & Z_{22} & \cdots & Z_{2n} \\ Z_{31} & Z_{32} & \cdots & Z_{3n} \\ \cdot & \cdot & & \cdot \\ \cdot & \cdot & & \cdot \\ \cdot & \cdot & & \cdot \\ Z_{n1} & Z_{n2} & \cdots & Z_{nn} \end{vmatrix} \tag{6-6}$$

When the quotient that represents I_k is simplified by the method of minors and cofactors, then I_k may be depicted as the sum:

$$I_k = \frac{(-1)^{k+1}E_1B_{1k}}{\Delta} + \frac{(-1)^{k+2}E_2B_{2k}}{\Delta} + \cdots + \frac{(-1)^{k+j}E_jB_{jk}}{\Delta} + \cdots + \frac{(-1)^{k+n}E_nB_{nk}}{\Delta} \tag{6-7}$$

where B_{jk} is the minor of the corresponding cofactor E_j, formed by canceling the jth row and kth column in Δ.

Mathematicians usually indicate the alternations in signs as shown above. However, for convenience in writing, it is common engineering practice to consider the sign $(-1)^{k+j}$ on the general term E_jB_{jk}/Δ to be incorporated into the minor B_{jk}, so that one may write

$$I_k = \frac{E_1B_{1k}}{\Delta} + \frac{E_2B_{2k}}{\Delta} + \cdots + \frac{E_jB_{jk}}{\Delta} + \cdots + \frac{E_nB_{nk}}{\Delta} \tag{6-8}$$

where the *sign has been absorbed* into the symbol B_{1k}, B_{2k}, . . . , B_{jk}, . . . , B_{nk}. The writing of Eq. (6-7) as (6-8) places the burden of responsibility for the proper algebraic sign of the respective minors on the engineer.

If the only generator is that present in mesh j, or E_j, all the terms in the above solution for I_k will drop out, and the current I_k in the kth mesh due to the voltage E_j acting in the jth mesh is

$$I_k = \frac{(-1)^{j+k}B_{jk}E_j}{\Delta} \tag{6-9}$$

If the algebraic sign is absorbed in B_{jk}, then

$$I_k = \frac{B_{jk}E_j}{\Delta} \tag{6-10}$$

Throughout the remainder of this chapter, the *algebraic sign* will in each case be absorbed and incorporated into the B_{jk} as in Eqs. (6-8) and (6-10).

When the voltage source and the current are in the same mesh, $j = k$, and Eq. (6-10) becomes

$$I_k = \frac{B_{kk}E_k}{\Delta} \tag{6-11}$$

The quotient B_{kk}/Δ has the dimensions of admittance. At the input or first mesh of a network, $j = k = 1$, and the short-circuit input admittance is

$$y_{11} = \frac{I_1}{E_1} = \frac{B_{11}}{\Delta} \tag{6-12}$$

The term *short circuit* refers to the fact that Eq. (6-12) and Eq. (6-9) presume that all the generators or voltage sources in all other meshes are zero, or, in other words, have been short-circuited.

If the output mesh of a network is designated as mesh 2, $j = k = 2$, and the short-circuit output admittance y_{22} is

$$y_{22} = \frac{I_2}{E_2} = \frac{B_{22}}{\Delta} \tag{6-13}$$

When a current I_k in mesh k is produced by a voltage E_j in mesh j, the short-circuit *transfer* admittance, from Eq. (6-10), is $y_{jk} = I_k/E_j = B_{jk}/\Delta$, and, in particular, the transfer admittance between input and output meshes is

$$y_{12} = \frac{I_2}{E_1} = \frac{B_{12}}{\Delta} \tag{6-14}$$

As an example of practical multimesh analysis, consider the parallel-T network analyzed by Cowles.[1] The network is shown in its usual form in Fig. 6-7*a*. In Fig. 6-7*b*, it has been redrawn to show the meshes

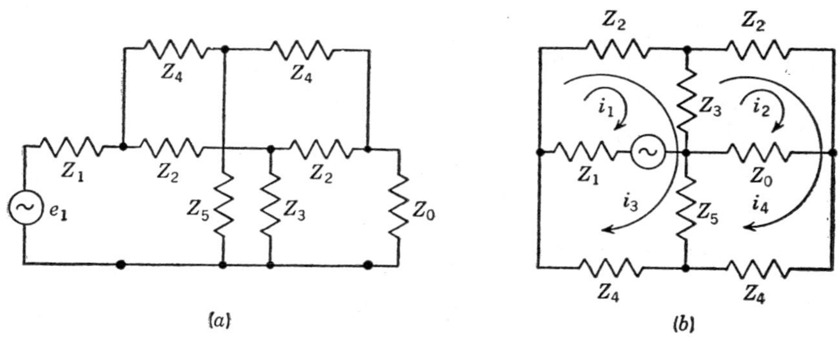

FIG. 6-7 General parallel-T network.

involved more clearly. Choosing the cyclic currents shown and taking the output current $i_0 = i_2$, the mesh equations for this network are

$$\begin{array}{lllll}
e_1 = +i_1(Z_1 + Z_2 + Z_3) & - i_0 Z_3 & + i_3(Z_2 + Z_3) & - i_4 Z_3 & \\
0 = -i_1 Z_3 & + i_0(Z_0 + Z_2 + Z_3) & - i_3 Z_3 & + i_4(Z_2 + Z_3) & \\
0 = +i_1(Z_2 + Z_3) & - i_0 Z_3 & + i_3(Z_2 + Z_3 + Z_4 + Z_5) & - i_4(Z_3 + Z_5) & \\
0 = -i_1 Z_3 & + i_0(Z_2 + Z_3) & - i_3(Z_3 + Z_5) & + i_4(Z_2 + Z_3 + Z_4 + Z_5) &
\end{array} \tag{6-15}$$

The fourth-order determinant Δ of the impedance coefficients, after adding the negative of row 3 to row 1, the negative of row 4 to row 2, and row 4 to row 3, is

$$\Delta = \begin{vmatrix}
Z_1 & 0 & (-Z_4 - Z_5) & Z_5 \\
0 & Z_0 & Z_5 & (-Z_4 - Z_5) \\
Z_2 & Z_2 & (Z_2 + Z_4) & (Z_2 + Z_4) \\
-Z_3 & (Z_2 + Z_3) & (-Z_3 - Z_5) & (Z_2 + Z_3 + Z_4 + Z_5)
\end{vmatrix} \tag{6-16}$$

The value of Δ is

$$\begin{aligned}
\Delta = {} & Z_1 Z_0 (Z_2 + Z_4)(Z_2 + 2Z_5 + Z_4 + 2Z_3) \\
& + (Z_1 + Z_0)[Z_2(Z_2 + 2Z_3)(Z_4 + Z_5) + Z_4(Z_4 + 2Z_5)(Z_2 + Z_3)] \\
& + Z_2 Z_4 (Z_2 + 2Z_3)(Z_4 + 2Z_5)
\end{aligned} \tag{6-17}$$

[1] L. G. Cowles, The Parallel-T Resistance Capacitance Network, *Proc. I.R.E.*, **40** (12): 1712–1717, December, 1952.

The input current i_1 is, therefore,

$$i_1 = \frac{Z_0(Z_2 + Z_4)(Z_2 + 2Z_3 + Z_4 + 2Z_5) + (Z_2 + Z_3)(Z_2 + Z_4)(Z_4 + 2Z_5)}{\Delta} e_1 + \frac{Z_2Z_3(Z_4 + 2Z_5) - Z_2Z_5(Z_2 + 2Z_3)}{\Delta} e_1 \tag{6-18}$$

and the output current i_0 is

$$i_0 = \frac{Z_2Z_5(Z_2 + 2Z_3) + Z_3Z_4(Z_4 + 2Z_5)}{\Delta} e_1 \tag{6-19}$$

Equation (6-19) indicates that the transfer admittance Y_{12} is symmetrical as regards the input and output impedances, and the existence of a null in the output voltage developed across Z_0 will be independent of the terminating impedances. The particular case of the parallel-T *RC* equivalent of the Wien bridge can be derived from this solution by making Z_2 and Z_5 pure resistances and Z_3 and Z_4 pure capacitances. Other worthwhile examples of mesh analysis which will be of interest to the reader are given by Spangenberg[1] and Dunn.[2]

6-4 Nodal Analysis

The second method of network analysis is based on Kirchhoff's second law and is described as the *nodal* method. This method has only recently begun to achieve popularity and has distinct advantage over the mesh analysis for certain applications such as electron-tube amplifier analysis. In nodal analysis, one node is selected as the reference node and the combination of the reference node with any other node is called an *independent* node pair. The selection of the reference node is arbitrary, but the grounded or common side of a network, if any, is usually so designated.

Since the nodal analysis is based on Kirchhoff's second law, the currents at the junctions or nodes are of interest rather than the voltage drops across elements, and the terms in the linear equations which are developed will have the dimensions of *current* rather than voltage. This means that wherever voltages are involved, they will appear in terms as factors with admittances, and constant-current sources with infinite internal impedance will be used as generator equivalents, as obtained by Norton's theorem.

The manner in which a nodal analysis is performed can best be explained in terms of a particular network. Consider Fig. 6-8, in which the nodes are numbered 1, 2, and 3, the third being selected as the reference node.

[1] K. R. Spangenberg, The Universal Characteristics of Triple-resonant-circuit Band-pass Filters, *Proc. I.R.E.*, **34** (9): 629–634, September, 1946.

[2] S. C. Dunn, A Useful Network Procedure, *Wireless Engr.*, **27** (6): 171–175, June, 1950.

Just as the direction of the cyclic currents was arbitrarily assumed to be clockwise, it will be assumed that the current directions indicated by arrows at each junction or node are *away* from the node. It will be noted that the voltage existing at each node with respect to the reference node is indicated as V_1, V_2 and that the conventional constant-voltage generators used in the mesh analysis have been replaced by constant-*current* generators by Norton's theorem. Thus a source with open-circuit voltage E and internal impedance Z is replaced with a constant-current, infinite-internal-impedance generator equivalent of EY amperes in parallel with an impedance Z or an admittance Y.

When the network has been set up, Kirchhoff's second law may be utilized to write a current equation for each node. First the contribution of the active element (the constant-current generator) at node 1 should

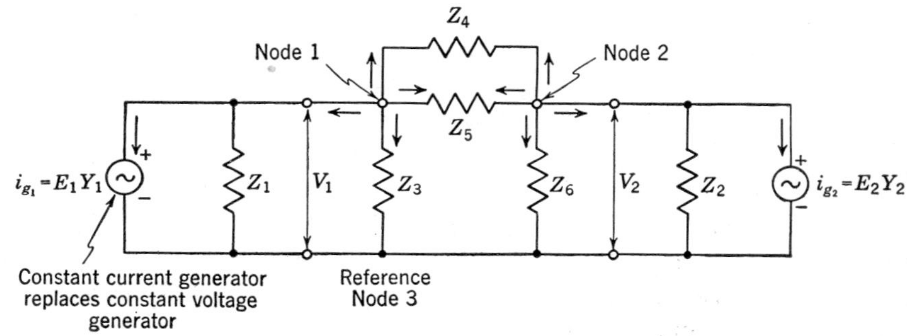

FIG. 6-8 Three-node network.

be considered. It produces a current assumed to be in a direction away from node 1 in value $-E_1Y_1$, where $Y_1 = 1/Z_1$. It is negative because its direction is opposed to the given polarity of the generator. The current leaving node 1 due to the node voltage V_1 must be determined *assuming that all remaining node voltages*, that is, V_2, are, for the moment, zero, or in other words short-circuited. Thus, V_1, acting at node 1 neglecting V_2, will produce currents away from node 1 of $V_1(Y_1 + Y_3 + Y_4 + Y_5)$. Second, the contributions at node 1 of the remaining node voltages are added based on the assumption for the moment that V_1 is equal to zero. In this case only V_2 contributes. Since the direction of the currents due to V_2 is assumed away from node 2, then the current which is due to V_2 will *approach* node 1 when V_1 is neglected. The component which is due to V_2 in this case is negative since it is approaching node 1 and is $-V_2(Y_4 + Y_5)$.

The complete current equation at node 1 is, therefore,

$$-E_1Y_1 + V_1(Y_1 + Y_3 + Y_4 + Y_5) - V_2(Y_4 + Y_5) = 0 \qquad (6\text{-}20)$$

and the corresponding equation for node 2 is

$$-E_2Y_2 - V_1(Y_4 + Y_5) + V_2(Y_2 + Y_4 + Y_5 + Y_6) = 0 \qquad (6\text{-}21)$$

It should be noted that in this particular instance the node voltage V_2 appeared in the equation for node 1. However, if a third node were situated between node 1 and node 2, a component due to V_2 would *not* appear in the node-1 equation, because, when the latter is written, the remaining nodes are considered short-circuited. Consider the four-node

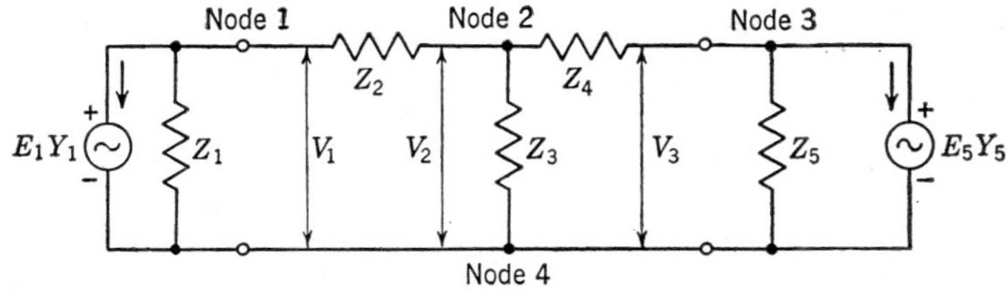

FIG. 6-9 Four-node network.

network shown in Fig. 6-9, the fourth being selected as the reference node. The nodal analysis at node 1 yields

$$-E_1Y_1 + V_1(Y_1 + Y_2) - V_2(Y_2) = 0 \qquad (6\text{-}22)$$

It can be seen that the node voltage V_3 does not appear in this equation because node 3 is not connected to node 1 except through node 2, and, when the contribution at node 1 due to V_3 is to be written, the short-circuiting of V_2 in the process will prevent any such component from appearing at node 1. It follows that, at node 2, components due to all the node voltages will appear and

$$-V_1(Y_2) + V_2(Y_2 + Y_3 + Y_4) - V_3(Y_4) = 0 \qquad (6\text{-}23)$$

while at node 3 no component due to V_1 will appear and

$$-V_2(Y_4) + V_3(Y_4 + Y_5) - E_5Y_5 = 0 \qquad (6\text{-}24)$$

Where mutual magnetic coupling exists between the network elements, it is impossible to perform the nodal analysis as just outlined without certain modifications. This is true because the voltage across a coil or winding coupled to another coil or winding is in most cases due to at least *two* different currents. Methods have been devised to modify the nodal analysis to accommodate such situations.[1] However, in general, where mutual induction exists in a network, it is better to resort to the mesh analysis rather than the nodal. In those cases where the relative d-c levels in the network are unimportant, transformers or coupled elements may be replaced by equivalent T or π networks as will be described in a later section. Such a transformation will resolve the difficulty except in the case where a network contains mutual induction

[1] M. B. Reed, Node Equations, *Proc. I.R.E.*, **32** (6), June, 1944.

and only the input currents are known rather than input voltages, in which case the mesh analysis cannot be used. Fortunately, the occurrence of such a problem is infrequent.

To summarize, the nodal analysis is performed by summing up and equating the current components at each node to zero, where one of the nodes is chosen and employed as a reference node. A sufficient number of nodes is used to obtain full information on all the network elements. The passive network elements are admittances, conductances, and susceptances. The active elements or generators are constant-current, infinite-internal-impedance generators which are placed in parallel with impedances which represent the internal impedances of the actual practical generators in accordance with Norton's theorem. The element connections in the nodal analysis are principally in parallel, and the basic concept is Kirchhoff's second law. The n-node network leads in general to $n - 1$ linear simultaneous equations, and each term in these equations has the dimensions of *current*.

6-5 The Multinode Network and the Determinant

If a linear bilateral network of n independent nodes is analyzed by the nodal analysis, then the node equations are

$$\begin{array}{l} I_1 = Y_{11}V_1 + Y_{12}V_2 + Y_{13}V_3 + \cdots + Y_{1n}V_n \\ I_2 = Y_{21}V_1 + Y_{22}V_2 + Y_{23}V_3 + \cdots + Y_{2n}V_n \\ I_3 = Y_{31}V_1 + Y_{32}V_2 + Y_{33}V_3 + \cdots + Y_{3n}V_n \\ \cdot \qquad \cdot \qquad \cdot \qquad \cdot \qquad \cdot \\ \cdot \qquad \cdot \qquad \cdot \qquad \cdot \qquad \cdot \\ I_j = Y_{j1}V_1 + Y_{j2}V_2 + Y_{j3}V_3 + \cdots + Y_{jn}V_n \\ \cdot \qquad \cdot \qquad \cdot \qquad \cdot \qquad \cdot \\ \cdot \qquad \cdot \qquad \cdot \qquad \cdot \qquad \cdot \\ I_n = Y_{n1}V_1 + Y_{n2}V_2 + Y_{n3}V_3 + \cdots + Y_{nn}V_n \end{array} \tag{6-25}$$

where $I_1, I_2, I_3, \ldots, I_n$ are the generator currents at the respective nodes, and $V_1, V_2, V_3, \ldots, V_n$ are the various node voltages referred to the reference node. The solution for any node voltage V_k in the determinant form is

$$V_k = \frac{\begin{vmatrix} Y_{11} & Y_{12} & \cdots & Y_{1,k-1} & I_1 & Y_{1,k+1} & \cdots & Y_{1n} \\ Y_{21} & Y_{22} & \cdots & Y_{2,k-1} & I_2 & Y_{2,k+1} & \cdots & Y_{2n} \\ Y_{31} & Y_{32} & \cdots & Y_{3,k-1} & I_3 & Y_{3,k+1} & \cdots & Y_{3n} \\ \cdot & \cdot & & \cdot & \cdot & \cdot & & \cdot \\ \cdot & \cdot & & \cdot & I_j & \cdot & & \cdot \\ \cdot & \cdot & & \cdot & \cdot & \cdot & & \cdot \\ Y_{n1} & Y_{n2} & \cdots & Y_{n,k-1} & I_n & Y_{n,k+1} & \cdots & Y_{nn} \end{vmatrix}}{\Delta} \tag{6-26}$$

where

$$\Delta = \begin{vmatrix} Y_{11} & Y_{12} & Y_{13} & \cdots & Y_{1n} \\ Y_{21} & Y_{22} & Y_{23} & \cdots & Y_{2n} \\ Y_{31} & Y_{32} & Y_{33} & \cdots & Y_{3n} \\ \cdot & \cdot & \cdot & & \cdot \\ \cdot & \cdot & \cdot & & \cdot \\ \cdot & \cdot & \cdot & & \cdot \\ Y_{n1} & Y_{n2} & Y_{n3} & \cdots & Y_{nn} \end{vmatrix} \tag{6-27}$$

By the method of cofactor-minor expansion, V_k may be expressed as follows:

$$V_k = \frac{B_{1k}I_1}{\Delta} + \frac{B_{2k}I_2}{\Delta} + \frac{B_{3k}I_3}{\Delta} + \cdots + \frac{B_{jk}I_j}{\Delta} + \cdots + \frac{B_{nk}I_n}{\Delta} \tag{6-28}$$

where B_{jk} is the minor of Δ with algebraic sign absorbed formed by canceling the jth row and kth column.

If there is only one current generator I_j, then the node voltage V_k at the node k due to current I_j is

$$V_k = I_j \frac{B_{jk}}{\Delta} \tag{6-29}$$

and the *open-circuit transfer impedance* is B_{jk}/Δ. The *open-circuit input impedance* of a multinode network analyzed on a node basis is

$$z_{11} = \frac{V_1}{I_1} = \frac{B_{11}}{\Delta} \tag{6-30}$$

while the open-circuit output impedance is

$$z_{22} = \frac{V_2}{I_2} = \frac{B_{22}}{\Delta} \tag{6-31}$$

The work required in the solution of a multimesh or multinode network can be further simplified by writing the determinant directly from the network. The elements along the principal diagonal are always the loop or mesh impedances or node admittances, that is, $Z_{11}, Z_{22}, Z_{33}, \ldots, Z_{nn}$; or $Y_{11}, Y_{22}, Y_{33}, \ldots, Y_{nn}$. The other elements that remain are all mutual impedances or admittances and may be taken directly from the network diagram with a correction for polarity when required.

When the ratio of two mesh currents is desired, it may be calculated by the evaluation of only two minors. If there is but one source voltage E_n, in mesh n, then the current in the jth mesh due to E_n is, from Eq. (6-10),

$$I_j = \frac{B_{nj}}{\Delta} E_n \tag{6-32}$$

and that in the kth mesh due to E_n is

$$I_k = \frac{B_{nk}}{\Delta} E_n \tag{6-33}$$

so that the ratio I_j/I_k is

$$\frac{I_j}{I_k} = \frac{B_{nj}}{B_{nk}} \tag{6-34}$$

Similarly, by the method of nodal analysis, the ratio of two node voltages E_k and E_j due to a generator current I_n at the nth node is, from Eq (6-29),

$$\frac{E_j}{E_k} = \frac{B_{nj}}{B_{nk}} \tag{6-35}$$

Of the two methods, the nodal method of network analysis is particularly important to the television and electronics engineer because electron-tube networks lend themselves so readily to this form of analysis. This would be expected since one side of an electron-tube network is quite often grounded, yielding a convenient reference node, and the various load impedances and interelectrode capacitances appear in parallel.

In conclusion it is important to point out that a paramount feature of the nodal method is that, as long as the number of nodes in a network is not altered, the number of nodal equations will *not* increase as additional branches are added in parallel to elements already contained in the network. These additional parallel branches will appear merely as additional admittance terms, so that a solution obtained by the nodal analysis may often be modified to take into account additional parallel branches such as interelectrode capacitances which were previously neglected, without requiring a reworking of the entire problem.

6-6 Comparison of Solutions by Mesh and Nodal Analysis: Example

Consider the bridged-T network shown in Fig. 6-10, a network configuration often used in audio and video applications as a selective network or isolation pad. At some frequency, the network impedances assume the values shown.

For the mesh analysis, assume mesh currents in meshes 1, 2, and 3, as indicated. The mesh equations are

$$\begin{aligned}
E &= (600 + 400 + j600)I_1 - (j600)I_2 - 400I_3\\
0 &= -(j600)I_1 + (900 + j600 - j600)I_2 - (-j600)I_3\\
0 &= -(400)I_1 - (-j600)I_2 + (600 + 400 - j600)I_3
\end{aligned}$$

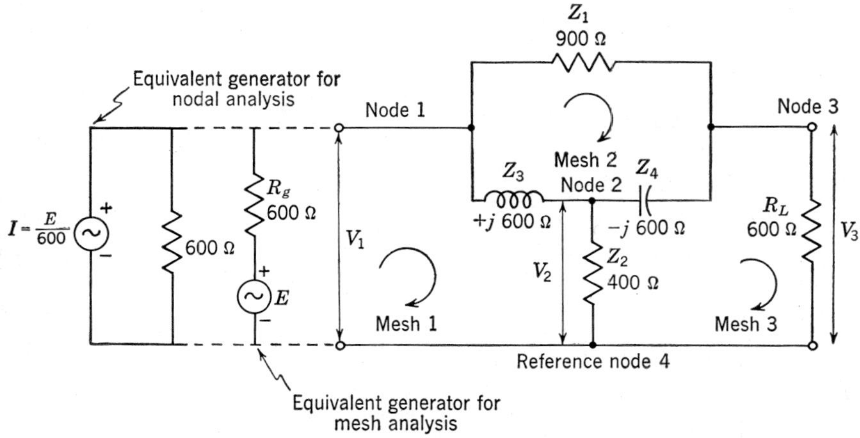

FIG. 6-10 Bridged-T network.

The denominator Δ, by Cramer's rule, for currents I_1, I_2, and I_3 will be

$$\Delta = \begin{vmatrix} (1{,}000 + j600) & -j600 & -400 \\ -j600 & 900 & j600 \\ -400 & j600 & (1{,}000 - j600) \end{vmatrix}$$

Factor 100 from each column, which gives

$$\Delta = (100)^3 \begin{vmatrix} (10 + j6) & -j6 & -4 \\ -j6 & 9 & j6 \\ -4 & j6 & (10 - j6) \end{vmatrix}$$

With the elements of the second column as cofactors,

$$\Delta = 10^6 \left[j6 \begin{vmatrix} -j6 & j6 \\ -4 & (10 - j6) \end{vmatrix} + 9 \begin{vmatrix} (10 + j6) & -4 \\ -4 & (10 - j6) \end{vmatrix} - j6 \begin{vmatrix} (10 + j6) & -4 \\ -j6 & j6 \end{vmatrix} \right]$$

$$= 10^6 \left[j6 \begin{vmatrix} -j6 & 0 \\ -4 & (6 - j6) \end{vmatrix} + 9 \begin{vmatrix} (10 + j6) & -4 \\ -4 & (10 - j6) \end{vmatrix} - j6 \begin{vmatrix} (10 + j6) & (6 + j6) \\ -j6 & 0 \end{vmatrix} \right]$$

$$= 10^6[216 - j216 + 1{,}080 + 216 + j216]$$

Therefore Δ when evaluated becomes

$$\Delta = 10^6(216 + 1{,}080 + 216) = 1.512 \times 10^9$$

Cramer's rule yields for numerators when solving for I_1, I_2, and I_3, respectively,

$$N_1 = (10^2)^2 \begin{vmatrix} E & -j6 & -4 \\ 0 & 9 & j6 \\ 0 & j6 & (10 - j6) \end{vmatrix} = E(10^4) \begin{vmatrix} 9 & j6 \\ j6 & (10 - j6) \end{vmatrix}$$

$$= E(10^4)(126 - j54)$$

$$N_2 = 10^4 \begin{vmatrix} (10 + j6) & E & -4 \\ -j6 & 0 & j6 \\ -4 & 0 & (10 - j6) \end{vmatrix} = -E(10^4) \begin{vmatrix} -j6 & j6 \\ -4 & (10 - j6) \end{vmatrix}$$

$$= E(10^4)(36 + j36)$$

$$N_3 = 10^4 \begin{vmatrix} (10 + j6) & -j6 & E \\ -j6 & 9 & 0 \\ -4 & j6 & 0 \end{vmatrix} = E(10^4) \begin{vmatrix} -j6 & 9 \\ -4 & j6 \end{vmatrix}$$

$$= E(10^4)(72)$$

I_1, I_2, and I_3 are then

$$I_1 = \frac{E(10^4)(126 - j54)}{1.512 \times 10^9} = E(83.3 - j35.7)(10^{-5})$$

$$I_2 = \frac{E(10^4)(36 + j36)}{1.512 \times 10^9} = E(23.8 + j23.8)(10^{-5})$$

$$I_3 = \frac{E(10)(72)}{1.512 \times 10^9} = E(47.6)(10^{-5})$$

For the nodal analysis, node voltages V_1, V_2, and V_3 for the three nodes indicated in Fig. 6-10 are assigned and the generator E is replaced with an equivalent constant-current source of $E/600$ amp in parallel with an impedance of 600 ohms according to Norton's theorem.

To simplify the analysis, the admittances of the network are tabulated first:

$$Y_G = \frac{1}{600} = 1.67 \times 10^{-3} \text{ mho}$$

$$Y_1 = \frac{1}{Z_1} = \frac{1}{900} = 1.1 \times 10^{-3} \text{ mho}$$

$$Y_2 = \frac{1}{400} = 2.5 \times 10^{-3} \text{ mho}$$

$$Y_3 = \frac{1}{j600} = -j1.67 \times 10^{-3} \text{ mho}$$

$$Y_4 = \frac{1}{-j600} = j1.67 \times 10^{-3} \text{ mho}$$

$$Y_L = \frac{1}{600} = 1.67 \times 10^{-3} \text{ mho}$$

The node equations can then be written in accordance with the procedure described previously:

$$\frac{E}{600} = V_1(1.67 \times 10^{-3} + 1.1 \times 10^{-3} - j1.67 \times 10^{-3}) - V_2(-j1.67 \times 10^{-3}) - V_3(1.1 \times 10^{-3})$$

$$0 = -V_1(-j1.67 \times 10^{-3}) + V_2(2.5 \times 10^{-3} - j1.67 \times 10^{-3} + j1.67 \times 10^{-3}) - V_3(j1.67 \times 10^{-3})$$

$$0 = -V_1(1.1 \times 10^{-3}) - V_2(j1.67 \times 10^{-3}) + V_3(1.67 \times 10^{-3} + 1.1 \times 10^{-3} + j1.67 \times 10^{-3})$$

The denominator Δ, by Cramer's rule, for V_1, V_2, and V_3 will be

$$\Delta = \begin{vmatrix} (2.77 - j1.67)10^{-3} & j1.67 \times 10^{-3} & -1.1 \times 10^{-3} \\ j1.67 \times 10^{-3} & 2.5 \times 10^{-3} & -j1.67 \times 10^{-3} \\ -1.1 \times 10^{-3} & -j1.67 \times 10^{-3} & (2.77 + j1.67)10^{-3} \end{vmatrix}$$

$$= (10^{-3})^3 \begin{vmatrix} (2.77 - j1.67) & j1.67 & (1.67 - j1.67) \\ j1.67 & 2.5 & 0 \\ -1.1 & -j1.67 & (1.67 + j1.67) \end{vmatrix}$$

$$= 10^{-9} \begin{vmatrix} (1.67 - j1.67) & 0 & 3.34 \\ j1.67 & 2.5 & 0 \\ -1.1 & -j1.67 & (1.67 + j1.67) \end{vmatrix}$$

$$= 2.5 \times 10^{-9} \begin{vmatrix} (1.67 - j1.67) & 3.34 \\ -1.1 & (1.67 + j1.67) \end{vmatrix} + j1.67 \times 10^{-9} \begin{vmatrix} (1.67 - j1.67) & 3.34 \\ j1.67 & 0 \end{vmatrix}$$

$$= 32.47 \times 10^{-9}$$

Again, by Cramer's rule, the numerators for V_1, V_2, and V_3, respectively, are

$$N_1 = \begin{vmatrix} \frac{E}{600} & j1.67 \times 10^{-3} & -1.1 \times 10^{-3} \\ 0 & 2.5 \times 10^{-3} & -j1.67 \times 10^{-3} \\ 0 & -j1.67 \times 10^{-3} & (2.77 \times 10^{-3} + j1.67 \times 10^{-3}) \end{vmatrix}$$

$$= \frac{E(10^{-3})^2}{600} \begin{vmatrix} 2.5 & -j1.67 \\ -j1.67 & (2.77 + j1.67) \end{vmatrix}$$

$$= \frac{E(10^{-6})}{600}(9.72 + j4.17)$$

$$N_2 = \begin{vmatrix} (2.77 - j1.67)10^{-3} & \dfrac{E}{600} & -1.1 \times 10^{-3} \\ j1.67 \times 10^{-3} & 0 & -j1.67 \times 10^{-3} \\ -1.1 \times 10^{-3} & 0 & (2.77 + j1.67)10^{-3} \end{vmatrix}$$

$$= \frac{-E(10^{-3})^2}{600} \begin{vmatrix} j1.67 & -j1.67 \\ -1.1 & (2.77 + j1.67) \end{vmatrix}$$

$$= \frac{E(10^{-6})}{600} (2.79 - j2.79)$$

$$N_3 = \begin{vmatrix} (2.77 \times 10^{-3} - j1.67 \times 10^{-3}) & j1.67 \times 10^{-3} & \dfrac{E}{600} \\ j1.67 \times 10^{-3} & 2.5 \times 10^{-3} & 0 \\ -1.1 \times 10^{-3} & -j1.67 \times 10^{-3} & 0 \end{vmatrix}$$

$$= \frac{E(10^{-6})}{600} \begin{vmatrix} j1.67 & 2.5 \\ -1.1 & -j1.67 \end{vmatrix} = \frac{E(5.54) \times 10^{-6}}{600}$$

V_1, V_2, and V_3 are evaluated, with slide-rule accuracy, to be

$$V_1 = \frac{E(9.72 + j4.17) \times 10^{-6}}{600(32.47 \times 10^{-9})} = E(0.499 + j0.214)$$

$$V_2 = \frac{E(2.79 - j2.79) \times 10^{-6}}{600(32.47 \times 10^{-9})} = E(0.143 - j0.143)$$

$$V_3 = \frac{E(5.54) \times 10^{-6}}{600(32.47 \times 10^{-9})} = E(0.284)$$

To check these results, the node voltages can also be computed from the values of I_1, I_2, and I_3 obtained previously by the mesh analysis.

From the circuit diagram it can be seen that, with respect to the reference node,

$$V_1 \text{ at node } 1 = E - I_1(R_g)$$
$$V_2 \text{ at node } 2 = Z_2(I_1 - I_3)$$
$$V_3 \text{ at node } 3 = I_3 Z_L$$

Substituting the appropriate values in these expressions, the node voltages are, with slide-rule accuracy, as follows:

$$V_1 = E - 600E(83.3 - j35.7)(10^{-5})$$
$$= E[1 - (50{,}000 - j21{,}400)(10^{-5})] = E(0.5 + j.214)$$
$$V_2 = 400[E(83.3 - j35.7)(10^{-5}) - E(47.6)(10^{-5})]$$
$$= E(10^{-5})(400)(35.7 - j35.7) = E(0.143 - j0.143)$$
$$V_3 = 600E(47.6)(10^{-5}) = E(0.285)$$

Thus the mesh analysis yields the identical information that the nodal analysis does, except that, where the end results desired are voltages across network branches, the nodal method is more direct.

6-7 Network Topology

Topology may be defined as a nonquantitative geometry which is concerned with those properties of geometric figures which would remain invariant if the figures were drawn on a rubber sheet or modeled in rubber and were then subjected to deformation. For example, the figures shown in Fig. 6-11 are topologically equivalent (or homeomorphic). Topology has been used in the past to solve such baffling problems as the Koenigsberg Bridge problem and to prove that no more than five different colors are required to color a map drawn on a plane so that neighboring states are distinguishable from each other by their different hues.[1]

The importance of topology to network analysis lies in the fact that, if all the electrical properties are abstracted from a network, there remains a

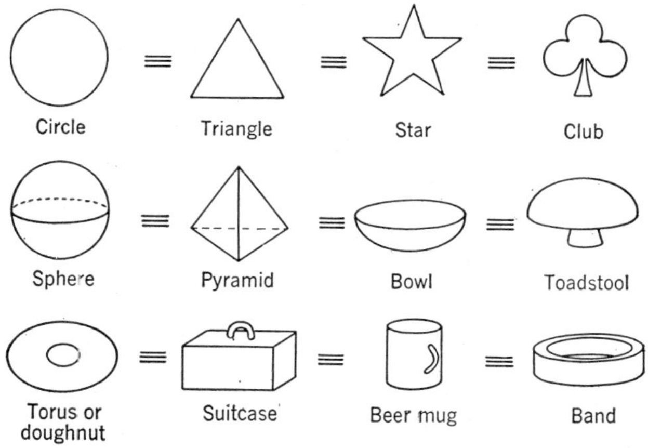

FIG. 6-11 Topological equivalents

geometrical circuit completely characterized by sets of branches terminating at various vertexes. From this geometrical circuit, it is possible to deduce many general properties of the network and to determine whether the network can be most easily solved by the mesh or by the nodal analysis.[2] The most important consideration, of course, is the question of which method will produce the smaller number of simultaneous equations and the resultant lower-order determinant.

[1] For further information on this interesting topic see A. Tucker and A. Bailey, Topology, *Sci. American*, January, 1950; E. Kasner and J. Newman, "Mathematics and the Imagination," Simon and Schuster, Inc., New York, 1950; H. Steinhaus, "Mathematical Snapshots," Oxford University Press, New York, 1950; S. Lefschetz, "Introduction to Topology," Princeton University Press, Princeton, N.J., 1949.

[2] R. M. Foster, Geometrical Circuits of Electrical Networks, *Trans. AIEE*, **51**: 309–317, June, 1932; J. L. Synge, The Fundamental Theorem of Electrical Networks, *Quart. Appl. Math.*, **9** (2): 113–127, July, 1951, and **11** (2): 215, July, 1953.

Consider the network shown in Fig. 6-12. In order to establish the nomenclature of network geometry, the outlines of a rectangular mesh have been drawn omitting the details of the network impedance elements. It is to be understood that some form of impedance element is connected between each pair of points 1–2, 2–3, 3–4, and 4–1. The points 1, 2, 3, and 4 are called nodes. A *node* is defined as a terminal of any branch

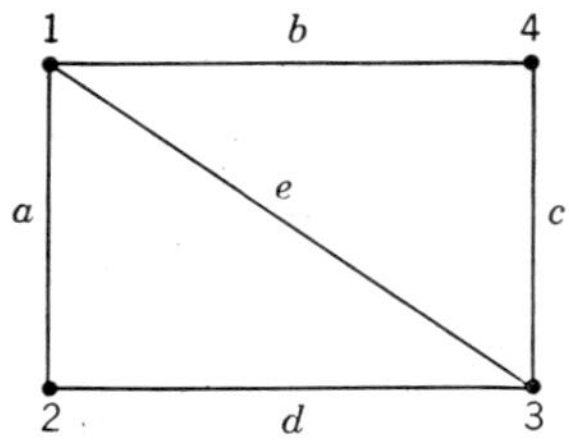

FIG. 6-12 Topological network.

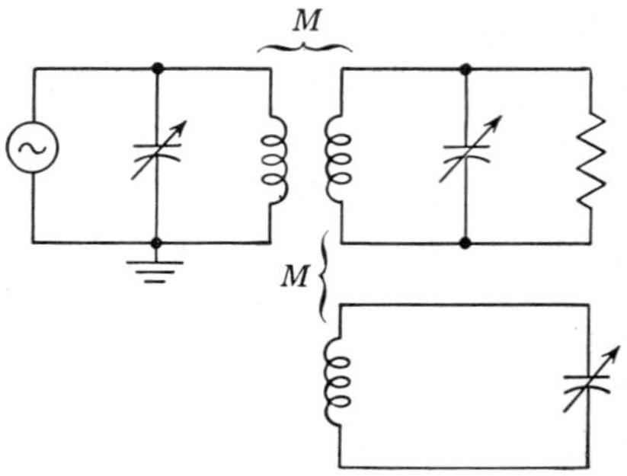

FIG. 6-13 Network consisting of three separate parts inductively coupled.

of a network or a terminal common to two or more branches of a network and is identical with a junction point, branch point, or vertex. The lines a, b, c, d, and e, which connect points 1, 2, 3, and 4, are called branches. A *branch* is defined as a portion of a network consisting of one or more two-terminal elements (electric devices such as impedances or generators) in series.

A *mesh* is a set of branches forming a closed path in a network, provided that, if any one branch is omitted from the set, the remaining branches

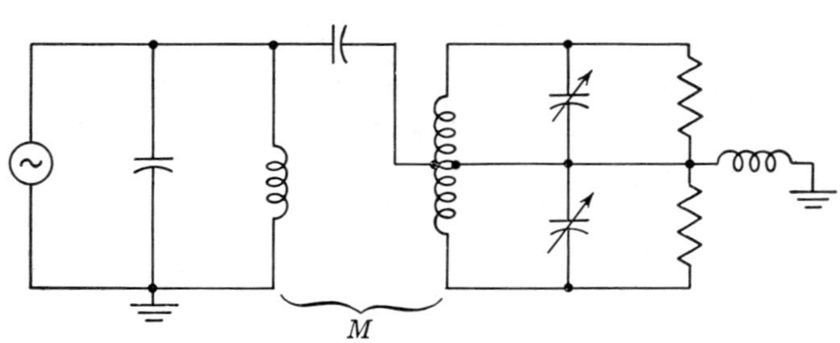

FIG. 6-14 Network consisting of one separate part.

of the set do not form a closed path. In Fig. 6-12, there are three sets of branches, *abcd*, *bce*, and *aed*, which form meshes.

A part of a network not directly connected, i.e., inductively coupled, is defined as a *separate part*. In Fig. 6-13, the network shown consists of three separate parts. The network in Fig. 6-14, on the other hand, consists of one separate part only.

Referring to Fig. 6-15, a set of connected branches which encloses no meshes, as shown, is defined as a *tree*. It can be seen that any network which consists of one or more meshes can be reduced to a tree by cutting or removing branches without disturbing the nodes. In other words, in a given connected network it is possible by cutting branches to construct a tree containing all the nodes of the network. Furthermore, the number of nodes V is

$$V = B + 1 \tag{6-36}$$

where B is the number of branches in the tree thus formed. For example, in Fig. 6-15, $B = 5$ and $V = 6$.

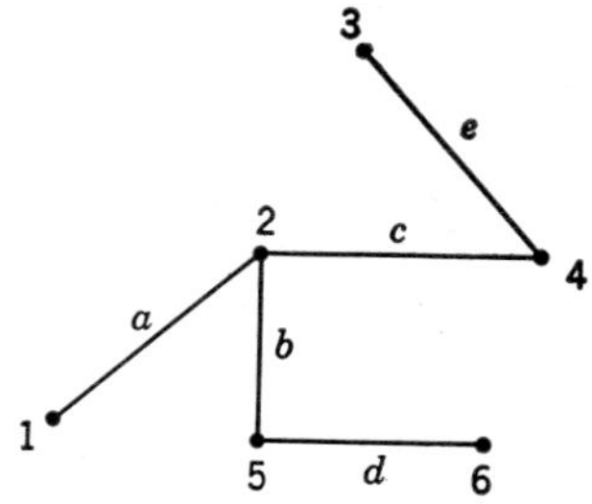

FIG. 6-15 Tree containing five branches and six nodes.

If a network has V nodes and B branches and if β branches must be removed or cut to convert it to a tree, it has been found that the relationship between V, B, and β is always

$$V = 1 + B - \beta \tag{6-37}$$

For the network in Fig. 6-12, $B = 5$, $\beta = 2$, and $V = 4$, which corresponds to the numbering of the nodes shown. The quantity β is known as the Betti number after Enrico Betti, an Italian mathematical physicist.[1]

Rewriting Eq. (6-37), an expression for the Betti number is obtained:

$$\beta = 1 + B - V \tag{6-38}$$

But the Betti number is also the number of meshes in a connected network drawn entirely in a plane with no intersecting paths, since β branches must be cut to convert a network which contains β meshes to a tree. Therefore, the number of meshes in a network N, which is also called the *nullity* of the network, must equal the Betti number, and, according to Eq. (6-38),

$$N = \beta = 1 + B - V \tag{6-39}$$

N, the nullity, is obviously also the number of independent simultaneous equations and the order of determinant which must be solved using the mesh analysis. The numerical constant 1 corresponds to a network which can be drawn in or confined to a single plane without intersecting paths, or a network which consists of one separate part.

[1] This relationship was originally established by Poincaré after earlier work by Betti and Riemann. The Betti number or its equivalent was used by Kirchhoff and Maxwell. For a rigorous proof of this relationship see O. Veblen, "Analysis Situs," 2d ed., American Mathematical Society, Providence, R. I., 1931.

A more general expression for the nullity N can be written as

$$N = P + B - V \tag{6-40}$$

where the constant 1 has been replaced by P, the number of separate parts in the network. For a network with no magnetically coupled sections, i.e., which has only one separate part or which can be drawn in a plane without intersecting branches, P is unity. For networks with more than one separate part, P is correspondingly greater than unity. For a network which cannot be drawn in a plane, but which can be drawn on a sphere without intersecting branches, such as a rounded cube or lattice configuration, $P = 2$.

In the nodal analysis, the number of independent nodal equations R, which is defined as the *rank* of the network, will obviously be one less than the number of nodes because no nodal equation is written for the reference node. In general, it can be shown that the number of independent nodal equations R required for the nodal analysis is

$$R = V - P \tag{6-41}$$

where P is the number of separate parts in the network.

It is now possible to compare the mesh and nodal methods of analysis as to economy of labor required. Assume a network in which the rank and nullity are the same. Then, the number of simultaneous equations required by either method will be identical, and

$$N = R \tag{6-42}$$

Substituting the values of N and R developed in Eqs. (6-40) and (6-41) gives

$$P + B - V = V - P \tag{6-43}$$

from which

$$B = 2(V - P) \tag{6-44}$$

If the mesh analysis requires *more* equations than the nodal, the left-hand side of the above equation will *exceed* the right and

$$B > 2(V - P) \tag{6-45}$$

Conversely, if the mesh analysis requires *fewer* equations than the nodal, the right-hand member of the inequality will exceed the left-hand member and

$$B < 2(V - P) \tag{6-46}$$

These results can be summarized as follows:

When $B > 2(V - P)$: nodal analysis is preferable.
When $B = 2(V - P)$: use either system.
When $B < 2(V - P)$: mesh analysis is preferable.

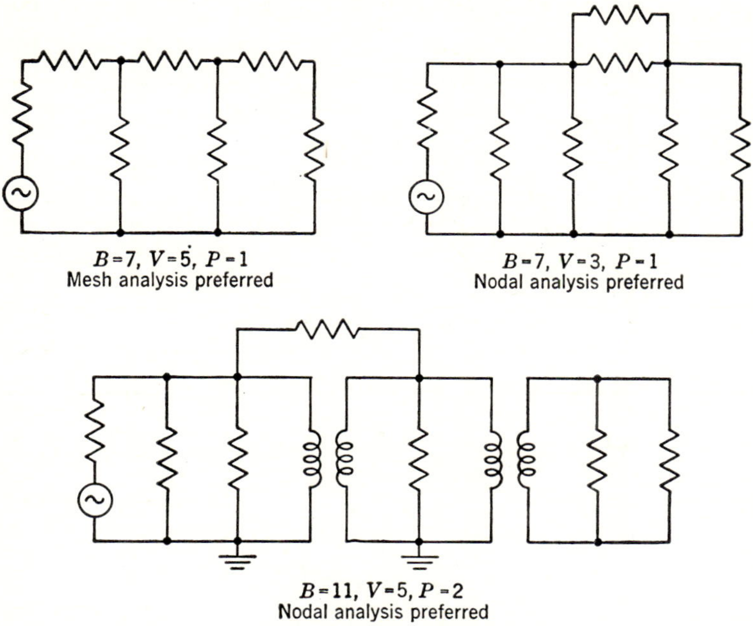

FIG. 6-16 Topological analysis of three networks.

In Fig. 6-16, three networks have been analyzed and the best method of analysis in each case is indicated. It should be noted that it is a topological fact that a network which contains five or more nodes cannot be drawn in a plane with a separate branch connecting each node with every other without the use of intersecting lines. For such cases, P must be greater than unity, and such networks will always contain at least one configuration such as is shown in Fig. 6-17.

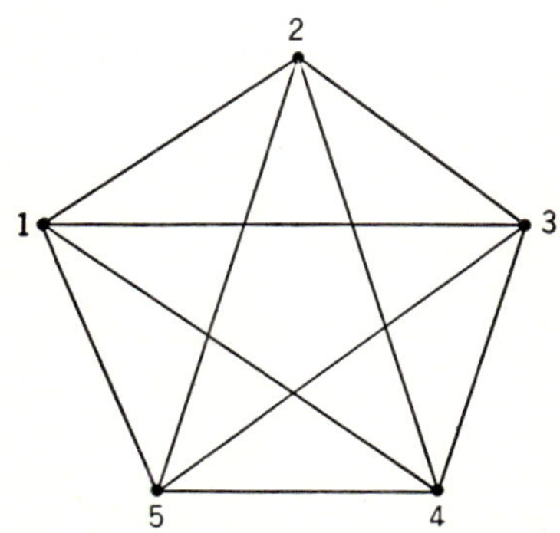

FIG. 6-17 Nonplanar network configuration in which V = 5.

Topology has also been employed in the development of a method of network analysis which avoids the use of simultaneous equations and determinants. In this method, developed by Percival,[1] the original network is replaced by trees and "two-trees," so that network solutions can be obtained directly

[1] W. S. Percival, The Solution of Passive Electrical Networks by Means of Mathematical Trees, *Proc. Inst. Elec. Engrs. (London)*, pt. III, **100**: 143–150, May, 1953; J. Lantieri, Method of Determining the Trees of a Network, *Ann. Telecommun.*, **5**: 204–208, May, 1950.

from the simplified diagrams by ordinary algebra. The methods involved are beyond the scope of this text, but Percival's work is recommended to all who are interested in a new approach to network analysis at a more advanced level.

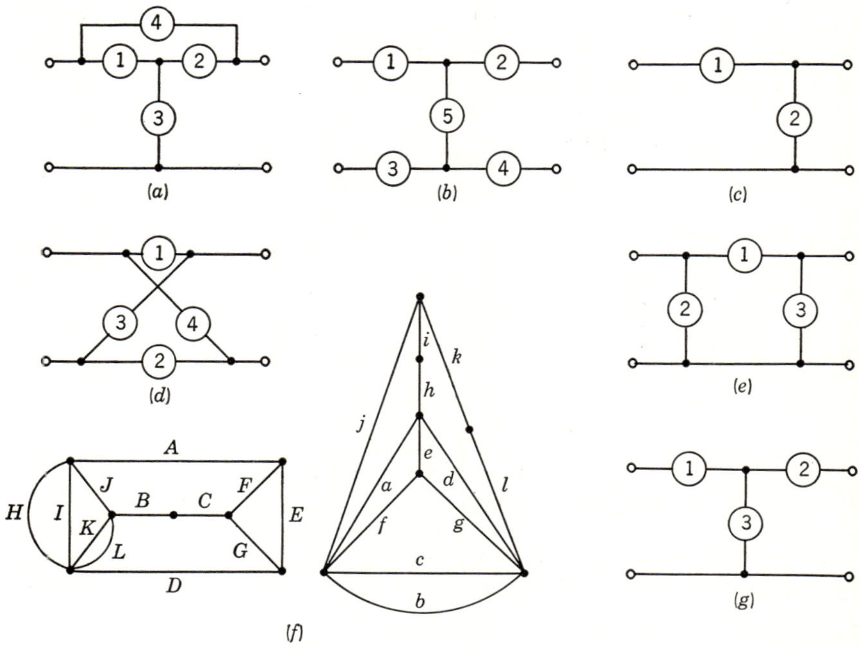

FIG. 6-18 Definitions of terms in network topology. (*a*) Bridged-T network. Branches 1, 2, and 3 comprise the T network, and branch 4 is the fourth branch. (*b*) H network. Branches 1 and 2 are the first two branches between an input and an output terminal; branches 3 and 4 are the second two branches; and branch 5 is the branch between the junction points. (*c*) L network. The free ends are the left-hand terminal pair, and the junction point and one free end are the right-hand terminal pair. (*d*) Lattice network. In the mesh 1, 2, 3, 4, the junction points between branches 4 and 1 and between branches 3 and 2 are the input terminals, and the junction points between branches 1 and 3 and between branches 2 and 4 are the output terminals. (*e*) π network. The junction point between branches 1 and 2 forms an input terminal, that between branches 1 and 3 forms an output terminal, and that between branches 2 and 3 forms a common input and output terminal. (*f*) Structurally dual networks. For example, the mesh *EFG* in the rectangular figure corresponds to the cut-set *efg* in the triangular figure, the mesh *bc* to the cut-set *BC*, and the mesh *JAEGCB* to the cut-set *jaegcb*. (*g*) T network. One end of each of the branches 1, 2, and 3 is connected to a common point. The other ends of branches 1 and 2 form, respectively, an input and an output terminal, and the other end of branch 3 forms a common input and output terminal.

In addition to those terms in network topology which have already been defined and those cited in Fig. 6-18, there are a number of others with which the reader should be familiar. These appear in the list as follows:

Definitions of Terms in Network Topology[1]

Accessible terminal. A network node that is available for external connections.

Arm. See Branch.

Branch (*arm*). A portion of a network consisting of one or more two-terminal elements in series.

Branch point. See Node.

Bridged-T network. A T network with a fourth branch connected across the two series arms of the T, between an input terminal and an output terminal.

Cascade. See Tandem.

Circuit. A network providing one or more closed paths.

Connected. A network is connected if there exists at least one path, composed of branches of the network, between every pair of nodes of the network.

Cut-set. A set of branches of a network such that the cutting of all the branches of the set increases the number of separate parts of the network, but the cutting of all the branches except one does not.

Degrees of freedom on a mesh basis. See Nullity.

Degrees of freedom on a node basis. See Rank.

Delta network. A set of three branches connected in series to form a mesh.

Dual networks. See Structurally dual networks.

Element. Any electrical device (such as inductor, resistor, capacitor, generator, line, electron tube) with terminals at which it may be directly connected to other electrical devices.

Four-pole. See Two-terminal-pair network.

H network. A network composed of five branches, two connected in series between an input terminal and an output terminal, two connected in series between another input terminal and output terminal, and the fifth connected from the junction point of the first two branches to the junction point of the second two branches.

Junction point. See Node.

L network. A network composed of two branches in series, the free ends being connected to one pair of terminals, and the junction point and one free end being connected to another pair of terminals.

Ladder network. A network composed of a sequence of H, L, T, or π networks connected in tandem.

Lattice network. A network composed of four branches connected in series to form a mesh, two nonadjacent junction points serving as input terminals, while the remaining two junction points serve as output terminals.

Loop. See note under Mesh.

Mesh. A set of branches forming a closed path in a network, provided that if any one branch is omitted from the set, the remaining branches of the set do not form a closed path. (NOTE: The term *loop* is sometimes used in the sense of *mesh.*)

Network. A combination of elements.

Node (*junction point*), *branch point, vertex.* A terminal of any branch of a network or a terminal common to two or more branches of a network.

Nonplanar network. A network which cannot be drawn on a plane without crossing of branches.

N-terminal network. A network with N accessible terminals.

N-terminal-pair network. A network with $2N$ accessible terminals grouped in pairs. In such a network one terminal of each pair may coincide with a network node.

[1] Standards on Circuits: Definitions of Terms in Network Topology, 1950, *Proc. I.R.E.*, **39** (1): 27–29, January, 1951.

Nullity (*degrees of freedom on a mesh basis*). The number of independent meshes that can be selected in a network. The nullity N is equal to the number of branches B minus the number of nodes V plus the number of separate parts P. $N = B - V + P$

Parallel elements. (*a*) Two-terminal elements are connected in parallel when they are connected between the same pair of nodes. (*b*) Two-terminal elements are connected in parallel when any cut-set including one must include the others.

Parallel two-terminal-pair networks. Two-terminal-pair networks are connected in parallel at the input or at the output terminals when their respective input or output terminals are in parallel.

π network. A network composed of three branches connected in series with each other to form a mesh, the three junction points forming an input terminal, an output terminal, and a common input and output terminal, respectively.

Planar network. A network which can be drawn on a plane without crossing of branches.

Quadripole. See Two-terminal-pair network.

Rank (*degrees of freedom on a node basis*). The number of independent cut-sets that can be selected in a network. The rank R is equal to the number of nodes V minus the number of separate parts P. $R = V - P$.

Separate parts of a network. The parts which are not connected.

Series elements. (*a*) Two-terminal elements are connected in series when they form a path between two nodes of a network such that only elements of this path, and no other elements, terminate at intermediate nodes along the path. (*b*) Two-terminal elements are connected in series when any mesh including one must include the others.

Series two-terminal-pair networks. Two-terminal-pair networks are connected in series at the input or at the output terminals when their respective input or output terminals are in series.

Star network. A set of three or more branches with one terminal of each connected at a common node.

Structurally dual networks. A pair of networks such that their branches can be marked in one-to-one correspondence so that any mesh of one corresponds to a cut-set of the other. Each network of such a pair is said to be the dual of the other.

Structurally symmetrical network. A network which can be arranged so that a cut through the network produces two parts that are mirror images of each other.

Symmetrical network. See Structurally symmetrical network.

Tandem (*cascade*). Two-terminal-pair networks are in tandem when the output terminals of one network are directly connected to the input terminals of the other network.

Terminal. A point at which any element may be directly connected to one or more other elements.

Terminal pair. An associated pair of accessible terminals, such as input pair, output pair, and the like.

T network. A network composed of three branches with one end of each branch connected to a common junction point, and with the three remaining ends connected to an input and output terminal, respectively.

Tree. A set of connected branches including no meshes.

Two-terminal-pair network (*quadripole, four-pole*). A network with four accessible terminals grouped in pairs. In such a network one terminal of each pair may coincide with a network node.

Vertex. See Node.

Y network. A star network of three branches.

6-8 Two-terminal-pair (Four-pole) Networks

The specialized network configuration which is probably of the greatest importance to the television and electronics engineer is the two-terminal-pair, or four-pole, network. This network possesses a pair of input and a pair of output terminals, which is a general description of much of the apparatus with which the engineer deals.

It can be shown that four independent parameters (which are generally all a function of frequency) can be employed to specify the performance of a linear four-pole structure, passive or active. If the structure is passive, i.e., contains no generators, then only three of the parameters are independent, and one may be eliminated if desired. Therefore, any linear passive four-pole network can be represented by T- or π-network

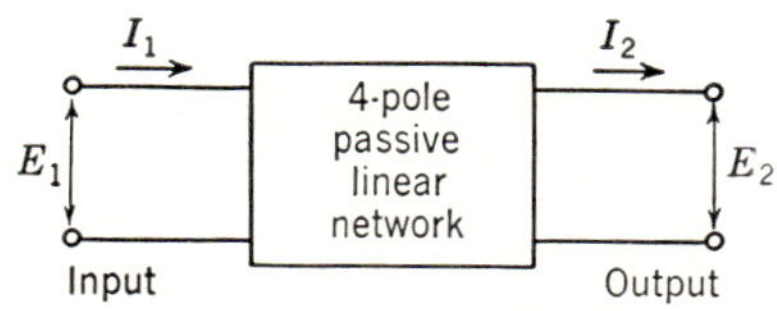

FIG. 6-19 Four-pole network.

equivalent at a given frequency if the *three* elements are properly proportioned; although all these elements may not always be physically realizable. It is the object of this section to summarize the relationships between the various network equivalents of the passive four-pole network.

Consider the four-pole network shown in Fig. 6-19. A number of sets of equations can be written representing the relations between V_1, I_1, V_2, and I_2. Assume that the input and output voltage polarities are arbitrarily assigned so that, from the mesh analysis,

$$\begin{aligned} E_1 &= Z_{11}I_1 + Z_{12}I_2 \\ E_2 &= Z_{21}I_1 + Z_{22}I_2 \end{aligned} \tag{6-47}$$

and, from the nodal analysis,

$$\begin{aligned} I_1 &= Y_{11}E_1 + Y_{12}E_2 \\ I_2 &= Y_{21}E_1 + Y_{22}E_2 \end{aligned} \tag{6-48}$$

If the voltage and current at the input end of the network are expressed in terms of the voltage and current at the output, then

$$\begin{aligned} E_1 &= AE_2 + BI_2 \\ I_1 &= CE_2 + DI_2 \end{aligned} \tag{6-49}$$

The impedance coefficients Z_{11}, Z_{12}, Z_{21}, and Z_{22} in Eq. (6-47) can be determined by measurement as follows, substituting each condition in Eq. (6-47):

$$
\begin{aligned}
Z_{11} &= \frac{E_1}{I_1} && \text{with } I_2 = 0 \text{ (output terminals open)} \\
Z_{21} &= \frac{E_2}{I_1} && \text{with } I_2 = 0 \text{ (output terminals open)} \\
Z_{12} &= \frac{E_1}{I_2} && \text{with } I_1 = 0 \text{ (input terminals open)} \\
Z_{22} &= \frac{E_2}{I_2} && \text{with } I_1 = 0 \text{ (input terminals open)}
\end{aligned}
\tag{6-50}
$$

The admittance coefficients Y_{11}, Y_{12}, Y_{21}, and Y_{33} in Eq. (6-48) can be determined by measurement as follows, substituting each condition in Eq. (6-48):

$$
\begin{aligned}
Y_{11} &= \frac{I_1}{E_1} && \text{with } E_2 = 0 \text{ (output terminals shorted)} \\
Y_{21} &= \frac{I_2}{E_1} && \text{with } E_2 = 0 \text{ (output terminals shorted)} \\
Y_{12} &= \frac{I_1}{E_2} && \text{with } E_1 = 0 \text{ (input terminals shorted)} \\
Y_{22} &= \frac{I_2}{E_2} && \text{with } E_1 = 0 \text{ (input terminals shorted)}
\end{aligned}
\tag{6-51}
$$

The general four-pole parameters A, B, C, and D in Eq. (6-49) can be determined by measurement as follows, substituting each condition in Eq. (6-49):

$$
\begin{aligned}
A &= \frac{E_1}{E_2} && \text{with } I_2 = 0 \text{ (output terminals open)} \\
B &= \frac{E_1}{I_2} && \text{with } E_2 = 0 \text{ (output terminals shorted)} \\
C &= \frac{I_1}{E_2} && \text{with } I_2 = 0 \text{ (output terminals open)} \\
D &= \frac{I_1}{I_2} && \text{with } E_2 = 0 \text{ (output terminals shorted)}
\end{aligned}
\tag{6-52}
$$

It can be seen from Eq. (6-52) that A and D are dimensionless constants, while B has the dimensions of impedance, and C the dimensions of admittance. Furthermore, since only three independent variables are necessary to define adequately a passive four-pole network, it can be shown that

$$AD - BC = 1 \tag{6-53}$$

If the four-pole network is lossless, then A and D will be real numbers and B and C imaginary. In a symmetrical network, $A = D$, and, if the network is terminated in $Z_C = \sqrt{B/C}$, then the input impedance will also be Z_C.

In addition to the three methods of specifying the properties of a four-pole network just treated, there are three others of importance:

(a) T-section equivalent impedances (see Fig. 6-20)

(b) π-section equivalent impedances (see Fig. 6-21)

(c) Image parameters

In the case of image parameters, W_1 is the image impedance at the input and W_2 is the image impedance at the output, so that, with W_2 as

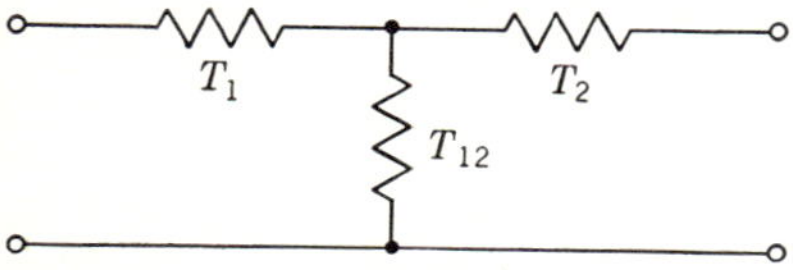

FIG. 6-20 T-section equivalent impedances.

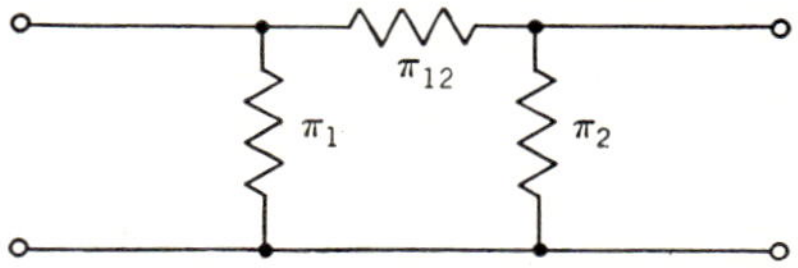

FIG. 6-21 π-section equivalent impedances.

the load, the input impedance of the network is W_1 and, with W_1 as the source impedance, the output impedance of the network is W_2.

The image transfer constant θ is defined by the equation

$$\frac{E_2}{E_1} = \left[\frac{W_2}{W_1}\right]^{1/2} \epsilon^{-\theta} \tag{6-54}$$

The real part of the image transfer constant is the image attenuation constant, and the imaginary part is the image phase constant.

The relationships of all these network equivalents to the general $ABCD$ network parameters are shown in Table 6-1. It will be noted that in the image-parameter case there is more than one relationship, as indicated by double algebraic signs ($\pm$) accompanying radicals and by

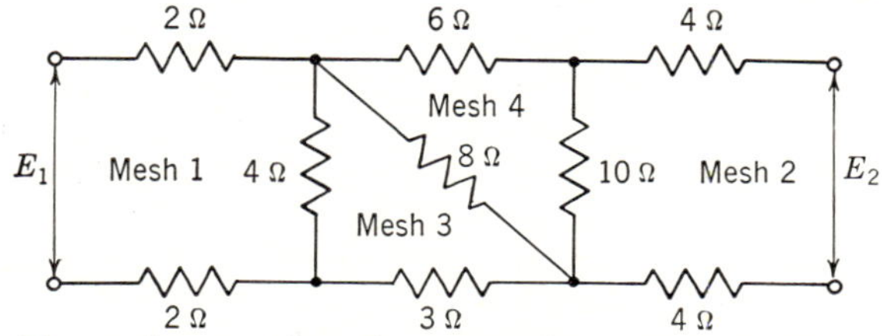

FIG. 6-22 Four-pole network to be analyzed in terms of $ABCD$ parameters.

the multiple-valued inverse hyperbolic function. However, if the Y, Z, T, or π parameters are specified, the network is unique and the relations between these sets are single-valued.

To illustrate exactly how the equivalents given in Table 6-1 are used, consider the network shown in Fig. 6-22. In this case, the internal nature of the network is known, and the $ABCD$ parameters can be evaluated by mesh analysis. If the network were unknown, however, these parameters could be evaluated by three measurements (in the case of a passive network) conducted according to the conditions specified for

TABLE 6-1. RELATIONSHIPS BETWEEN VARIOUS FOUR-POLE BILATERAL PASSIVE-NETWORK PARAMETERS AND *ABCD* PARAMETERS

	ABCD parameters
Z coefficients: Z_{11}	$\frac{A}{C}$
Z_{21}	$\frac{1}{C}$
Z_{22}	$-\frac{D}{C}$
Y coefficients: Y_{11}	$\frac{D}{B}$
Y_{21}	$\frac{1}{B}$
Y_{22}	$-\frac{A}{B}$
T equivalents: T_1	$\frac{A-1}{C}$
T_{12}	$\frac{1}{C}$
T_2	$\frac{D-1}{C}$
π equivalents: π_1	$\frac{B}{D-1}$
π_{12}	B
π_2	$\frac{B}{A-1}$
Image parameters: W_1	$\pm\sqrt{\frac{AB}{CD}}$
W_2	$\pm\sqrt{\frac{BD}{AC}}$
θ	$\cosh^{-1}(\pm\sqrt{AD})$

NOTE: The relationships between the Z and Y coefficients and the *ABCD* parameters can be derived by interrelating Eqs. (6-47) and (6-48) with Eq. (6-49). Because of the conventions chosen (polarity of upper terminals in Fig. 6-19 positive for all cases), Z_{12}, Z_{22}, Y_{12}, and Y_{22} are negative when related to the *ABCD* parameters. A complete summary of these relationships is also given in Table 8-1. Note also that the admittance and impedance coefficients Y_{22} and Z_{22} are negative when related to the short-circuit output admittance y_{22} as given by Eq. (6-13) and the open-circuit output impedance z_{22} as given by Eq. (6-31).

Eq. (6-52). If the upper input and output terminals of the network are taken as positive and cyclic currents are assigned as shown *with mesh* 1 *at the input and mesh* 2 *at the output*, the network equations are

$$\begin{aligned} E_1 &= 8I_1 + 0 - 4I_3 + 0 \\ E_2 &= 0 + 18I_2 + 0 - 10I_4 \\ 0 &= -4I_1 + 0 + 15I_3 - 8I_4 \\ 0 &= 0 - 10I_2 - 8I_3 + 24I_4 \end{aligned} \tag{6-55}$$

The value of the network determinant Δ is

$$\Delta = \begin{vmatrix} 8 & 0 & -4 & 0 \\ 0 & 18 & 0 & -10 \\ -4 & 0 & 15 & -8 \\ 0 & -10 & -8 & 24 \end{vmatrix} = 25{,}312 \qquad (6\text{-}56)$$

The values of the minors B_{11}, B_{22}, and B_{21} are

$$B_{11} = \begin{vmatrix} 18 & 0 & -10 \\ 0 & 15 & -8 \\ -10 & -8 & 24 \end{vmatrix} = 3{,}828 \qquad (6\text{-}57)$$

$$B_{22} = \begin{vmatrix} 8 & -4 & 0 \\ -4 & 15 & -8 \\ 0 & -8 & 24 \end{vmatrix} = 1{,}984 \qquad (6\text{-}58)$$

$$B_{21} = -\begin{vmatrix} 0 & -4 & 0 \\ 0 & 15 & -8 \\ -10 & -8 & 24 \end{vmatrix} = 320 \qquad (6\text{-}59)$$

The admittance coefficients in Table 6-1 are related to the short-circuit admittances given by Eqs. (6-12), (6-13), and (6-14) (taking into account the note appended to Table 6-1) as follows: $y_{11} = Y_{11}$, $y_{21} = Y_{21}$, and $y_{22} = -Y_{22}$. On this basis, the admittance coefficients for this network are

$$\begin{aligned} Y_{11} &= \frac{B_{11}}{\Delta} = \frac{3{,}828}{25{,}312} = 0.1513 \text{ mho} \\ Y_{21} &= \frac{B_{21}}{\Delta} = \frac{320}{25{,}312} = 0.01263 \text{ mho} \\ Y_{22} &= \frac{-B_{22}}{\Delta} = \frac{-1{,}984}{25{,}312} = -0.0784 \text{ mho} \end{aligned} \qquad (6\text{-}60)$$

From the relationships in Table 6-1, the $ABCD$ parameters are

$$\begin{aligned} A &= \frac{-Y_{22}}{Y_{21}} = \frac{B_{22}}{B_{21}} = \frac{1{,}984}{320} = 6.2 \\ B &= \frac{1}{Y_{12}} = \frac{\Delta}{B_{21}} = \frac{25{,}312}{320} = 79.1 \text{ ohms} \\ D &= \frac{Y_{11}}{Y_{21}} = \frac{B_{11}}{B_{21}} = \frac{3{,}828}{320} = 11.96 \end{aligned} \qquad (6\text{-}61)$$

If $AD - BC = 1$, from Eq. (6-53),

$$C = \frac{AD - 1}{B} = 0.925 \text{ mho} \qquad (6\text{-}62)$$

From Table 6-1, the Z coefficients are

$$\begin{aligned} Z_{11} &= \frac{A}{C} = 6.7 \text{ ohms} \\ Z_{22} &= \frac{-D}{C} = -12.93 \text{ ohms} \\ Z_{21} &= \frac{1}{C} = 1.08 \text{ ohms} \end{aligned} \tag{6-63}$$

The T-network equivalents are

$$\begin{aligned} \mathrm{T}_1 &= \frac{A-1}{C} = \frac{6.2-1}{0.925} = 5.65 \text{ ohms} \\ \mathrm{T}_{12} &= \frac{1}{C} = \frac{1}{0.925} = 1.08 \text{ ohms} \\ \mathrm{T}_2 &= \frac{D-1}{C} = \frac{11.96-1}{0.925} = 11.85 \text{ ohms} \end{aligned} \tag{6-64}$$

The π-network equivalents are

$$\begin{aligned} \pi_1 &= \frac{B}{D-1} = \frac{79}{11.96-1} = 7.2 \text{ ohms} \\ \pi_{12} &= B = 79 \text{ ohms} \\ \pi_2 &= \frac{B}{A-1} = \frac{79}{6.2-1} = 15.2 \text{ ohms} \end{aligned} \tag{6-65}$$

The image parameters are

$$\begin{aligned} W_1 &= \sqrt{\frac{AB}{CD}} = \sqrt{\frac{(6.2)(79)}{(0.925)(11.96)}} = 6.65 \text{ ohms} \\ W_2 &= \sqrt{\frac{DB}{AC}} = \sqrt{\frac{(79)(11.96)}{(0.925)(6.2)}} = 12.83 \text{ ohms} \\ \theta &= \cosh^{-1}\sqrt{AD} = \cosh^{-1}\sqrt{(6.2)(11.96)} = 2.845 \end{aligned} \tag{6-66}$$

Thus the source impedance for this network must be 6.65 ohms and the load impedance 12.83 ohms for image-impedance operation. The equivalent T and π networks are shown in Fig. 6-23.

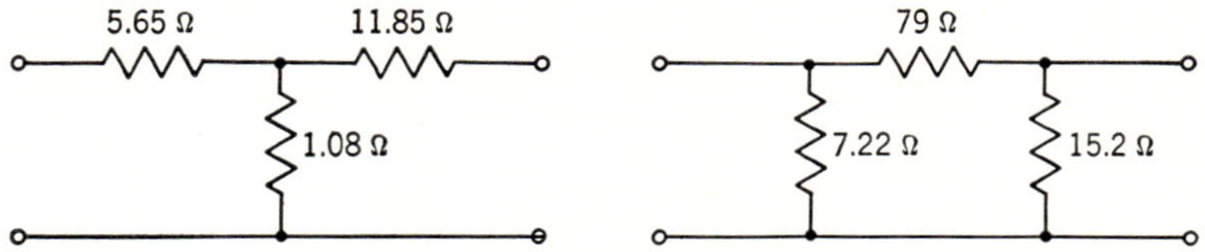

FIG. 6-23 Equivalent T and π networks for network of Fig. 6-22.

In closing this section, it should be noted that the use of the four-pole configuration in network analysis is not confined to passive circuits and has been applied to circuits containing both electron tubes and transistors as well.[1] In addition, the I.R.E. Electron Tube Committee has adopted the four-pole viewpoint and has proposed methods of tests for experimentally determining the four-pole parameters of electron tubes.[2] This approach will be considered in Chap. 8.

6-9 Electron-tube Networks

Generally, it is preferable to use the nodal method in the analysis of networks which contain electron-tube amplifiers. This is because there are usually many more branches than nodes in such circuits, and $B > 2(V - P)$.

In order to represent the electron tube in an equivalent network, it is customary to replace the tube with an equivalent constant-voltage or

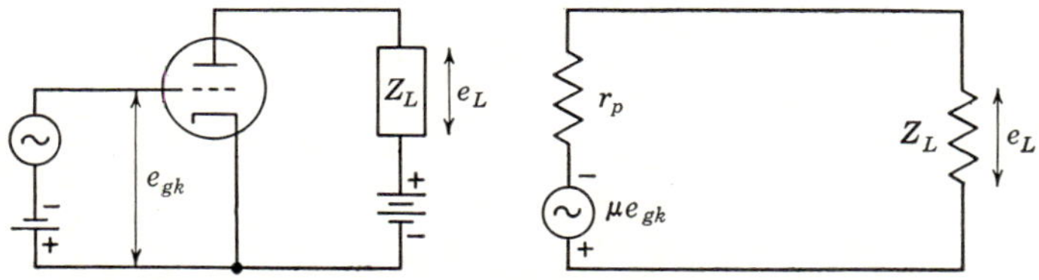

FIG. 6-24 A-C network equivalent of electron-tube amplifier.

constant-current generator. In Fig. 6-24 is shown the series, or constant-voltage, equivalent of the triode amplifier. This equivalent network is obtained on the basis of Thévenin's theorem so that, on the assumption that the operation of the tube is *linear*, it may be replaced by an open-circuit voltage of $-\mu e_{gk}$ in series with the internal plate resistance of the tube r_p where e_{gk} is the *grid-to-cathode* voltage. For example, for one section of a 6SL7 twin triode, the plate resistance r_p is 44,000 ohms and the amplification factor μ is 70 for conditions under which the quiescent (no-signal) plate voltage is 250 volts, the plate current is 2.3 ma, and the grid bias is -2 volts. The equivalent network for this tube is shown in Fig. 6-25 provided that the peak value of $e_{gk} < 0$ volts.

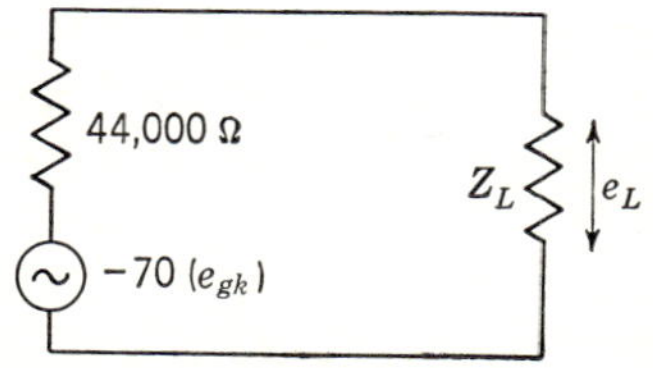

FIG. 6-25 Equivalent network of 6SL7 triode section.

From the equivalent network, the network voltage gain $\alpha = e_L/e_g$ will be

[1] L. C. Peterson, Equivalent Circuits of Linear Active Four-terminal Networks, *Bell System Tech. J.*, **27** (4): 593–622, October, 1948; L. J. Giacoletto, Terminology and Equations for Linear Active Four-terminal Networks, Including Transistors, *RCA Rev.*, **14** (1): 28–46, March, 1953.

[2] Standards on Electron Tubes: Methods of Testing, 1950, *Proc. I.R.E.*, **38** (8): 917–948, August, 1950.

$$\frac{e_L}{e_{gk}} = \frac{-\mu Z_L}{r_p + Z_L} \tag{6-67}$$

and if $Z_L = R_L = 0.5$ megohm, then the gain α is

$$\alpha = \frac{e_L}{e_{gk}} = \frac{-(70)(0.5 \times 10^6)}{0.544 \times 10^6} = -64.3 \tag{6-68}$$

where the negative sign indicates a phase reversal of 180°.

In Fig. 6-26 is shown the parallel, or constant-current, equivalent of the triode amplifier. This representation is especially valuable in the nodal analysis in which constant-current sources are required and is obtained by an application of Norton's theorem. Assuming that the operation of the tube is *linear*, it may then be replaced by a constant-current generator of $-g_m e_{gk}$ in parallel with the plate resistance r_p, where g_m is the transconductance of the tube.

The voltage e_L developed across Z_L and r_p in this network is

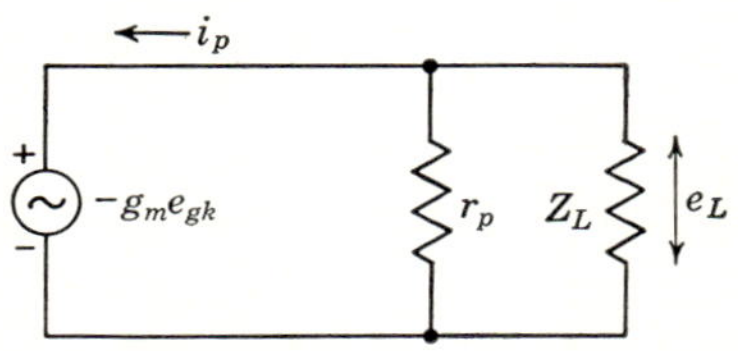

Fig. 6-26 Constant-current equivalent of triode.

$$e_L = \frac{-(g_m e_{gk})(r_p Z_L)}{r_p + Z_L} \tag{6-69}$$

and the gain α is

$$\alpha = \frac{e_L}{e_{gk}} = \frac{-g_m(r_p Z_L)}{r_p + Z_L} \tag{6-70}$$

Since in any electron tube,

$$\mu = g_m r_p \tag{6-71}$$

this expression for α is identical with that previously obtained, and the gain calculated for the 6SL7 triode can also be obtained from the above expression using a g_m of 1,600 μmho, which is the transconductance for the operating conditions specified.[1]

For pentode tubes, $r_p \gg Z_L$ in many cases, and the parallel plate resistance in the equivalent network may be omitted. This leads to a simplified expression for the gain of a pentode:

$$\alpha = -g_m Z_L \tag{6-72}$$

A brief comparison of triode characteristics with those of the pentode is of interest. Table 6-2 shows such a comparison for the type 6SJ7, which may be operated as a triode *or* a pentode. It will be noted that the g_m of the triode-connected 6SJ7 is about the same as that for pentode operation, but that the plate resistances in the two modes of operation differ by a factor of about 150:1. Thus the advantage gained by pentode operation, aside from the improved isolation between grid and plate

[1] It should be noted that these operating conditions were chosen for the sake of discussion and are considerably different from those which would probably obtain in a practical resistance-coupled 6SL7-stage design.

circuits, is the enormous increase in plate resistance possible which permits much higher amplifier-stage gains. This can be most clearly seen from the constant-current equivalent of the amplifier circuit, in which, for a fixed g_m, the gain will be a maximum when the r_p which shunts the load impedance is made as large as possible. For this reason, the order of magnitude of gains obtainable in amplifier stages will be higher with the use of tube types having larger plate resistances, which, on the basis of the constant-voltage equivalent circuit, may seem somewhat contradictory, but actually is not.

Once the equivalent plate circuits of the vacuum tube as a generator have been established, the equivalent networks of amplifiers and certain types of controlled-amplitude oscillators may be set up and solved by either the mesh or nodal analysis exactly as were previous networks

TABLE 6-2

Characteristic	6SJ7 (triode-connected)	6SJ7 (pentode-connected)
g_m	2,500 μmho	1,650 μmho
r_p	7,600 ohms	10^6 ohms (approx)
μ	19	1,650 (approx)
Screen current	....	0.8 ma
Plate current	9.2 ma	3.0 ma

which were passive rather than active, i.e., did not contain any internal generators. As was pointed out previously, the nodal method will generally be the most convenient for the analysis of such active equivalent networks, and it is interesting to note that it can also be adapted to an analysis of the *internal* electron-stream behavior of the tube in order to take into account the effects of finite electron transit time on amplifier operation.[1]

The simple equivalent plate networks developed above are adequate where the sole object of the network analysis is the behavior of the plate circuit alone. However, where some feedback path exists from the plate to the control grid, whether it be due to internal interelectrode capacitance or to an external feedback path deliberately provided, the grid voltage e_{gk} will no longer be independent of the plate circuit and the equivalent network of the electron tube must be elaborated upon.

Consider a four-terminal-network equivalent of the interelectrode capacitances of a triode operated with the cathode effectively grounded as shown in Fig. 6-27. C_{gk} represents the capacitance from grid to cathode, C_{gp} that from grid to plate, and C_{pk} the capacitance from plate to

[1] F. B. Llewellyn and L. C. Peterson, Vacuum Tube Networks, *Proc. I.R.E.*, **32** (3), March, 1944.

cathode. If the triode is to be operated at high audio or radio frequencies, these capacitances cannot be ignored and must be incorporated into the equivalent constant-current circuit of the triode as shown in Fig. 6-28.

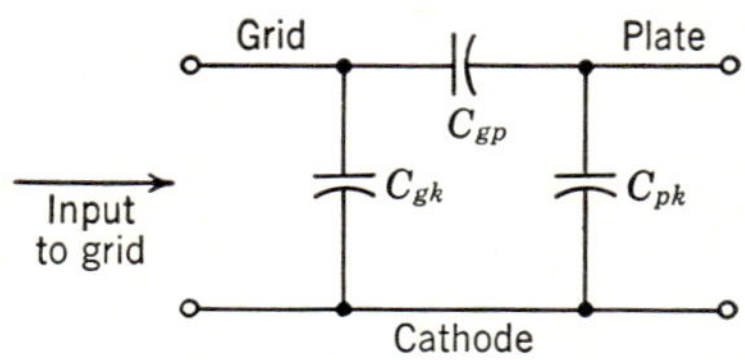

FIG. 6-27 Interelectrode capacitances of a triode.

The inclusion of these capacitances converts the constant-current equivalent of the triode into a *two*-node network. The equivalent network of the pentode would be approximately the same as that of the triode, except that r_p might be omitted if it were much greater than the magnitude of Z_L. C_{gp} would still be present, because even in a pentode C_{gp} is not completely eliminated but merely reduced.

The difference between the triode and the pentode with regard to interelectrode capacitances is shown in Table 6-3 for the 6SJ7. The pentode

TABLE 6-3

Capacitance	6SJ7	
	Triode-connected, μμf	Pentode-connected, μμf
Grid 1 to plate	2.8	0.005
Input	3.4	6.0
Output	11.0	7.0

is considerably superior to the triode in the size of C_{gp}. The input capacitance of the triode C_{gk} is smaller than that of the pentode because, in the pentode connection, capacitances exist between the control grid and the screen and suppressor grids which are effectively connected to the cathode in normal operation.

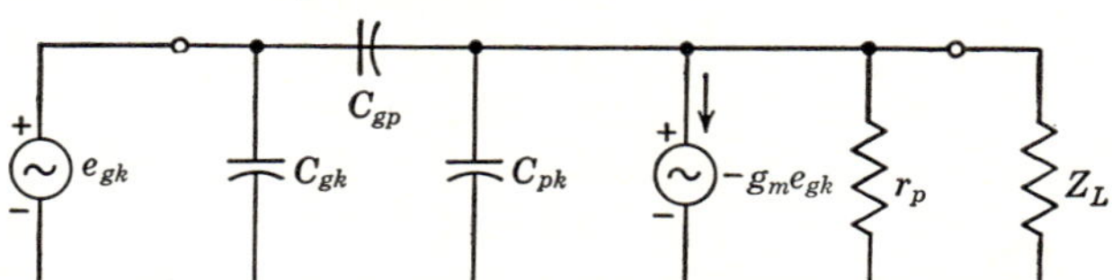

FIG. 6-28 Network equivalent of grounded-cathode triode.

Of course, if the operation of an amplifier stage is to be analyzed at 100 cps, the reactances of these internal capacitances will be so great that they can be considered open circuits as far as the remainder of the equivalent network is concerned. This obviously is not the case in an amplifier which is to be operated at video or radio frequencies or, for that matter, even at 15 kcs.

The equivalent network shown in Fig. 6-28 can be generalized by replacing the capacitances with admittances. In this form, the generalized two-node equivalent for a vacuum tube operated with a grounded cathode will be as shown in Fig. 6-29. In this network Y_s is the admittance of the signal source e_s; Y_1 is the total admittance of all paths

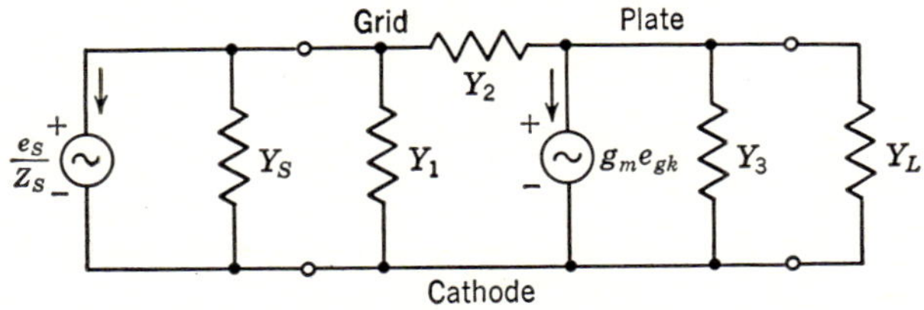

FIG. 6-29 Generalized equivalent network of grounded-cathode amplifier.

which exist between grid and cathode; Y_2 is the total admittance of all paths which exist from grid to plate; Y_3 is the reciprocal of the tube plate resistance combined with the total admittance of all parallel paths from plate to cathode exclusive of Y_L; and Y_L is the admittance of the load. It should be remembered that these admittances are perfectly general in nature and that Y_s, Y_1, Y_2, Y_3, and Y_L may be resistive, capacitive, or

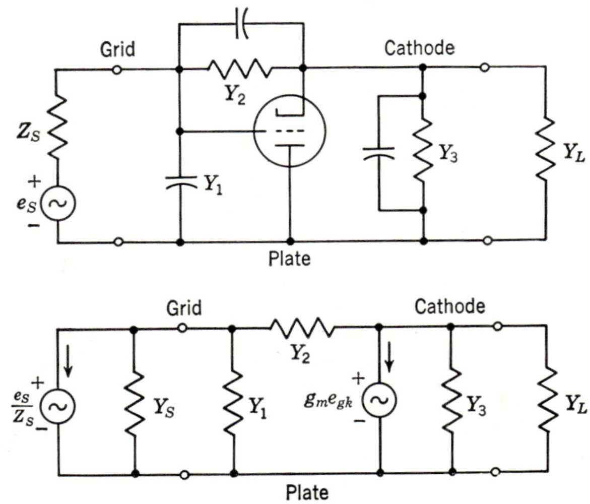

FIG. 6-30 Equivalent network of the grounded-plate amplifier.

inductive depending on the design of the particular network under analysis. The value of such a generalized network is that an analysis performed on the basis of such a generalized approach can be used to predict the performance of seemingly widely diverse types of circuits.

In modern electronic techniques, the grounded-cathode amplifier is but one of three amplifier configurations commonly employed. A second type is the grounded-plate or cathode-follower amplifier stage. The generalized two-node equivalent network for this amplifier is shown in Fig. 6-30. In this case, Y_1 is the total admittance of all paths from grid to plate, Y_2 is the total admittance of all paths from grid to cathode, Y_3 is

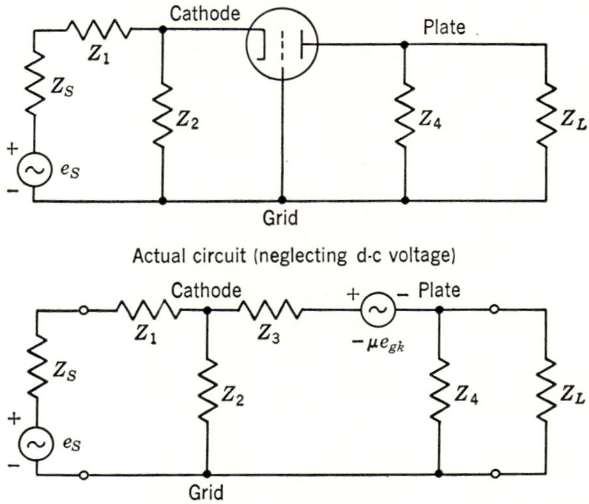

FIG. 6-31 Equivalent network of the grounded-grid amplifier using constant-voltage generator.

the total admittance of all paths from cathode to plate including the plate resistance of the tube, and Y_s and Y_L are as before.

The third amplifier configuration is that of the grounded-grid amplifier, which can be analyzed by either the mesh or nodal analysis. In Fig. 6-31 is shown the network equivalent suitable for the mesh analysis. In

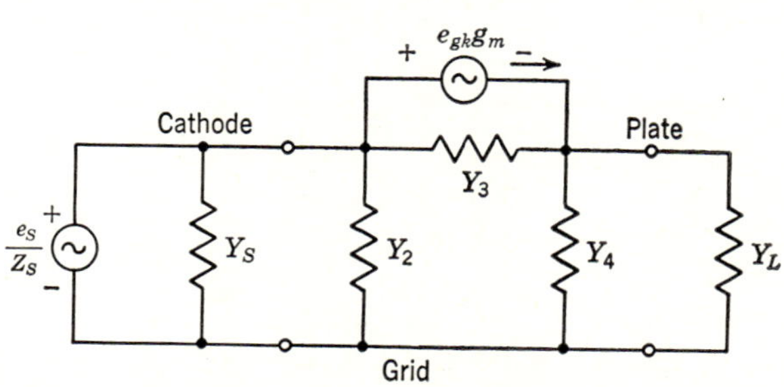

FIG. 6-32 Equivalent network of grounded-grid amplifier using constant-current generator.

this network, Z_1 is the total impedance between the cathode and the signal source, if any; Z_2 is the total impedance from cathode to grid; Z_3 is the total impedance from cathode to plate including the internal plate resistance of the tube; and Z_4 is the total impedance from the plate to the grid exclusive of the load impedance Z_L. The constant-current equivalent of this circuit is shown in Fig. 6-32, where Y_2 corresponds to Z_2, Y_3 to Z_3, Y_4 to Z_4, and Y_L to Z_L.

Before some of the practical applications of these equivalent networks are considered, some note should be made of those amplifier circuits in which the nodal analysis cannot be used with convenience because they contain some form of mutual coupling, either direct or magnetic.

For these cases, Dishal[1] has compiled a useful set of equivalent networks which may be employed to convert any given case to a form suitable for analysis by the nodal method.

Figure 6-33 shows one basic form of a two-node network consisting of two resonant circuits coupled together both inductively and capacitively. By virtue of π and T transformations and the use of transformers, i.e.,

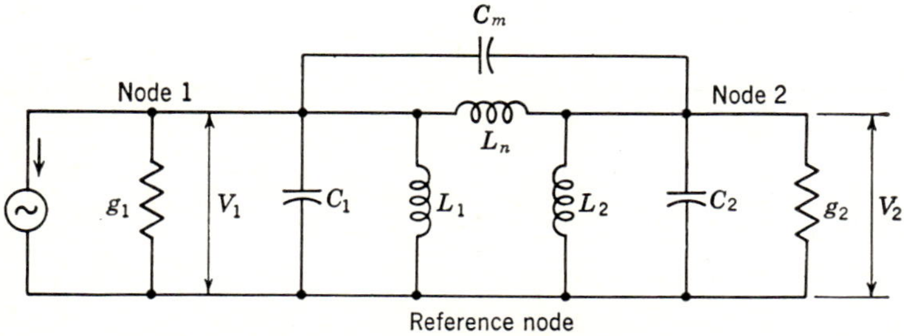

FIG. 6-33 Basic double-tuned two-node wide-band-pass circuit with both inductive and capacitive coupling.

magnetic coupling, a nodal analysis of the network in Fig. 6-33 will also apply to every one of the additional 10 two-node networks shown in Fig. 6-34. The manner in which the values of the equivalent elements are derived is shown in Fig. 6-35 for the most frequently encountered cases of inductive and capacitive coupling. Lattice or bridged-T equivalents are also possible.

If the analysis of two separate networks produces equations of exactly the same form so that the set obtained by the mesh analysis for one network can be converted to the set obtained by the nodal analysis of the second network by a simple interchange of symbols, the networks are said to be *duals*, or inverse structures. Thus, the solution of one can be made applicable to the other in many cases by the simple substitution of capacitance terms for all the self- and mutual-inductance terms, inductive terms for all the capacitive terms, and conductance terms for all the resistance terms in the original solution. In other words, the determinants obtained for dual networks are identical in form, and only the

[1] M. Dishal, Exact Design and Analysis of Double- and Tripled-tuned Band-pass Amplifiers, *Proc. I.R.E.*, **35** (6): 606–626, June, 1947.

dimensions of the quantities need be altered for the solution to be valid in either case.

In Fig. 6-36 is shown a dual of the original circuit in Fig. 6-33. The mathematics of the solution obtained for the network of Fig. 6-36 will

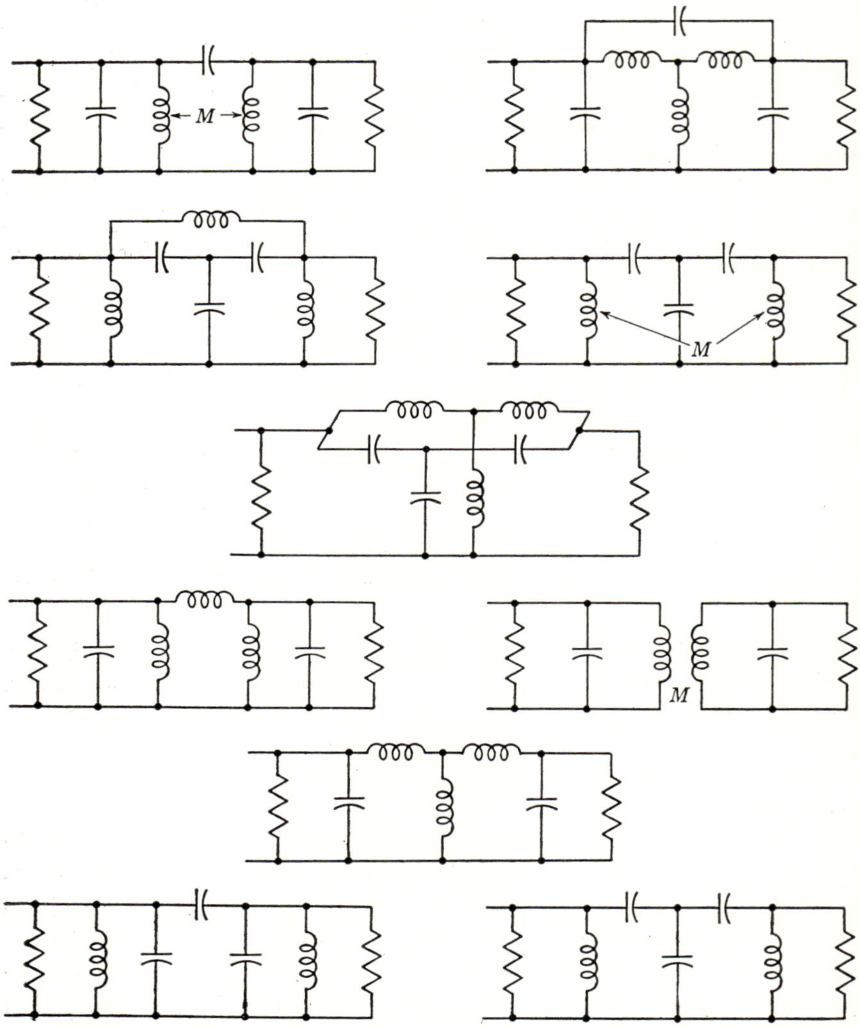

FIG. 6-34 Ten two-node circuits. The circuit of Fig. 6-33 is exactly equivalent to these circuits. (*After Dishal.*)

be identical to that obtained for the two-node network of Fig. 6-33 except that e, R, L, and C terms are respectively substituted in place of the i, G, C, and L terms used in the latter case.

The advantage of this network duality is that the two-mesh network

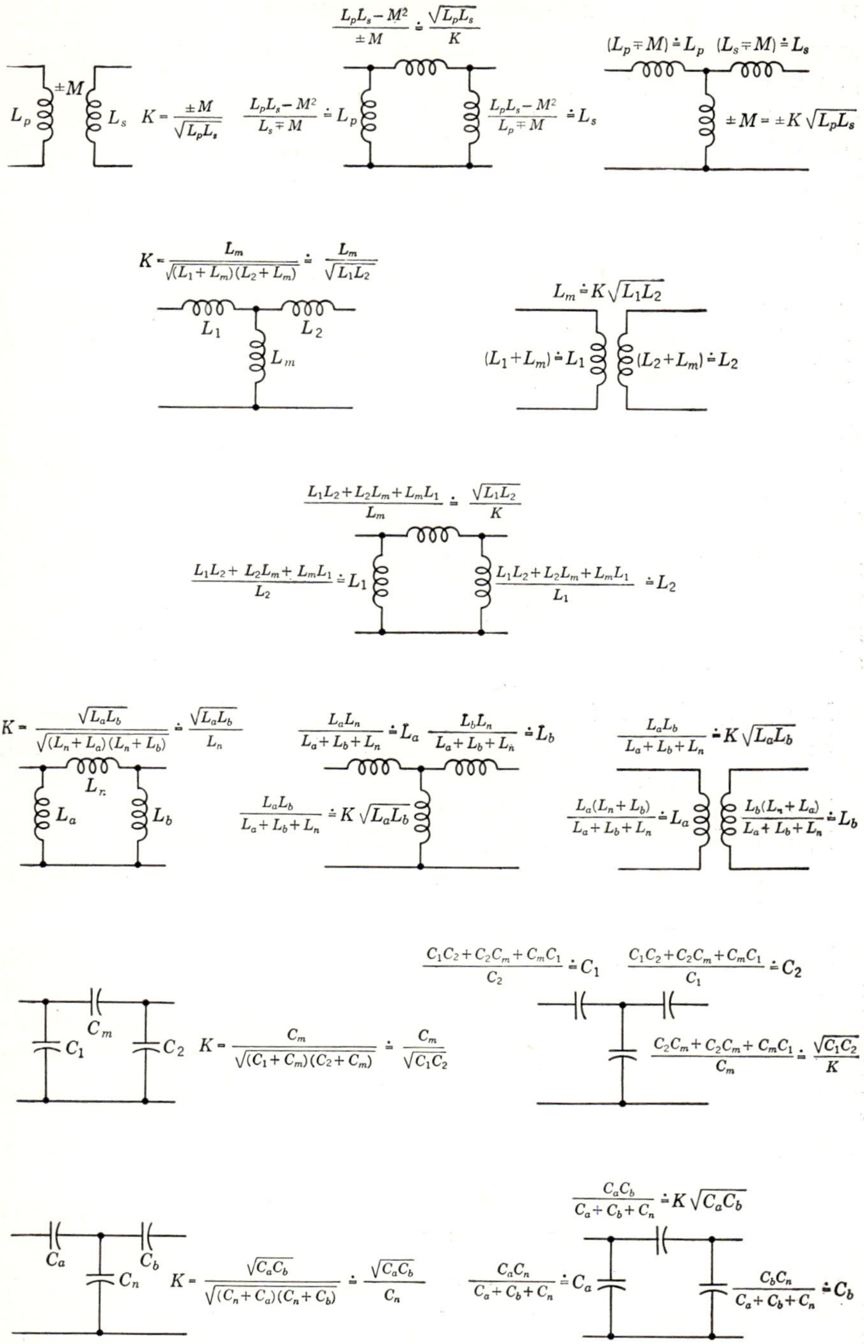

FIG. 6-35 Coefficient of coupling used in the analysis of the π, T, and transformer exact equivalents, and approximations for small couplings. (*After Dishal.*)

of Fig. 6-36 also possesses a set of equivalents which can be secured by π and T transformations and the use of transformers. Ten of these are shown in Fig. 6-37 so that the solution of the one two-node network shown in Fig. 6-33 leads to results which can be applied also to any of the 22 band-pass structures illustrated in Figs. 6-34, 6-36, and 6-37. For the complete results of such an analysis and information on the design of such structures, the reader is referred to the excellent paper by Dishal previously cited.[1]

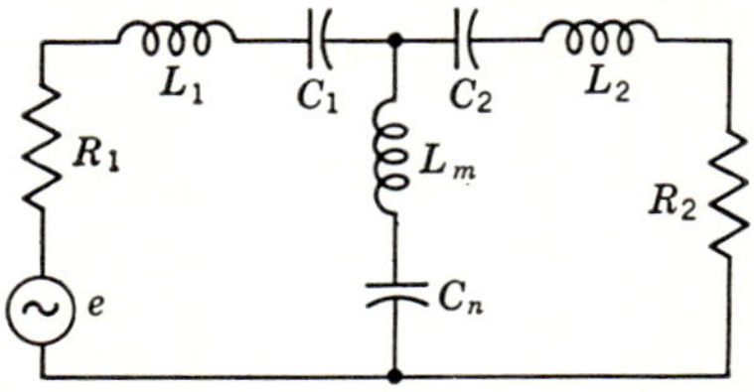

FIG. 6-36 Dual of network in Fig. 6-33.

As an example of the application of nodal analysis, it is worthwhile to undertake a complete analysis of a resistance-capacitance-coupled pentode amplifier stage shown in Fig. 6-38. The equivalent network for this circuit is shown in Fig. 6-39.

It has been assumed that C_{gp} can be neglected, and therefore the equivalent network does not include the control grid as a node, as it did in the case of the generalized network for the grounded-cathode amplifier in Fig. 6-29. The component parts of the admittances shown in the equivalent network are as follows:

$Y_p = 1/r_p$ where r_p is the plate resistance of the tube
$Y_1 =$ the conductance of R_L and the susceptance of C_{pk} in parallel, or $1/R_L + j\omega C_{pk}$
$Y_2 =$ the susceptance of the coupling capacitor C_c, or $j\omega C_c$
$Y_3 =$ the conductance of R_g, the grid resistor of the succeeding amplifier stage, and the susceptance of the C_{gk} of the next stage in parallel, or $1/R_g + j\omega C_{gk}$

Y_1 and Y_3 should properly include any distributed capacitances which may be contributed by the stage wiring if the gain of the stage at the higher audio frequencies is to be determined accurately, and such provision is usually made by increasing C_{pk} and C_{gk} accordingly. It should be noted that the direction of the current for the constant-current generator is taken as away from the node to which it is attached.

The nodal equations are

$$\begin{aligned} (Y_1 + Y_p + Y_2)e_1 - Y_2e_2 &= -g_m e_{gk} \\ -Y_2e_1 + (Y_2 + Y_3)e_2 &= 0 \end{aligned} \tag{6-73}$$

[1] For additional applications of the principle of duality see A. Block, On Methods for the Construction of Networks Dual to Non-Planar Networks, *Proc. Phys. Soc.*, **58**: 677–694, November, 1946; R. L. Wallace and G. Raisbeck, Duality as a Guide in Transistor Circuit Design, *Bell System Tech. J.*, **30** (2), April, 1951; G. R. Harris, Duality and Balance Equations, *Proc. I.R.E.*, **37** (8): 882, August, 1949.

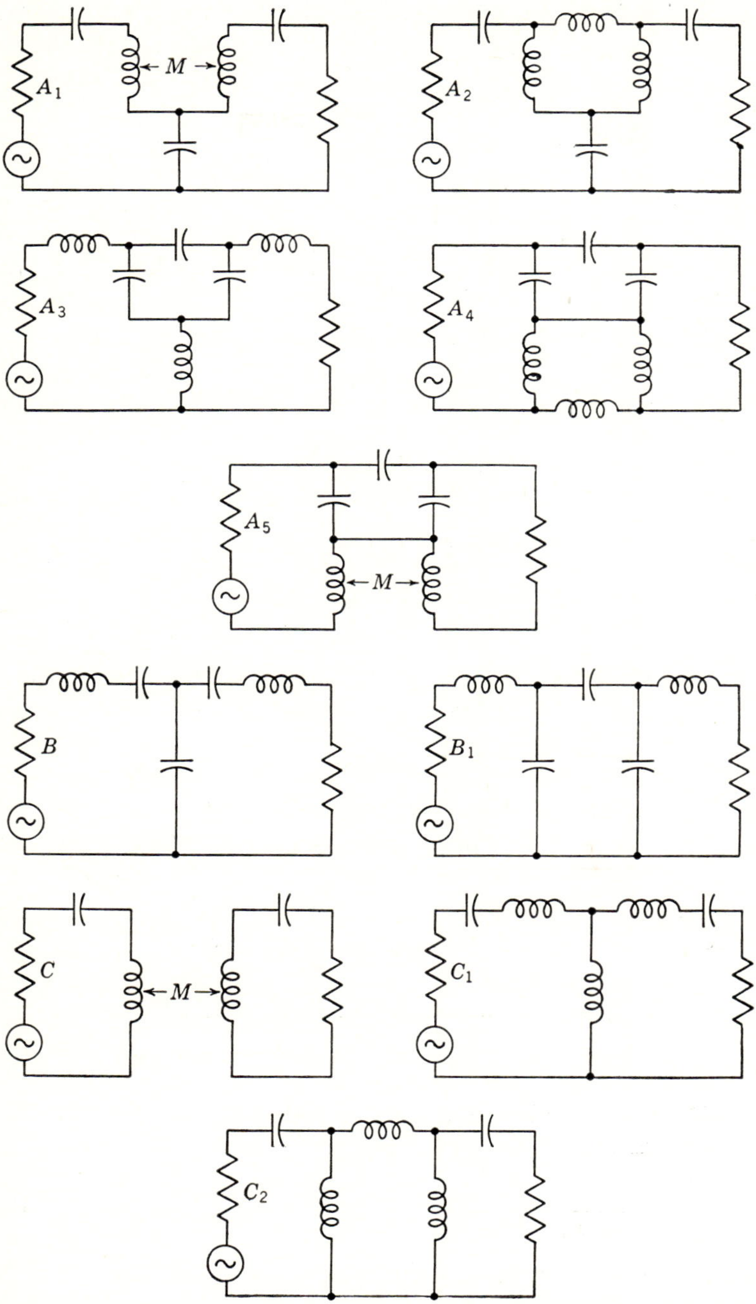

FIG. 6-37 Ten two-mesh circuits. The circuit of Fig. 6-36 is exactly equivalent to these circuits. (*After Dishal.*)

The solution for e_2, the output voltage, is given by

$$e_2 = \frac{\begin{vmatrix} (Y_1 + Y_p + Y_2) & -g_m e_{gk} \\ -Y_2 & 0 \end{vmatrix}}{\begin{vmatrix} (Y_1 + Y_p + Y_2) & -Y_2 \\ -Y_2 & (Y_2 + Y_3) \end{vmatrix}} \tag{6-74}$$

and the gain α of the amplifier is

$$\alpha = \frac{e_2}{e_{gk}} = \frac{-g_m Y_2}{(Y_1 + Y_p + Y_2)(Y_2 + Y_3) - Y_2^2} \tag{6-75}$$

By substituting the values of the admittances,

$$\alpha = \frac{-g_m(j\omega C_c)}{(1/R_L + j\omega C_{pk} + 1/r_p + j\omega C_c)(j\omega C_c + 1/R_g + j\omega C_{gk}) + \omega^2 C_c^2} \tag{6-76}$$

At the lower audio frequencies, the reactances of C_{pk} and C_{gk} will be exceedingly high, so that the $j\omega C_{pk}$ and $j\omega C_{gk}$ terms will be much smaller

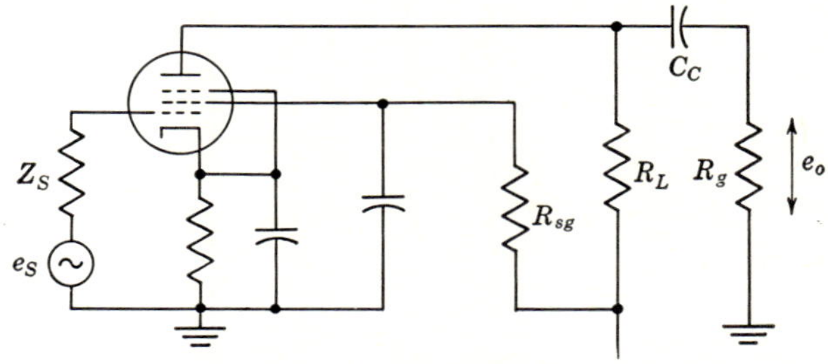

FIG. 6-38 Resistance-coupled pentode amplifier stage.

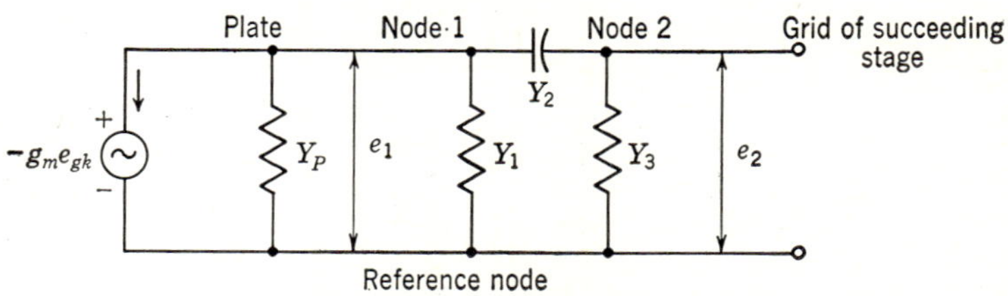

FIG. 6-39 Equivalent two-node network for Fig. 6-38.

than all the others in the denominator and can be neglected. Under these conditions the low-frequency gain α_l is

$$\alpha_l = \frac{-g_m(j\omega C_c)}{(1/R_L + 1/r_p + j\omega C_c)(j\omega C_c + 1/R_g) + \omega^2 C_c^2} \tag{6-77}$$

Expanding the denominator and factoring out the quantity

$$\left(\frac{1}{r_p} + \frac{1}{R_L} + \frac{1}{R_g}\right)$$

the low-frequency gain expression becomes

$$\alpha_l = \frac{-g_m(j\omega C_c)}{\left(\frac{1}{r_p} + \frac{1}{R_L} + \frac{1}{R_g}\right)\left[j\omega C_c + \frac{1}{R_g}\left(\frac{1/R_L + 1/r_p}{1/r_p + 1/R_L + 1/R_g}\right)\right]} \tag{6-78}$$

The combined resistance of R_p, R_L, and R_g in parallel is

$$R_{par} = \frac{1}{1/r_p + 1/R_L + 1/R_g} \tag{6-79}$$

or

$$\frac{1}{R_{par}} = \frac{1}{r_p} + \frac{1}{R_L} + \frac{1}{R_g} \tag{6-80}$$

The admittance around the loop formed by R_g on one side and r_p and

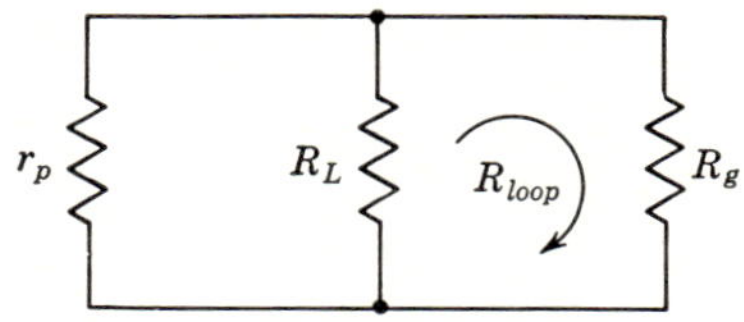

FIG. 6-40 Calculation of R_{loop}.

R_L in parallel on the other is (see Fig. 6-40)

$$Y_{\text{loop}} = \frac{1}{R_{\text{loop}}} = \frac{(1/R_g)(1/r_p + 1/R_L)}{1/R_g + 1/r_p + 1/R_L} \tag{6-81}$$

Substituting R_{par} and R_{loop} in the expression for α_l and dividing numerator and denominator by $j\omega C_c$ yield the following for the low-frequency gain:

$$\alpha_l = \frac{-g_m R_{par}}{1 + \frac{1}{R_{\text{loop}}(j\omega C_c)}} \tag{6-82}$$

or

$$\alpha_l = \frac{-g_m R_{par}}{1 - j/\omega C_c R_{\text{loop}}} \tag{6-83}$$

The absolute value of the low-frequency gain α_l is

$$|\alpha_l| = \frac{g_m R_{par}}{\sqrt{1 + (1/\omega C_c R_{\text{loop}})^2}} \tag{6-84}$$

To determine the amplifier-stage gain at high audio frequencies α_h, expand the denominator of the original expression for α [Eq. (6-76)] and obtain for the denominator the expression

$$\frac{(j\omega C_{pk} + j\omega C_c) + (1/r_p + 1/R_L)}{R_g} + \left(\frac{1}{r_p} + \frac{1}{R_L}\right)(j\omega C_c + j\omega C_{gk}) + j\omega C_c(j\omega C_{pk} + j\omega C_{gk}) + (j\omega C_{pk})(j\omega C_{gk}) \tag{6-85}$$

At high audio frequencies, the reactance of the coupling capacitor C_c will be negligible, so that $j\omega C_c \gg j\omega C_{pk}$ or $j\omega C_{gk}$. Simplifying the expression for the denominator of α_h on this basis, it becomes

$$\frac{(j\omega C_c) + (1/r_p + 1/R_L)}{R_g} + \left(\frac{1}{r_p} + \frac{1}{R_L}\right) j\omega C_c + j\omega C_c(j\omega C_{pk} + j\omega C_{gk}) + (j\omega C_{pk})(j\omega C_{gk}) \tag{6-86}$$

Dividing the numerator $-g_m(j\omega C_c)$ and denominator by $j\omega C_c$ and noting that

$$1 \gg \frac{(1/r_p + 1/R_L)}{j\omega C_c}$$

and that the term $(j\omega C_{pk})(j\omega C_{gk})/(j\omega C_c)$ is much smaller than the other terms in the denominator, the expression for α_h becomes

$$\alpha_h = \frac{-g_m}{(1/r_p + 1/R_g + 1/R_L) + (j\omega C_{pk} + j\omega C_{gk})} \tag{6-87}$$

Substituting

$$\frac{1}{R_{par}} = \left(\frac{1}{r_p} + \frac{1}{R_g} + \frac{1}{R_L}\right) \tag{6-88}$$

and

$$C_{par} = (C_{pk} + C_{gk}) \tag{6-89}$$

the high-frequency gain α_h is

$$\alpha_h = \frac{-g_m R_{par}}{1 + R_{par}(j\omega C_{par})} \tag{6-90}$$

The absolute value of the high-frequency gain is

$$|\alpha_h| = \frac{g_m R_{par}}{\sqrt{1 + (\omega C_{par} R_{par})^2}} \tag{6-91}$$

The maximum gain of an amplifier stage in which the reactances of C_c, C_{pk}, and C_{gk} are neglected (i.e., at the middle of the audio-frequency range) is

$$\alpha_m = -g_m R_{par} \tag{6-92}$$

which has an absolute value of

$$|\alpha_m| = g_m R_{par} \tag{6-93}$$

The absolute ratio of the low-frequency gain α_l to the maximum obtainable at mid-band is then

$$\left|\frac{\alpha_l}{\alpha_m}\right| = \frac{1}{\sqrt{1 + (1/\omega C_c R_{\text{loop}})^2}} \tag{6-94}$$

and for the high-frequency case

$$\left|\frac{\alpha_h}{\alpha_m}\right| = \frac{1}{\sqrt{1 + (\omega C_{par} R_{par})^2}} \tag{6-95}$$

At some low frequency f_l, the reactance of C_c will equal the value of R_{loop}, and half the power will be developed in R_{loop} that would have been developed at mid-band. Also, at some high frequency f_h, the reactance of C_{par} will equal the resistance R_{par}, and half the power will be developed in R_{par} that would have been developed at mid-band. These half-power frequencies are

$$f_l = \frac{1}{2\pi C_c R_{\text{loop}}} \tag{6-96}$$

and

$$f_h = \frac{1}{2\pi C_{par} R_{par}} \tag{6-97}$$

If the power developed at f_l and f_h is one-half the maximum possible, then the voltage gain will be $\sqrt{1/2}$, or 70.7 per cent of the maximum. In other words, the amplifier voltage gain will be $-20 \log \sqrt{2}$, or 3 db, below the maximum obtained at mid-band for the frequencies f_l and f_h.

Substituting these values for f_l and f_h in the expressions for $|\alpha_l/\alpha_m|$ and $|\alpha_h/\alpha_m|$ gives the following relationships for the gain at any frequency f compared to the maximum gain at mid-band:

For the low-frequency range:

$$\left|\frac{\alpha_f}{\alpha_m}\right| = \frac{1}{\sqrt{1 + (f_l/f)^2}} \tag{6-98}$$

For the high-frequency range:

$$\left|\frac{\alpha_h}{\alpha_m}\right| = \frac{1}{\sqrt{1 + (f/f_h)^2}} \tag{6-99}$$

where $|\alpha_m| = g_m R_{par}$.

With regard to high-frequency response, the behavior of the network is similar to that of a broadly tuned amplifier with a relatively low plate tank Q which has a band width of $1/(2\pi C_{par} R_{par})$. The product of mid-frequency gain and band width f_h from Eq. (6-97) is therefore a constant, for

$$\alpha_m f_h = \frac{g_m R_{par}}{2\pi C_{par} R_{par}} = \frac{g_m}{2\pi C_{par}} \tag{6-100}$$

Thus the maximum gain of a wideband amplifier varies directly with the quotient g_m/C_{par} and inversely with the band width or upper half-power frequency f_h. This indicates why tubes with high transconductances and low input and output capacitances are so desirable as video amplifiers.

As a second application of the nodal method, it is desired to calculate the input admittance of a triode based on the generalized equivalent network of the grounded-cathode amplifier shown in Fig. 6-29. The exact circuit to be analyzed is also shown in Fig. 6-28.

In this case let the voltage at the first, or grid, node be e_1 and the voltage at the plate node be e_2, with

$$\begin{aligned} Y_1 &= j\omega C_{gk} \\ Y_2 &= j\omega C_{gp} \\ Y_3 &= \frac{1}{r_p} + j\omega C_{pk} \\ Y_L &= \frac{1}{Z_L} \end{aligned} \tag{6-101}$$

The current which enters the first node due to the input signal is i_1, and the current at the second node due to the tube as a generator is $-g_m e_1$.

The nodal equations are

$$\begin{aligned} i_1 &= (Y_1 + Y_2)e_1 - Y_2 e_2 \\ -g_m e_1 &= -Y_2(e_1) + (Y_2 + Y_3 + Y_L)e_2 \end{aligned} \tag{6-102}$$

The corresponding determinant for E_1 is

$$e_1 = \frac{\begin{vmatrix} i_1 & -Y_2 \\ -g_m e_1 & (Y_2 + Y_3 + Y_L) \end{vmatrix}}{\begin{vmatrix} (Y_1 + Y_2) & -Y_2 \\ -Y_2 & (Y_2 + Y_3 + Y_L) \end{vmatrix}} \tag{6-103}$$

$$= \frac{i_1(Y_2 + Y_3 + Y_L) - (g_m e_1) Y_2}{(Y_1 + Y_2)(Y_2 + Y_3 + Y_L) - Y_2^2} \tag{6-104}$$

$$e_1[(Y_1 + Y_2)(Y_2 + Y_3 + Y_L) - Y_2^2 + g_m Y_2] = i_1(Y_2 + Y_3 + Y_L) \tag{6-105}$$

The input admittance Y_{in} is i_1/e_1, or

$$Y_{\text{in}} = \frac{(Y_1 + Y_2)(Y_2 + Y_3 + Y_L) - Y_2^2 + g_m Y_2}{(Y_2 + Y_3 + Y_L)} \tag{6-106}$$

Substituting the appropriate values, Y_{in} is

$$Y_{\text{in}} = \frac{(j\omega C_{gk} + j\omega C_{gp})(j\omega C_{gp} + 1/r_p + 1/Z_L) + \omega^2 C_{gp}^2 + g_m(j\omega C_{gp})}{j\omega C_{gp} + 1/r_p + j\omega C_{pk} + 1/Z_L} \tag{6-107}$$

Now, for triodes, in many cases, $1/j\omega C_{gp}$ and $1/j\omega C_{pk}$ are much greater than r_p or Z_L, so that $j\omega C_{gp}$ and $j\omega C_{pk}$ are negligible as compared to the remainder of the denominator. The conductance ωC_{gp} is normally quite small and may also be neglected.

Under these circumstances, the input admittance becomes

$$Y_{\text{in}} = \frac{(j\omega C_{gk} + j\omega C_{gp})(1/r_p + 1/Z_L) + g_m(j\omega C_{gp})}{1/r_p + 1/Z_L} \tag{6-108}$$

and, separating components,

$$Y_{\text{in}} = \underbrace{(j\omega C_{gk} + j\omega C_{gp})}_{\substack{\text{due to static}\\ \text{interelectrode}\\ \text{capacitance}}} + \underbrace{\frac{g_m}{1/r_p + 1/Z_L}(j\omega C_{gp})}_{\substack{\text{additional susceptance}\\ \text{due to stage amplifi-}\\ \text{cation}}} \tag{6-109}$$

The first two terms in the final expression for the admittance Y_{in} are the susceptances due to the static interelectrode capacitances C_{gk} and C_{gp}, which are always present. The third term represents a susceptance which varies directly with the gain of the amplifier stage, since the gain α is $g_m/(1/r_p + 1/Z_L)$.

In other words, in a tube such as a triode where C_{gp} is not negligible, there will be an apparent increase in the input capacitance of the grid circuit, over and above the normal input capacitance, which varies in magnitude depending on the amplification of the stage. The additional capacitance is αC_{gp} where α is the stage gain. This effect is described as the *Miller effect* and is not the least of the disadvantages of the triode, since it tends to limit the high-frequency amplification possible with cascaded triode amplifiers. There is a considerable advantage in the solution of this problem by the nodal rather than the mesh analysis. One published solution by the latter method required no less than *seven* branch equations.[1]

The increment in the input admittance because of the Miller effect can also be represented in the equivalent network of an amplifier as an additional shunt path across the input. Vallese[2] has tabulated in useful form the formulas for the input and output impedances of the grounded-cathode, grounded-plate, and grounded-grid amplifiers along with equivalent network representations of these using this method.

The nodal method may also be used in the analysis of multistage amplifiers. As an example of this, in Fig. 6-41 is shown the circuit of a low-noise cascode radio-frequency amplifier. The cascode amplifier consists of a grounded-cathode amplifier followed by a grounded-grid stage and is found to produce the least internal noise of any of the nine possible ways in which two triodes can be cascaded. Consequently, it is frequently encountered in television-receiver tuner circuits. The cascode amplifier shown in Fig. 6-41 provides the stability of a pentode, the

[1] R. S. Glasgow, "Principles of Radio Engineering," 1st ed., p. 228, McGraw-Hill Book Company, Inc., New York, 1936.

[2] L. M. Vallese, Network Representation of Input and Output Admittances of Amplifiers, *Proc. I.R.E.*, **37** (4): 407–408, April, 1949.

amplification and gain of a pentode, and the low noise factor of the first triode.

The equivalent network of the cascode amplifier is shown in Fig. 6-42. The nodal equations for this network are

$$
\begin{aligned}
I_s &= (Y_{gk1} + Y_{gp1})E_{g1} - (Y_{gp1})E_{p1} + 0 \\
0 &= (g_{m1} - Y_{gp1})E_{g1} + (g_{m2} + Y_{gp1} + Y_{p1} + Y_{L1} + Y_{pk2})E_{p1} \\
&\qquad - Y_{pk2}E_{p2} \qquad (6\text{-}110) \\
0 &= 0 - (g_{m2} + Y_{pk2})E_{p1} + (Y_{pk2} + Y_{gp2})E_{p2}
\end{aligned}
$$

The significant thing about these equations is that the contributions of the current generators $-g_{m1}E_{g1}$ and $-g_{m2}E_{g2}$ are incorporated within

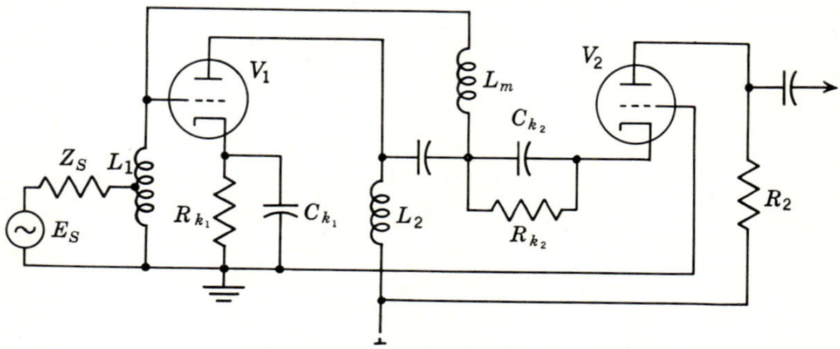

FIG. 6-41 Diagram of cascode low-noise amplifier.

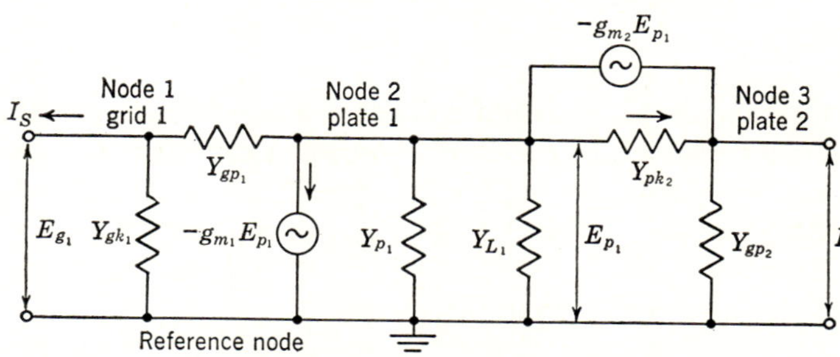

FIG. 6-42 Equivalent network of cascode amplifier.

the admittance expressions instead of being transposed to the left side as is I_s. This general method facilitates a solution for amplifier gains by permitting easy access to the ratio of output to input voltage, in this case E_{p2}/E_{g1}. A solution for this ratio shows that the gain of the cascode amplifier is approximately $g_{m1}R_2$. The cascode circuit is thus equivalent to a pentode of transconductance g_{m1}. For the details of the proof which leads to this result, the reader is referred to the original paper on the subject.[1]

[1] H. Wallman, A. B. Macnee, and C. P. Gadsden, A Low-noise Amplifier, *Proc. I.R.E.*, **36** (7), July, 1948.

In conclusion, the advantages of the nodal method in the analysis of electron-tube networks are as follows:

(*a*) Less work is required for the nodal method because, in most cases, $B > 2(V - P)$, particularly in multistage amplifiers.

(*b*) The performance of amplifiers is generally specified in terms of voltage ratios, which the nodal analysis gives directly.

(*c*) In the equivalent networks of electron tubes, many elements occur in parallel, and parallel components may be taken into account merely by adding additional terms to the nodal equations without disturbing their basic form.

Problems

Prob. 6-1 Write the mesh equations for the network shown in Fig. 6-43, and determine the current in the 5-ohm branch.

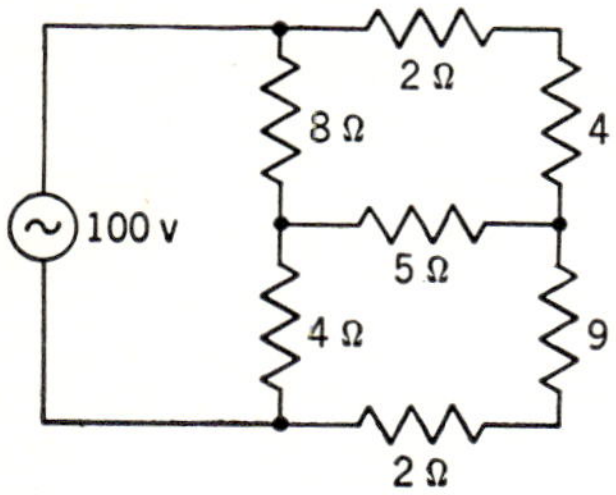

FIG. 6-43

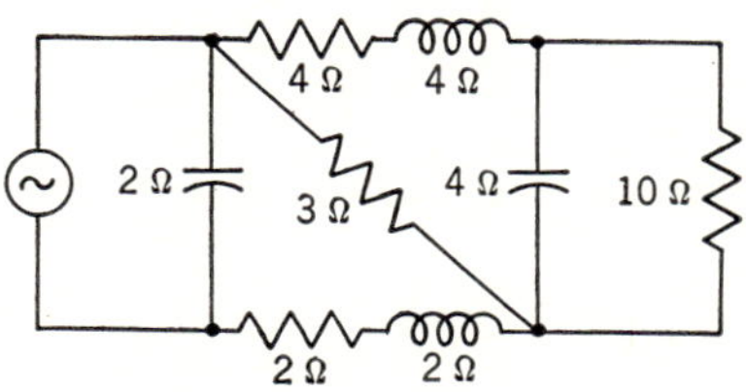

FIG. 6-44

Prob. 6-2 Find y_{11}, y_{12}, and y_{22} for the network shown in Fig. 6-44.

Prob. 6-3 Write the nodal equations for the network shown in Fig. 6-45 and determine z_{11}, z_{13}, and z_{33}. What is the value of E_0? The values shown are impedance values in ohms.

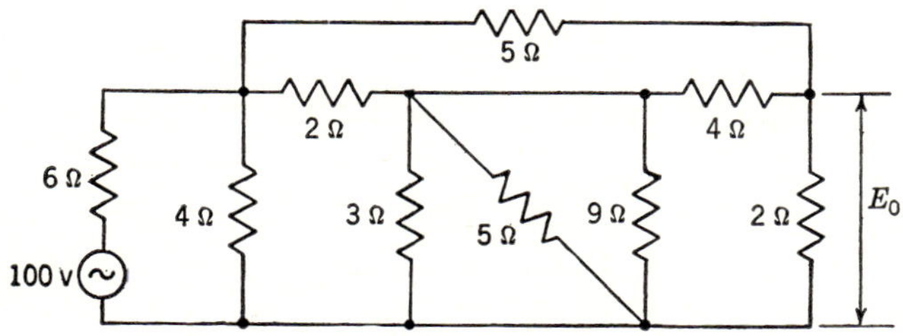

FIG. 6-45

Prob. 6-4 Determine whether the mesh or the nodal analysis would be preferred in the analysis of the networks shown in Fig. 6-46*a* and *b*.

Prob. 6-5 In a certain network, currents in certain meshes and voltages across certain node pairs are measured as follows:

	E_1	E_2	E_3	E_4
When $I_1 = 4$	1	2	3	4
$I_2 = 1$	4	1	2	3
$I_3 = -2$	2	4	2	1
$I_4 = 2$	8	2	4	0

What value will I_4 have when $E_4 = 10$ and all the remaining Es are equal to zero? Also find the value of I_3 when $E_3 = 0$ under the same conditions.

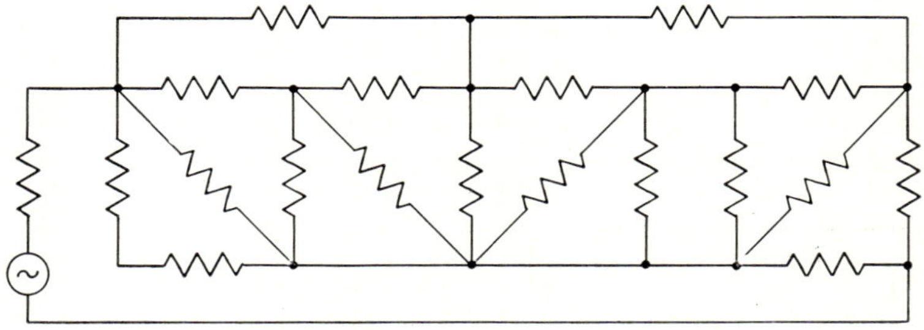

(a)

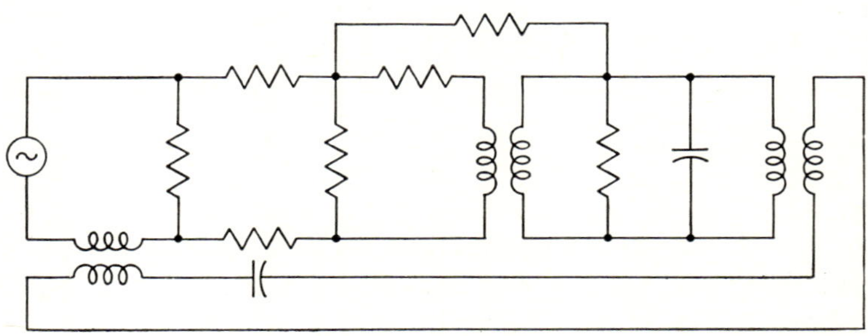

(b)

FIG. 6-46

Prob. 6-6 Write the mesh equations for the network shown in Fig. 6-47.

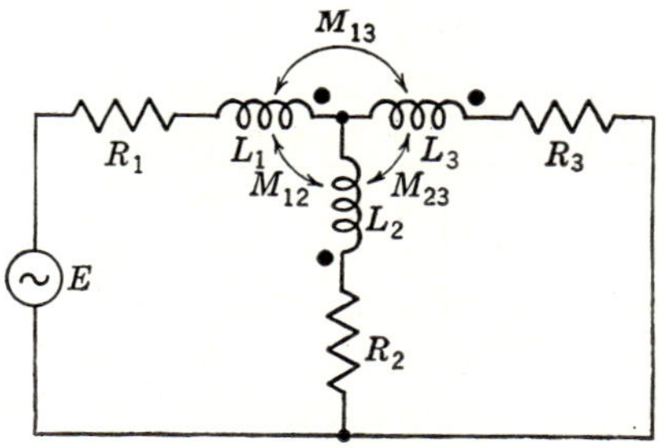

FIG. 6-47

Prob. 6-7 Write the mesh equations for the network shown in Fig. 6-48. Determine the input or driving-point admittance presented to the generator. Assume that $L_1 = L_2 = L_3$ and that all inductances are part of the same winding.

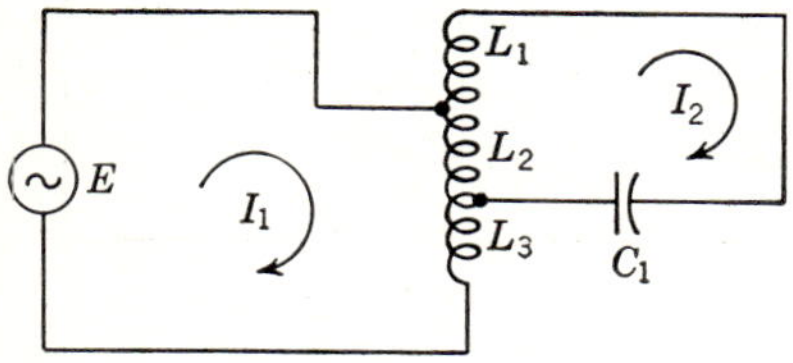

FIG. 6-48

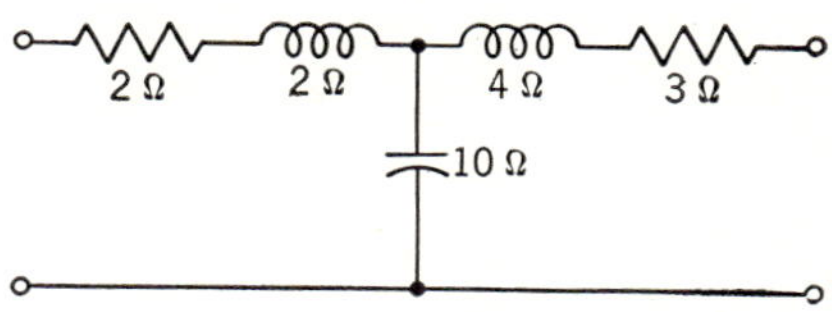

FIG. 6-49

Prob. 6-8 Determine the $ABCD$ parameters and the equivalent π network for the network in Fig. 6-49.

Prob. 6-9 Show by network analysis that, if in Fig. 6-50 $\omega L = 1/2\omega C$ and $R_p = 1/R(\omega C)^2$, transmission through the network will be zero. Is there a π-network equivalent for this network?

Prob. 6-10 Develop a full set of equivalent network constants if the general parameters of a passive network are (a) $A = 1.5$, $B = 11$ ohms, $D = 2.5$; (b) $A = 1.66$, $B = 4$ ohms, $C = 1$ mho.

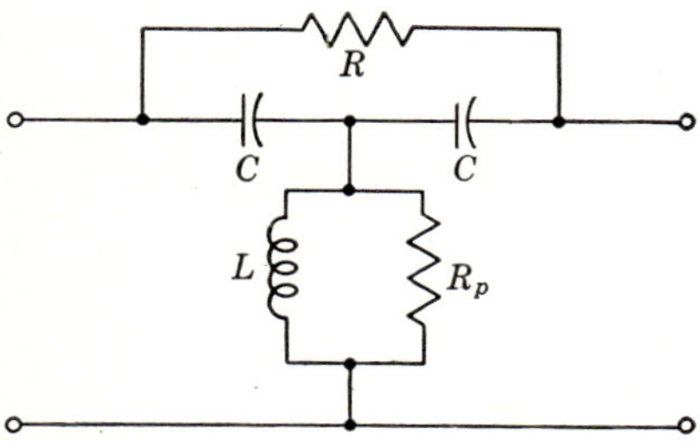

FIG. 6-50

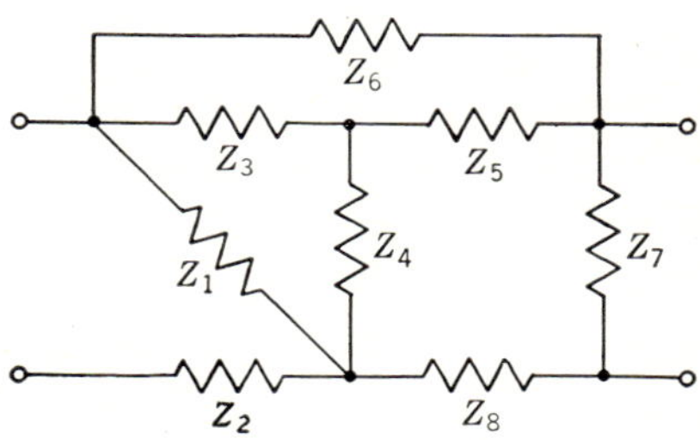

FIG. 6-51 $Z_1 = 2 + j2$, $Z_2 = 3 + j4$, $Z_3 = 3 - j4$, $Z_4 = 5 - j5$, $Z_5 = 8 + j6$, $Z_6 = 6 - j8$, $Z_7 = 1 - j1$, and $Z_8 = 2 - j4$.

Prob. 6-11* Find equivalent T and π networks for the network shown in Fig. 6-51 and determine whether they are physically realizable in a passive network.

NOTE: As a check, the mesh analysis should yield $\Delta = 4{,}704 - j21{,}472$.

Prob. 6-12 In Fig. 6-52 is shown the circuit of a T-network attenuator which is part of a standard volume indicator in a television-studio audio-

* M. B. Reed, General Formulas for "T" and "π" Network Equivalents, *Proc. I.R.E.*, **33** (12), December, 1945.

control console. If $T_1 = T_2$, determine the values of T_1 and T_{12} if the attenuator is set for a 5-db loss.

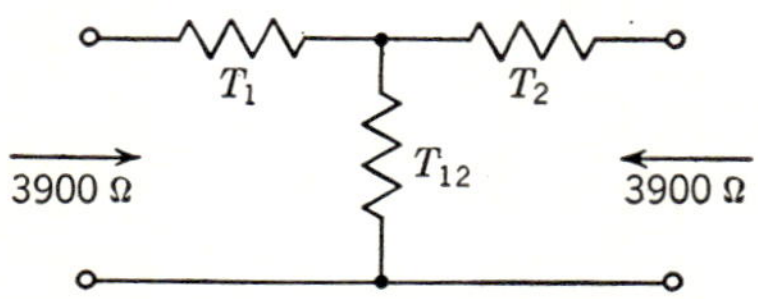

FIG. 6-52 For 5 db loss, $1{,}093\,\Omega = T_1 = T_2$ and $6{,}415\,\Omega = T_{12}$.

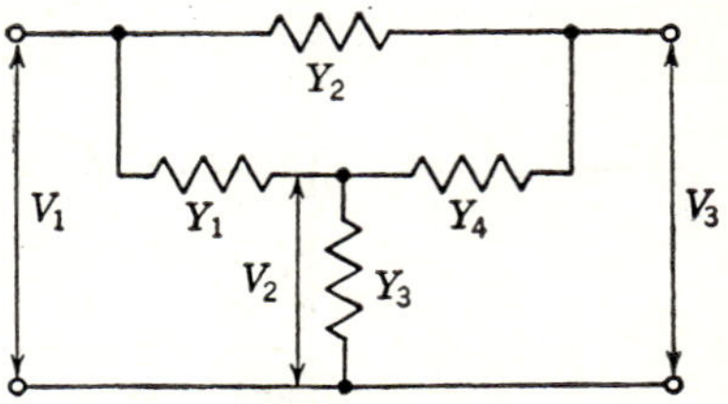

FIG. 6-53

Prob. 6-13 (*a*) Write the nodal equations for the bridged-T network shown in Fig. 6-53.

(*b*)* Show that the voltage ratio V_3/V_1 is

$$\frac{V_3}{V_1} = \frac{Y_1 + Y_2 + Y_1Y_2/Y_4 + Y_2Y_3/Y_4}{Y_1 + Y_2 + Y_1Y_2/Y_4 + Y_2Y_3/Y_4 + Y_3}$$

(*c*) If $Y_1 = Y_4 = j\omega C$, $Y_2 = 1/aR$, and $Y_3 = a/R$, where a is a constant, show that

$$\frac{V_3}{V_1} = \frac{2/a + j\mu}{2/a + a + j\mu}$$

where $\mu = (\omega/\omega_0) - (\omega_0/\omega)$
$\omega_0 = 1/RC$

(*d*) Show that a voltage-ratio minimum, accompanied by zero phase shift, is obtained at ω_0.

(*e*) The equivalent Q of the network is defined as

$$Q = \left(\frac{\omega_0}{\Delta\omega}\right) = \frac{1}{\mu}$$

where the limits of $\Delta\omega$ occur when the network response is 3 db higher than its minimum value. By equating the expression for the magnitude $|V_3/V_1|$ obtained in (*c*) to $\sqrt{2}$ times its minimum value, show that

$$\mu = \frac{2}{a\sqrt{1 - 8/(2 + a^2)^2}} \qquad \text{and} \qquad Q \cong \frac{a}{2} \qquad \text{for } a \gg 1$$

Prob. 6-14 Find the elements of the L network, the lattice network, and the bridged-T network shown in Fig. 6-54 in terms of the $ABCD$ parameters.

* P. G. Sulzer, A Note on a Bridged-T Network, *Proc. I.R.E.*, **39**(7): 819–821, July, 1951.

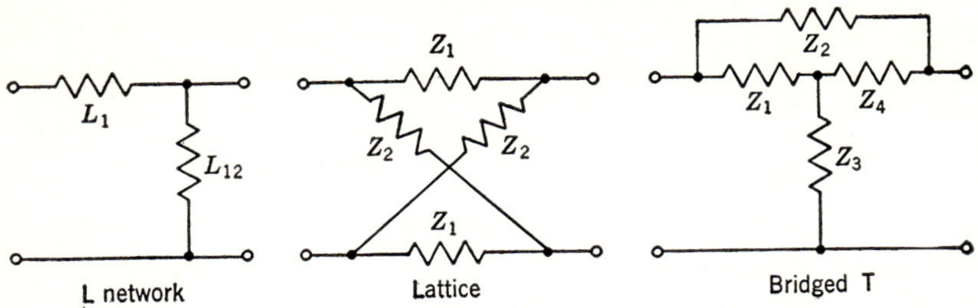

FIG. 6-54

Prob. 6-15 Find the resonant frequency of the lattice network shown in Fig. 6-55 by converting it to an equivalent bridge network.

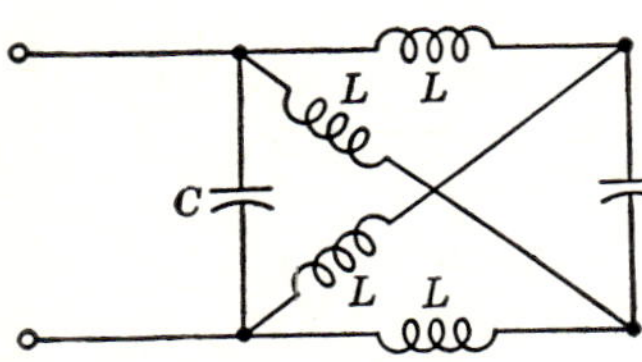

FIG. 6-55

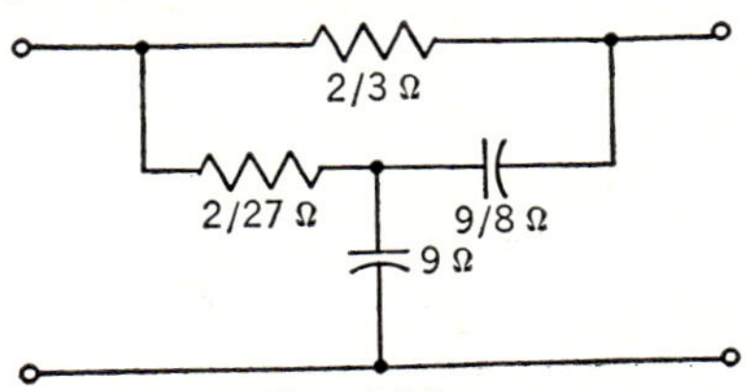

FIG. 6-56

Prob. 6-16* The network in Fig. 6-56 is supposed to have a voltage gain of *greater* than unity and, specifically, 1.14. Determine whether or not this is true.

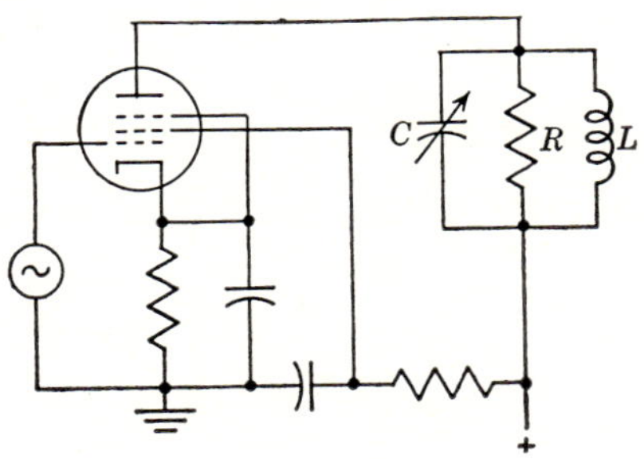

FIG. 6-57

Prob. 6-17 Analyze the tuned r-f amplifier shown in Fig. 6-57, and determine the stage gain at the resonant frequency of the plate-circuit elements. Neglect C_{gp} and C_{pk}.

NOTE: r_p cannot be neglected.

Prob. 6-18 Show that (*a*) the input resistance of a grounded-cathode amplifier will be negative with an inductive plate load and provided that $\omega^2 C_{gp} L < 1$; (*b*) the input conductance of a cathode follower with a capacitive load is negative.

How could these effects be used to obtain oscillator action?

* A. D. Fialkow and I. Gerst, The Maximum Gain of an *RC* Network, *Proc. I.R.E.*, **41** (3): 392–395, March, 1953.

Prob. 6-19* Show that at ultrahigh frequencies, where the cathode lead inductance is not negligible, the input admittance of a grounded-cathode pentode amplifier is

$$Y_{\text{in}} = j\omega(C_{gk} + C_{gs}) + \omega^2 L_k C_{gk} g_m$$

where C_{gs} is the capacitance between the control grid and the screen grid, which is grounded electrically, and L_k is the cathode lead inductance.

Hint: Set up a two-node equivalent network with the grid at one node and the cathode at the other, the two nodes interconnected by C_{gk}. Note also that in most cases $\omega^2 L_k{}^2 g_m{}^2 \ll 1$.

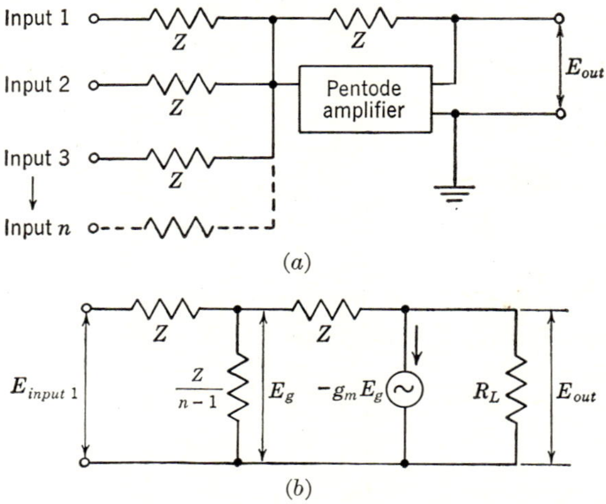

FIG. 6-58 (*a*) Feedback-mixer block diagram. (*b*) Equivalent network of feedback mixer.

Prob. 6-20† Analyze by the nodal method the circuit of the feedback mixer shown in Fig. 6-58*a*. Using the equivalent circuit in Fig. 6-58*b*, show that the gain of the circuit with all inputs but input 1 grounded is approximately

$$\frac{E_{\text{out}}}{E_{\text{input 1}}} \cong -\frac{1}{1 + (n+1)/g_m R_L}$$

for $Y \ll G_L \ll g_m$, where $Y = 1/Z$ and $G_L = 1/R_L$; so that if $g_m R_L \gg n + 1$ the output voltage is equal to the negative of the input voltage at any input, independent of g_m.

* M. J. O. Strutt, The Causes for the Increase of the Admittances of Modern High-frequency Amplifier Tubes on Short Waves, *Proc. I.R.E.*, **26** (8), August, 1938; E. W. Herold and L. Malter, Some Aspects of Radio Reception at Ultra-high Frequency, pt. II, *Proc. I.R.E.*, **31** (9), September, 1943.

† G. E. Valley and H. Wallman, "Vacuum Tube Amplifiers," Radiation Laboratory Series, vol. 18, pp. 99–100, McGraw-Hill Book Company, Inc., New York, 1948.

Chapter 7

MATRIX ALGEBRA

7-1 Introduction

The method and notation of determinants which were presented in Chaps. 5 and 6 provided a systematic means for the solution of the linear simultaneous equations encountered in the analysis of networks. A more flexible and compact means for the solution of problems which involve such equations is the method of matrices which was devised by Cayley in 1857. This chapter is devoted to a study of the operations of matrix algebra. Chapter 8, which follows, deals with the applications of matrix algebra to network analysis.

7-2 Basic Definitions and Notation

A system of $m \cdot n$ elements which are arranged to form a rectangular array is termed a *matrix*. There are several notations used to represent a matrix. Those commonly found in the technical literature are

$$\begin{bmatrix} a_{11} & a_{12} & \cdots & a_{1m} \\ a_{21} & a_{22} & \cdots & a_{2m} \\ \cdot & \cdot & & \cdot \\ \cdot & \cdot & & \cdot \\ \cdot & \cdot & & \cdot \\ a_{n1} & a_{n2} & \cdots & a_{nm} \end{bmatrix} \tag{7-1}$$

or

$$\begin{Vmatrix} a_{11} & a_{12} & \cdots & a_{1m} \\ a_{21} & a_{22} & \cdots & a_{2m} \\ \cdot & \cdot & & \cdot \\ \cdot & \cdot & & \cdot \\ \cdot & \cdot & & \cdot \\ a_{n1} & a_{n2} & \cdots & a_{nm} \end{Vmatrix} \tag{7-2}$$

or

$$\begin{pmatrix} a_{11} & a_{12} & \cdots & a_{1m} \\ a_{21} & a_{22} & \cdots & a_{2m} \\ \cdot & \cdot & & \cdot \\ \cdot & \cdot & & \cdot \\ \cdot & \cdot & & \cdot \\ a_{n1} & a_{n2} & \cdots & a_{nm} \end{pmatrix} \tag{7-3}$$

Of these, (7-1) is the form which will be used in this text. The corresponding abbreviated notations for Eqs. (7-1) to (7-3) are $[a_{ij}]$, $\|a_{ij}\|$, (a_{ij}), respectively, for $i = 1, 2, 3, \ldots, n$; and $j = 1, 2, 3, \ldots, m$. The *order* of the rectangular matrix described above is $\{n,m\}$.

If, in the matrix cited above, the number of rows (n) and the number of columns (m) are made equal, that is, $n = m$, then the matrix that results is a *square matrix* of order simply equal to n.

EXAMPLE 1. The rectangular matrix

$$\begin{bmatrix} 3 & 5 \\ 2 & 1 \\ 1 & 3 \end{bmatrix} \tag{7-4}$$

has three rows and two columns; hence it is of order $\{3,2\}$, whereas

$$\begin{bmatrix} 3 & 5 & -1 \\ 2 & 1 & 2 \\ 1 & 3 & 4 \end{bmatrix} \tag{7-5}$$

is a square matrix of *third order*, or $\{3,3\}$.

Any rectangular matrix can be made into a square matrix by the augmentation with zero elements of the rectangular array to make up a square matrix of order n, that is, the addition of rows or columns of zeros.

EXAMPLE 2. The rectangular matrix

$$\begin{bmatrix} 3 & 5 \\ 2 & 1 \\ 1 & 3 \end{bmatrix} \tag{7-6}$$

may be made into the square matrix

$$\begin{bmatrix} 3 & 5 & 0 \\ 2 & 1 & 0 \\ 1 & 3 & 0 \end{bmatrix} \tag{7-7}$$

The *rank of a matrix* is said to be equal to r if the matrix contains at least one nonsingular, i.e., nonzero, determinant of order r and all its determinants of order greater than r are zero.

EXAMPLE 3. The rank of the square matrix below is equal to 2.

$$\begin{bmatrix} 5 & 8 & 7 \\ 13 & 11 & -2 \\ 18 & 19 & 5 \end{bmatrix} \tag{7-8}$$

Note that in this case the matrix contains one third-order determinant which is zero, nine second-order determinants, and nine first-order determinants (elements).

EXAMPLE 4. The rank of the rectangular matrix below is equal to 2.

$$\begin{bmatrix} 3 & 2 & 1 \\ 6 & 5 & 4 \end{bmatrix} \tag{7-9}$$

If the rows and columns of a matrix are interchanged, then the *transpose of the matrix*, or *transposed matrix*, is obtained. The element a_{ij}, which is in the ith row and jth column, is interchanged with a_{ji}, the element which is located in the jth row and ith column, for all i, j to form the transpose of the matrix.

The third-order matrix

$$[A] = \begin{bmatrix} a_{11} & a_{12} & a_{13} \\ a_{21} & a_{22} & a_{23} \\ a_{31} & a_{32} & a_{33} \end{bmatrix} \tag{7-10}$$

has for a transposed matrix

$$\begin{bmatrix} a_{11} & a_{21} & a_{31} \\ a_{12} & a_{22} & a_{32} \\ a_{13} & a_{23} & a_{33} \end{bmatrix} \tag{7-11}$$

The usual notation for the transposed matrix of $[a_{ij}]$ is $[a_{ij}]'$, where

$$[a_{ij}]' = [a_{ji}] \tag{7-12}$$

and

$$[a_{ji}]' = [a_{ij}] \tag{7-13}$$

EXAMPLE 5. If

$$[A] = \begin{bmatrix} 3 & 5 & 1 & -1 \\ 2 & 6 & 2 & -2 \\ 1 & 7 & 3 & -8 \\ 4 & 9 & 4 & 5 \end{bmatrix} \tag{7-14}$$

then its transposed matrix is

$$[A]' = \begin{bmatrix} 3 & 2 & 1 & 4 \\ 5 & 6 & 7 & 9 \\ 1 & 2 & 3 & 4 \\ -1 & -2 & -8 & 5 \end{bmatrix} \tag{7-15}$$

A matrix which has unit elements along the principal diagonal and in which all other elements are zero is called the *unit matrix* $[U]$.[1]

$$[U] = \begin{bmatrix} 1 & 0 & . & . & . & . & 0 \\ 0 & 1 & 0 & . & . & . & 0 \\ 0 & 0 & 1 & . & . & . & 0 \\ . & & & & . & & . \\ . & & & & & . & . \\ . & & & & & . & . \\ 0 & . & . & . & . & . & 1 \end{bmatrix} \tag{7-16}$$

$[U]$ is often written in another form as

$$[U] = [\delta_{ij}] \tag{7-17}$$

where δ_{ij} is identified as the *Kronecker delta*. It has the properties

$$\delta_{ij} = \begin{cases} 1 & \text{for } i = j \\ 0 & \text{for } i \neq j \end{cases} \tag{7-18}$$

EXAMPLE 6. A unit matrix $[U]$ of order 3 is represented by

$$[U] = \begin{bmatrix} 1 & 0 & 0 \\ 0 & 1 & 0 \\ 0 & 0 & 1 \end{bmatrix} \tag{7-19}$$

In a *symmetrical matrix*, the given matrix and its transposed matrix are identical, or

$$[a_{ij}] = [a_{ij}]' \tag{7-20}$$

EXAMPLE 7. The matrix

$$\begin{bmatrix} 2 & 3 & 1 \\ 3 & 5 & -2 \\ 1 & -2 & 5 \end{bmatrix} \tag{7-21}$$

and its transposed matrix are identical.

It is evident that a symmetrical matrix is symmetrical about the principal diagonal.

If the elements of matrix $a_{ij} = -a_{ji}$, but the elements a_{ii} are not *all* zero, then the matrix is called a *skew* matrix. If, however, $a_{ij} = -a_{ji}$ and as an additional property $a_{ii} = 0$, the matrix is *skew-symmetric*. In other words, if a matrix is skew-symmetric, then the transpose is equal to the negative of the given matrix, or, in symbols,

[1] $[I]$ or $[E]$ is generally used to denote the unit matrix, but in this text the notation indicated above will be used to avoid confusion with current or voltage matrices.

$$[a_{ij}] = [-a_{ij}]' \tag{7-22}$$

Symmetrical and skew-symmetric matrices are always square.

EXAMPLE 8. The matrix

$$\begin{bmatrix} 3 & 2 & 1 & 5 \\ -2 & 4 & 3 & 6 \\ -1 & -3 & 0 & -8 \\ -5 & -6 & 8 & -2 \end{bmatrix} \tag{7-23}$$

is skew, but

$$\begin{bmatrix} 0 & 2 & 1 & 5 \\ -2 & 0 & 3 & 6 \\ -1 & -3 & 0 & -8 \\ -5 & -6 & 8 & 0 \end{bmatrix} \tag{7-24}$$

is skew-symmetric.

A matrix which has all its elements a_{ij} equal to zero is known as a *zero*, or *null*, matrix and is designated by [0], where

$$[0] = \begin{bmatrix} 0 & 0 & . & . & . & 0 \\ 0 & 0 & . & . & . & 0 \\ . & & . & & & . \\ . & & & . & & . \\ . & & & & . & . \\ 0 & . & . & . & . & 0 \end{bmatrix} \tag{7-25}$$

A singular matrix is a square matrix which has a determinant equal to zero. All other square matrices are identified as nonsingular. All rectangular matrices must be classified as singular, since it will be recalled that nonsquare matrices are made square by adding the necessary rows or columns of zero elements.

Singular matrices cannot always be identified as singular simply by inspection. One important form of a singular matrix is that in which the elements of the various rows are numerically proportional to the elements in the top row.

EXAMPLE 9

$$[A] = \begin{bmatrix} a_{11} & a_{12} & a_{13} \\ a_{21} & a_{22} & a_{23} \\ \omega a_{11} & \omega a_{12} & \omega a_{13} \end{bmatrix} \tag{7-26}$$

is a singular matrix, because of the fact that the value of its determinant is equal to zero.

It is now possible to compare determinants and matrices. The determinant always consists of a square array which must be combined in a prescribed manner (e.g., Laplace's development or an adaptation thereof) to yield a unique value known as the value of the determinant (symbol). On the other hand, the matrix represents a rectangular array (or its special case, a square array) of elements which may be used to represent a set of linear operations, that is, linear transformations actually are expressed through the particular elements that are employed to make up the matrix. The matrix does not have any specifiable value, although, as is expected, the determinant of a matrix, just like any other determinant, does have a unique value. Another important fact to be stressed is that the elements of a determinant always have the same dimensions, whereas the elements of a matrix are not required to be dimensionally homogeneous.

7-3 Operations of Matrix Algebra

Two matrices $[a_{ij}]$ and $[b_{ij}]$ are equal if for every element a_{ij} in the first matrix there corresponds an element b_{ij} in the second matrix such that $a_{ij} = b_{ij}$ for all i, j, that is, two matrices are equal if and only if the corresponding elements of the matrices are identically equal.

The operation of *matrix addition* may be represented as follows. Consider the matrices $[a_{ij}]$ and $[b_{ij}]$, where

$$[a_{ij}] = \begin{bmatrix} a_{11} & a_{12} & a_{13} & \cdots & a_{1n} \\ a_{21} & a_{22} & a_{23} & \cdots & a_{2n} \\ a_{31} & a_{32} & a_{33} & \cdots & a_{3n} \\ \cdot & \cdot & \cdot & & \cdot \\ \cdot & \cdot & \cdot & & \cdot \\ \cdot & \cdot & \cdot & & \cdot \\ a_{n1} & a_{n2} & a_{n3} & \cdots & a_{nn} \end{bmatrix} \tag{7-27}$$

and

$$[b_{ij}] = \begin{bmatrix} b_{11} & b_{12} & b_{13} & \cdots & b_{1n} \\ b_{21} & b_{22} & b_{23} & \cdots & b_{2n} \\ b_{31} & b_{32} & b_{33} & \cdots & b_{3n} \\ \cdot & \cdot & \cdot & & \cdot \\ \cdot & \cdot & \cdot & & \cdot \\ \cdot & \cdot & \cdot & & \cdot \\ b_{n1} & b_{n2} & b_{n3} & \cdots & b_{nn} \end{bmatrix} \tag{7-28}$$

The *sum of matrices* $[a_{ij}] + [b_{ij}]$ is defined by the matrix

$$[a_{ij} + b_{ij}] = \begin{bmatrix} (a_{11} + b_{11}) & (a_{12} + b_{12}) & (a_{13} + b_{13}) & \cdots & (a_{1n} + b_{1n}) \\ (a_{21} + b_{21}) & (a_{22} + b_{22}) & (a_{23} + b_{23}) & \cdots & (a_{2n} + b_{2n}) \\ (a_{31} + b_{31}) & (a_{32} + b_{32}) & (a_{33} + b_{33}) & \cdots & (a_{3n} + b_{3n}) \\ \cdot & \cdot & \cdot & & \cdot \\ \cdot & \cdot & \cdot & & \cdot \\ \cdot & \cdot & \cdot & & \cdot \\ (a_{n1} + b_{n1}) & (a_{n2} + b_{n2}) & (a_{n3} + b_{n3}) & \cdots & (a_{nn} + b_{nn}) \end{bmatrix} \tag{7-29}$$

In a similar manner, *subtraction of matrices* is defined in such a manner that

$$[a_{ij}] - [b_{ij}] = [a_{ij} - b_{ij}] \tag{7-30}$$

EXAMPLE 1. Given the matrices $[A]$ and $[B]$, where

$$[A] = \begin{bmatrix} 3 & 2 & 1 & 5 \\ 1 & 4 & -8 & 1 \\ 2 & 5 & 0 & 8 \\ 3 & 6 & 2 & 7 \end{bmatrix} \tag{7-31}$$

and

$$[B] = \begin{bmatrix} 1 & 2 & 3 & 4 \\ 8 & 7 & 6 & 5 \\ 9 & 0 & 1 & 2 \\ 3 & 4 & 5 & 6 \end{bmatrix} \tag{7-32}$$

then the sum is

$$[A] + [B] = \begin{bmatrix} (3 + 1) & (2 + 2) & (1 + 3) & (5 + 4) \\ (1 + 8) & (4 + 7) & (-8 + 6) & (1 + 5) \\ (2 + 9) & (5 + 0) & (0 + 1) & (8 + 2) \\ (3 + 3) & (6 + 4) & (2 + 5) & (7 + 6) \end{bmatrix} \tag{7-33}$$

which, when simplified, becomes

$$[A] + [B] = \begin{bmatrix} 4 & 4 & 4 & 9 \\ 9 & 11 & -2 & 6 \\ 11 & 5 & 1 & 10 \\ 6 & 10 & 7 & 13 \end{bmatrix} \tag{7-34}$$

However, the difference between two matrices $[A]$ and $[B]$, or $[A] - [B]$, is expressed in matrix form as follows:

$$[A] - [B] = \begin{bmatrix} (3 - 1) & (2 - 2) & (1 - 3) & (5 - 4) \\ (1 - 8) & (4 - 7) & (-8 - 6) & (1 - 5) \\ (2 - 9) & (5 - 0) & (0 - 1) & (8 - 2) \\ (3 - 3) & (6 - 4) & (2 - 5) & (7 - 6) \end{bmatrix} \tag{7-35}$$

which reduces to the matrix

$$[A] - [B] = \begin{bmatrix} 2 & 0 & -2 & 1 \\ -7 & -3 & -14 & -4 \\ -7 & 5 & -1 & 6 \\ 0 & 2 & -3 & 1 \end{bmatrix} \tag{7-36}$$

If a matrix $[A] = [a_{ij}]$ is *multiplied by a scalar quantity* k, constant or variable, then

$$k[A] = [ka_{ij}] \tag{7-37}$$

EXAMPLE 2.

$$3 \cdot \begin{bmatrix} 2 & 3 & -1 \\ 1 & 8 & 0 \\ 4 & -7 & 13 \end{bmatrix} = \begin{bmatrix} 6 & 9 & -3 \\ 3 & 24 & 0 \\ 12 & -21 & 39 \end{bmatrix} \tag{7-38}$$

The *product* of matrix $[A] = [a_{ij}]$ and matrix $[B] = [b_{ij}]$ is

$$[A] \cdot [B] = [C],$$

where matrix $[C]$ has for its elements c_{ij}, where

$$c_{ij} = \sum_{k=1}^{s} a_{ik}b_{kj} \tag{7-39}$$

and the orders of matrices $[A]$, $[B]$, and $[C]$ are $\{m,s\}$, $\{s,n\}$, and $\{m,n\}$, respectively. This operation is expressed in words as follows: The product of the matrices $[A]$ and $[B]$ is a matrix $[C]$ in which the element which is located in the ith row and the jth column is obtained by multiplying each element of the ith row of $[A]$ by the corresponding element of the jth column of $[B]$ and then adding the results.

For example, consider the second-order matrices $[A]$ and $[B]$. Let

$$[A] = \begin{bmatrix} a_{11} & a_{12} \\ a_{21} & a_{22} \end{bmatrix} \tag{7-40}$$

and

$$[B] = \begin{bmatrix} b_{11} & b_{12} \\ b_{21} & b_{22} \end{bmatrix} \tag{7-41}$$

Then

$$[A] \cdot [B] = \begin{bmatrix} (a_{11}b_{11} + a_{12}b_{21}) & (a_{11}b_{12} + a_{12}b_{22}) \\ (a_{21}b_{11} + a_{22}b_{21}) & (a_{21}b_{12} + a_{22}b_{22}) \end{bmatrix} \tag{7-42}$$

Again, consider the matrix $[A]$ of order 3 and the matrix $[B]$ of order 2.

$$[A] = \begin{bmatrix} a_{11} & a_{12} & a_{13} \\ a_{21} & a_{22} & a_{23} \\ a_{31} & a_{32} & a_{33} \end{bmatrix} \tag{7-43}$$

and

$$[B] = \begin{bmatrix} b_{11} & b_{12} \\ b_{21} & b_{22} \\ b_{31} & b_{32} \end{bmatrix} \tag{7-44}$$

Then

$$[A] \cdot [B] = \begin{bmatrix} (a_{11}b_{11}+a_{12}b_{21}+a_{13}b_{31}) & (a_{11}b_{12}+a_{12}b_{22}+a_{12}b_{32}) \\ (a_{21}b_{11}+a_{22}b_{21}+a_{23}b_{31}) & (a_{21}b_{12}+a_{22}b_{22}+a_{23}b_{23}) \\ (a_{31}b_{11}+a_{32}b_{21}+a_{33}b_{31}) & (a_{31}b_{12}+a_{32}b_{22}+a_{33}b_{32}) \end{bmatrix} \tag{7-45}$$

Matrices obey the usual algebraic rules, e.g., associative and distributive, with the exception of the multiplicative commutative law. Thus if $[A]$, $[B]$, and $[C]$ denote matrices and μ and ν are scalars (in the real or complex domain):

(*a*) In addition of matrices,

$$\begin{aligned} [A] + [B] &= [B] + [A] \\ \{[A] + [B]\} + [C] &= \{[A] + [C]\} + [B] \\ &= \{[B] + [C]\} + [A] \\ &= \{[A] + [B]\} + [C] \end{aligned} \tag{7-46}$$

(*b*) In the multiplication of matrices by scalars μ and ν,

$$\begin{aligned} \mu[A] &= [A]\mu \\ \mu\{[A] + [B]\} &= \mu[A] + \mu[B] \\ \{\mu + \nu\}[A] &= \mu[A] + \nu[A] \end{aligned} \tag{7-47}$$

But note that, in general,

$$[A] \cdot [B] \neq [B] \cdot [A] \tag{7-48}$$

and

$$\begin{aligned} \{[A] + [B]\} \cdot [C] &= [A] \cdot [C] + [B] \cdot [C] \\ [C]\{[A] + [B]\} &= [C] \cdot [A] + [C] \cdot [B] \\ \{[A] \cdot [B]\} \cdot [C] &= [A] \cdot \{[B] \cdot [C]\} \\ &= [A] \cdot [B] \cdot [C] \\ [A] \cdot [B] \cdot [C] &\neq [A] \cdot [C] \cdot [B] \neq [C] \cdot [B] \cdot [A] \end{aligned} \tag{7-49}$$

The last expression, which represents the multiplication of more than two matrices, indicates that the order of indicated multiplication must be preserved. The practical consequences of an incorrect order of multiplication of the matrix factors will become evident in Chap. 8.

The noncommutative law of matrix algebra is void for diagonal matrices of the same order, where the same product is obtained regardless of the order in which the factors are multiplied. A special case worthy of mention here is the fact that any matrix is commutable with the unit matrix $[U]$.

To specify the order in which the matrix multiplication is to be performed, special mathematical terminology is used. The operation $[A] \cdot [B]$ is described as premultiplication of $[B]$ by $[A]$. The operation $[B] \cdot [A]$ is postmultiplication of $[B]$ by $[A]$.

On occasion, the product of two matrices $[A]$ and $[B]$ may produce the null matrix $[0]$, or

$$[A] \cdot [B] = [0] \tag{7-50}$$

It can be shown that this is possible only under the following sets of conditions:

$$[A] = [0] \quad \text{and} \quad [B] \neq [0] \tag{7-51}$$
$$[A] \neq [0] \quad \text{and} \quad [B] = [0] \tag{7-52}$$
$$[A] = [0] \quad \text{and} \quad [B] = [0] \tag{7-53}$$
$$[A] \neq [0] \quad \text{and} \quad [B] \neq [0] \quad \text{but } [A] \text{ and } [B] \text{ are singular matrices} \tag{7-54}$$

It may also be shown that, if the product of two square matrices is the null matrix, the sum of their ranks cannot exceed their orders.

If a square matrix $[A]$ is raised to a power, e.g., multiplied by itself n times, the product is represented as $[A]^n$. In addition, the matrix $[A]^0$ is equivalent to the unit matrix $[U]$.

The reciprocal matrix $[A]^{-1}$ exists provided that $[A]$ is nonsingular. To summarize, by the associative law of multiplication,

$$[A]^n \cdot [A]^m = [A]^{m+n} \tag{7-55}$$
$$\{[A]^n\}^m = \{[A]^m\}^n = [A]^{mn} \tag{7-56}$$
$$\{[A]^{-1}\}^n = [A]^{-n} \tag{7-57}$$

The division of matrices can be explained in terms of the adjoint matrix. The adjoint of a square matrix is formed by replacing each element of the matrix by its corresponding determinant cofactor (as determined in the manner described in Chap. 5) and then transposing the result.

EXAMPLE 3. If

$$[A] = \begin{bmatrix} a_{11} & a_{12} & a_{13} \\ a_{21} & a_{22} & a_{23} \\ a_{31} & a_{32} & a_{33} \end{bmatrix}$$

then the cofactors of the elements are

$$A_{11} = \begin{vmatrix} a_{22} & a_{23} \\ a_{32} & a_{33} \end{vmatrix} = a_{22}a_{33} - a_{23}a_{32} \tag{7-58}$$

$$A_{21} = -\begin{vmatrix} a_{12} & a_{13} \\ a_{32} & a_{33} \end{vmatrix} = a_{13}a_{32} - a_{12}a_{33} \tag{7-59}$$

a_{ij} has for its cofactor A_{ij} with proper algebraic sign prefixed (see Chap. 5).

The cofactors are then placed in the matrix in the positions originally held by the corresponding elements, which produces the matrix

$$\begin{bmatrix} A_{11} & A_{12} & A_{13} \\ A_{21} & A_{22} & A_{23} \\ A_{31} & A_{32} & A_{33} \end{bmatrix} \tag{7-60}$$

The transpose of this matrix gives the adjoint of $[A]$, or

$$\text{adj } [A] = \begin{bmatrix} A_{11} & A_{21} & A_{31} \\ A_{12} & A_{22} & A_{32} \\ A_{13} & A_{23} & A_{33} \end{bmatrix} \tag{7-61}$$

It can be shown that the product of an adjoint with the matrix itself will, regardless of the order of multiplication, be equal to the product of a unit matrix and the determinant of the matrix expressed symbolically:

$$[A] \cdot \text{adj } [A] = \text{adj } [A] \cdot [A] = |A| \cdot [U] \tag{7-62}$$

where $|A|$ is the determinant of the matrix. Now if the inverse or reciprocal of a matrix $[A]^{-1}$ is such that, by definition, the product of the matrix and its inverse yields the unit matrix, or

$$[A] \cdot [A]^{-1} = [U] \tag{7-63}$$

then from the relationship above,

$$[A] \cdot [A]^{-1} = [U] = \frac{\text{adj } [A] \cdot [A]}{|A|} \tag{7-64}$$

and an expression for the inverse matrix in terms of the adjoint matrix and the determinant of the matrix is obtained:

$$[A]^{-1} = \frac{\text{adj } [A]}{|A|} = \text{adj } [A] \cdot |A|^{-1} \tag{7-65}$$

It is obvious that only nonsingular matrices possess finite inverse matrices, because for singular matrices, by definition, $|A|$, the determinant of the matrix, is zero, so the expression for the inverse becomes infinite.

Once the inverse matrix has been calculated, the division of a matrix is a simple as well as an obvious operation, since division of a matrix may be performed by multiplication by its inverse.

If the quotient of matrix $[B]$ by matrix $[A]$ is to be obtained, then

$$\frac{[B]}{[A]} = [B] \cdot [A]^{-1} \tag{7-66}$$

where $[A]^{-1}$ is the inverse of matrix $[A]$ if $|A| \neq 0$. (Since matrix multiplication is in general noncommutative, there are two types of division, predivision and postdivision, $[A]^{-1} \cdot [B]$ and $[B] \cdot [A]^{-1}$, respectively.)

EXAMPLE 4. If the determinant of the matrix cited in Example 3 is defined as $|A|$, then the inverse of the matrix will be expressible in the following manner:

$$[A]^{-1} = \text{adj } [A]|A|^{-1} \tag{7-67}$$

$$[A]^{-1} = \frac{\begin{bmatrix} A_{11} & A_{21} & A_{31} \\ A_{12} & A_{22} & A_{32} \\ A_{13} & A_{23} & A_{33} \end{bmatrix}}{|A|} \tag{7-68}$$

where $|A|$ is the determinant of the matrix $[A]$.

By the property expressed previously of a matrix multiplied by a scalar, it follows that

$$[A]^{-1} = \begin{bmatrix} \frac{A_{11}}{|A|} & \frac{A_{21}}{|A|} & \frac{A_{31}}{|A|} \\ \frac{A_{12}}{|A|} & \frac{A_{22}}{|A|} & \frac{A_{32}}{|A|} \\ \frac{A_{13}}{|A|} & \frac{A_{23}}{|A|} & \frac{A_{33}}{|A|} \end{bmatrix} \tag{7-69}$$

For a two-by-two matrix, a special simplified procedure may be utilized. The inverse is obtained simply by the following steps:

(*a*) Interchange the elements on the principal diagonal ($i = j$).

(*b*) Change the algebraic sign of the elements in the remaining diagonal ($i \neq j$).

(*c*) Divide each element of the matrix obtained after 1 and 2 by the determinant of the matrix.

EXAMPLE 5. Consider the square matrix of order 2:

$$\begin{bmatrix} A & B \\ C & D \end{bmatrix} \tag{7-70}$$

The determinant symbol of the matrix is

$$\Delta = \begin{vmatrix} A & B \\ C & D \end{vmatrix} \tag{7-71}$$

and the value of the determinant Δ is

$$\Delta = AD - BC \tag{7-72}$$

In accord with the rules cited above,

$$\begin{bmatrix} A & B \\ C & D \end{bmatrix}^{-1} = \frac{1}{AD - BC}\begin{bmatrix} D & -B \\ -C & A \end{bmatrix} \tag{7-73}$$

if $AD - BC \neq 0$.

If the inverse of a matrix which is the product of several other matrices is to be obtained, it can be shown that the inverse of a product itself composed of matrices is equal to the product of the inverses of the individual matrices taken in the reverse order, or

$$[\{[A] \cdot [B]\}]^{-1} = [B]^{-1} \cdot [A]^{-1} \qquad (7\text{-}74)$$

and

$$[\{[A] \cdot [B] \cdot [C]\}]^{-1} = [C]^{-1}[\{[A] \cdot [B]\}]^{-1} = [C]^{-1}[B]^{-1}[A]^{-1} \qquad (7\text{-}75)$$

Sufficient matrix theory has now been presented to permit application to the more elementary operations which will be required in the discussion of the various aspects of network analysis treated in Chap. 8. It should be recognized that what has been presented here represents only the rudiments of matrix theory. For more advanced phases of the topic, the following references are suggested:

Bôcher, M.: "An Introduction to Higher Algebra," The Macmillan Company, New York, 1930.

Ferrar, N. L.: "Algebra—A Textbook of Determinants, Matrices and Algebraic Forms," Oxford University Press, London, 1941.

Frazer, R. A., W. J. Duncan, and A. R. Collar: "Elementary Matrices and Some Applications to Dynamics and Differential Equations," Cambridge University Press, London, 1938.

Guillemin, E. A.: "The Mathematics of Circuit Analysis," John Wiley & Sons, Inc., New York, 1949.

Kron, G.: "Tensor Analysis and Networks," John Wiley & Sons, Inc., New York, 1939.

LeCorbeiller, P.: "Matrix Analysis for Electric Networks" (Harvard University Press), John Wiley & Sons, Inc., New York, 1950.

Nodelman, H. M.: "Similarity of Matrices with Elements over the Commutative Field and over the Real Quaternion Field," M. A. thesis, Yale University, New Haven, Conn., 1937.

Pipes, L. A.: Matrices in Engineering, *Elec. Eng.*, **56**: 1177–1190, September, 1937.

Schreier, O., and E. Sperner: "Vorlesungen über Matrizen," Verlag B. G. Teubner, Leipzig, 1932.

Wedderburn, J. H. M.: "Lectures on Matrices," American Mathematical Society, Providence, R. I., 1934.

Problems

Prob. 7-1 If

$$A = \begin{bmatrix} 1 & 9 & 7 \\ 2 & 6 & 1 \\ 3 & 5 & 8 \end{bmatrix} \quad \text{and} \quad B = \begin{bmatrix} 0 & 1 & 2 \\ 1 & 5 & 8 \\ 2 & 7 & 6 \end{bmatrix}$$

find the following:

$$\begin{array}{llll} A \cdot B & B \cdot A & A' & B' \\ A^{-1} & B^{-1} & A + B & A - B \end{array}$$

Prob. 7-2 Find the number of determinants of third order contained in the matrix A, where

$$A = \begin{bmatrix} 1 & 2 & 9 & 1 \\ 2 & 0 & 19 & 0 \\ 1 & 3 & 4 & 6 \end{bmatrix}$$

Prob. 7-3 Determine the rank of the matrix B, where

$$B = \begin{bmatrix} 0 & 1 & 1 & 0 \\ 0 & 2 & 0 & 2 \\ 3 & 0 & 0 & 3 \\ 0 & 4 & 4 & 0 \end{bmatrix}$$

Prob. 7-4 Find the inverse of the matrix

$$\begin{bmatrix} 1 & 3 \\ 2 & 5 \end{bmatrix}$$

Prob. 7-5 Compute the fourth power of the matrix

$$A = \begin{bmatrix} 1 & -1 & 2 & 1 \\ -1 & 3 & 0 & -3 \\ 2 & 0 & 9 & -6 \\ 1 & -3 & -6 & 19 \end{bmatrix}$$

Prob. 7-6* Find the product

$$\begin{bmatrix} 1 & 1 & 1 \\ 1 & a^2 & a \\ 1 & a & a^2 \end{bmatrix} \cdot \begin{bmatrix} I_0 \\ I_1 \\ I_2 \end{bmatrix}$$

Prob. 7-7 Show by matrix algebra that if

$$\begin{aligned} Y_1 &= a_{11}X_1 + a_{12}X_2 \\ Y_2 &= a_{21}X_1 + a_{22}X_2 \end{aligned} \qquad \text{and} \qquad \begin{aligned} Z_1 &= b_{11}Y_1 + b_{12}Y_2 \\ Z_2 &= b_{21}Y_1 + b_{22}Y_2 \end{aligned}$$

then $\qquad Z_1 = c_{11}X_1 + c_{12}X_2 \qquad$ and $\qquad Z_2 = c_{21}X_1 + c_{22}X_2$

where

$$\begin{aligned} c_{11} &= b_{11}a_{11} + b_{12}a_{21} & \qquad c_{21} &= b_{21}a_{11} + b_{22}a_{21} \\ c_{12} &= b_{11}a_{12} + b_{12}a_{22} & \qquad c_{22} &= b_{21}a_{12} + b_{22}a_{22} \end{aligned}$$

Prob. 7-8 Find the inverse of the matrix

$$\begin{bmatrix} 3 & 1 & 0 & 0 \\ 1 & 2 & 0 & 0 \\ 0 & 0 & 2 & 0 \\ 0 & 0 & 0 & 4 \end{bmatrix}$$

* R. E. Vowels, Matrix Methods in the Solution of Ladder Networks, *J. Inst. Elec. Engrs. (London)*, pt. III, **95**: 40–50, January, 1948.

Prob. 7-9 Determine the rank of the matrix $[M]$, where

$$[M] = \begin{bmatrix} 2 & 3 & 3 & 16 & 22 & 9 \\ 4 & 3 & 2 & 2 & 4 & 1 \\ 6 & 6 & 7 & 6 & 18 & 5 \\ 12 & 12 & 10 & 36 & 52 & 20 \\ 10 & 9 & 11 & 2 & 22 & 6 \end{bmatrix}$$

Prob. 7-10 Prove that the determinant of a matrix formed by the multiplication of two or more matrices is equal to the product of the determinants of the individual matrices.

Prob. 7-11 If the matrix equation

$$\begin{bmatrix} a_{11} & a_{12} \\ a_{21} & a_{22} \end{bmatrix} \cdot \begin{bmatrix} Y_1 \\ Y_2 \end{bmatrix} = \begin{bmatrix} b_{11} & b_{12} \\ b_{21} & b_{22} \end{bmatrix} \cdot \begin{bmatrix} Y_1 \\ Y_2 \end{bmatrix}$$

is identically true, prove that $a_{ij} = b_{ij}$ for corresponding i,j.

Prob. 7-12 If the matrix equation

$$[Y_1 \quad Y_2] \begin{bmatrix} a_{11} & a_{12} & a_{13} \\ a_{21} & a_{22} & a_{23} \end{bmatrix} = [Y_1 \quad Y_2] \begin{bmatrix} b_{11} & b_{12} & b_{13} \\ b_{21} & b_{22} & b_{23} \end{bmatrix}$$

is identically true, prove that $a_{ij} = b_{ij}$ for corresponding i,j.

Prob. 7-13 Perform the indicated operations:

$$\begin{bmatrix} 1 & 2 \\ 3 & 4 \end{bmatrix}^2 \qquad \begin{bmatrix} 1 & 2 \\ 3 & 4 \end{bmatrix}^3 \qquad \begin{bmatrix} 1 & 2 \\ 3 & 4 \end{bmatrix}^{-1} \qquad \begin{bmatrix} 1 & 2 \\ 3 & 4 \end{bmatrix}^{-2}$$

Prob. 7-14 Show that, if $[U]$ is defined as the unit matrix, then

$$[U] \cdot [U]^{-1} = [U]^{-1} \cdot [U]$$

Prob. 7-15 Find the resultant matrix $[P]$, where

$$[P] = [A] \cdot [B]$$

The matrices $[A]$ and $[B]$ are given as

$$[A] = \begin{bmatrix} a_{11} & a_{12} & 0 \\ a_{21} & a_{22} & 0 \\ a_{31} & a_{32} & 0 \end{bmatrix} \qquad [B] = \begin{bmatrix} 0 & 0 & 0 \\ 0 & 0 & 0 \\ b_{31} & b_{32} & b_{33} \end{bmatrix}$$

Prob. 7-16 Find the transpose of the matrix

$$\begin{bmatrix} 6 & 5 & 1 \\ 3 & 4 & 2 \\ 2 & 2 & 1 \end{bmatrix}$$

Prob. 7-17 Prove that, if the product of two square matrices is the null matrix, then the sum of their ranks cannot exceed their orders.

Prob. 7-18 Find the adjoint of the matrix

$$\begin{bmatrix} 3 & 4 & 1 \\ 1 & 2 & 8 \\ 0 & 5 & 7 \end{bmatrix}$$

Prob. 7-19* If

$$Z = \begin{bmatrix} Z_0 & Z_1 & Z_2 \\ Z_2 & Z_0 & Z_1 \\ Z_1 & Z_2 & Z_0 \end{bmatrix} \quad \text{and} \quad A = \begin{bmatrix} 1 & 1 & 1 \\ 1 & a^2 & a \\ 1 & a & a^2 \end{bmatrix}$$

find $[Z] \cdot [A]$ and $[A] \cdot [Z]$.

Prob. 7-20* If

$$A = \begin{bmatrix} (5 + j2) & 0 & 5 \\ -3 & (2 - j7) & 2 \\ (3 + j2) & 6 & (3 + j4) \end{bmatrix}$$

and

$$B = \begin{bmatrix} (1 - j2) & 1 & 3 \\ 4 & (4 - j2) & (5 + j4) \\ 2 & 3 & (2 + j5) \end{bmatrix}$$

find $A + B$, $A - B$.

Prob. 7-21 Show that matrices $[A]$ and $[B]$, which are such that

$$[A] \cdot [B] \neq [B] \cdot [A]$$

have determinants $|A|$ and $|B|$ such that

$$|A| \cdot |B| = |B| \cdot |A|$$

Prob. 7-22 Show that if the matrices P, Q, and R are given as follows:

$$P = \begin{bmatrix} 6 & 0 & 7 \\ 0 & -3 & 4 \\ -5 & 2 & 0 \end{bmatrix} \quad Q = \begin{bmatrix} -7 & 21 & 4 \\ 8 & 5 & 1 \\ -2 & -9 & 17 \end{bmatrix} \quad R = \begin{bmatrix} 1 & 2 \\ 0 & 1 \\ 3 & -1 \end{bmatrix}$$

then

$$[P] \cdot [R] \neq [Q] \cdot [R]$$

Prob. 7-23 If a matrix N_1 is composed of elements from the complex domain, i.e., its elements are complex numbers, and the matrix N_2 is composed of elements which are complex conjugates of the corresponding elements of N_1, and if N_2' is the transpose of matrix N_2, then, if $N_1 N_2' = 0$, show that N_1 and N_2 are both null matrices.

Prob. 7-24 If the matrix representation of the equations

$$E_1 = AE_2 + BI_2$$
$$I_1 = CE_2 + DI_2$$

* R. E. Vowels, *loc. cit.*

is given as

$$\begin{bmatrix} E_1 \\ I_1 \end{bmatrix} = \begin{bmatrix} A & B \\ C & D \end{bmatrix} \begin{bmatrix} E_2 \\ I_2 \end{bmatrix}$$

show that

$$\begin{bmatrix} E_2 \\ I_2 \end{bmatrix} = \begin{bmatrix} A & B \\ C & D \end{bmatrix}^{-1} \begin{bmatrix} E_1 \\ I_1 \end{bmatrix}$$

Prob. 7-25 Show that if the matrix $[Y]$ is given as

$$[Y] = \begin{bmatrix} \cosh y & A \sinh y \\ \dfrac{\sinh y}{A} & \cosh y \end{bmatrix}$$

then

$$[Y]^r = \begin{bmatrix} \cosh ry & A \sinh ry \\ \dfrac{\sinh ry}{A} & \cosh ry \end{bmatrix}$$

for r defined as a positive or negative integer.

Prob. 7-26 If the elements of a matrix A are real, and if the product of A and its transpose A' yields the null matrix, prove that the matrix A is itself a null matrix.

Prob. 7-27 Prove that a matrix in which elements of the various rows are numerically proportional to the elements in the first row is singular.

Prob. 7-28 Prove that if a unit matrix is modified by interchanging the first and second rows (or the first and second columns) then pre- or postmultiplication of any other matrix $[X]$ by the unit matrix so altered produces the same change in $[X]$.

Prob. 7-29 Solve for matrices X_1 and X_2, respectively, where X_1 and X_2 are described by the equations

$$\begin{bmatrix} 1 & 5 & 1 \\ -1 & -5 & 1 \\ 5 & 25 & 2 \end{bmatrix} \cdot X_1 = 0 \qquad \text{and} \qquad X_2 \cdot \begin{bmatrix} 1 & 5 & 1 \\ -1 & -5 & 1 \\ 5 & 25 & 2 \end{bmatrix} = 0$$

Chapter 8

NETWORK SOLUTIONS BY MATRICES

8-1 Introduction

A considerable length of time elapsed before it was realized that Cayley's matrix algebra could be employed in the solution of the linear simultaneous equations which usually appear in electric-network problems. In 1929, however, Strecker and Feldtkeller[1] made the first application of matrix algebra to the study of four-terminal passive networks which operate under steady-state conditions. Since that time, the application of matrices to network analysis has become increasingly popular, and they are frequently encountered in the technical literature. The use of matrix algebra provides an enormous simplification in the solution of networks and supplants the need, in the case of the four-pole network, for the constant rewriting of network equations and their solution by the method of determinants.

8-2 General Matrix Forms for Four-pole Networks

Consider the simple system of linear equations

$$Y_1 = a_{11}X_1 + a_{12}X_2 \tag{8-1}$$
$$Y_2 = a_{21}X_1 + a_{22}X_2 \tag{8-2}$$

In terms of the products of matrices, (8-1) and (8-2) may be expressed in matrix form as

$$\begin{bmatrix} Y_1 \\ Y_2 \end{bmatrix} = \begin{bmatrix} a_{11} & a_{12} \\ a_{21} & a_{22} \end{bmatrix} \cdot \begin{bmatrix} X_1 \\ X_2 \end{bmatrix} \tag{8-3}$$

It will be recalled that similar equations were obtained for the relations between the input voltage and current and the output voltage and current of a linear passive four-pole network in Chap. 6. These were

[1] F. Strecker and R. Feldtkeller, Theory of the Generalized Quadripole, *Elek. Nachr. Tech.*, **6**: 93–112, March, 1929.

$$
\begin{aligned}
E_1 &= AE_2 + BI_2 \\
I_1 &= CE_2 + DI_2
\end{aligned} \tag{8-4}
$$

where A, B, C, and D are the general network parameters. It will also be recalled that these constants completely define the properties of a network and that the entire series of network equivalents (open-circuit impedances, short-circuit admittances, π and T equivalents, etc.) can be derived directly provided the A, B, C, and D parameters are known.

In matrix notation, according to the procedure followed for Eqs. (8-1) and (8-2), Eq. (8-4) is

$$
\begin{bmatrix} E_1 \\ I_1 \end{bmatrix} = \begin{bmatrix} A & B \\ C & D \end{bmatrix} \cdot \begin{bmatrix} E_2 \\ I_2 \end{bmatrix} \tag{8-5}
$$

where

$$
\begin{bmatrix} A & B \\ C & D \end{bmatrix}
$$

is defined as the *transfer* matrix of the network and is written $[A]$. Equation (8-5) is usually abbreviated thus:

$$
\begin{bmatrix} E_1 \\ I_1 \end{bmatrix} = [A] \cdot \begin{bmatrix} E_2 \\ I_2 \end{bmatrix} \tag{8-6}
$$

The other form of Eq. (8-6) is obviously

$$
\begin{bmatrix} E_2 \\ I_2 \end{bmatrix} = [A]^{-1} \cdot \begin{bmatrix} E_1 \\ I_1 \end{bmatrix} \tag{8-7}
$$

In addition to Eqs. (8-6) and (8-7), there are four other possible forms of the input-output equation for the four-pole network. These are

$$
\begin{bmatrix} E_1 \\ E_2 \end{bmatrix} = [Z] \cdot \begin{bmatrix} I_1 \\ I_2 \end{bmatrix} \tag{8-8}
$$

$$
\begin{bmatrix} I_1 \\ I_2 \end{bmatrix} = [Y] \cdot \begin{bmatrix} E_1 \\ E_2 \end{bmatrix} \tag{8-9}
$$

$$
\begin{bmatrix} E_1 \\ I_2 \end{bmatrix} = [H] \cdot \begin{bmatrix} I_1 \\ E_2 \end{bmatrix} \tag{8-10}
$$

and

$$
\begin{bmatrix} I_1 \\ E_2 \end{bmatrix} = [H]^{-1} \cdot \begin{bmatrix} E_1 \\ I_2 \end{bmatrix} \tag{8-11}
$$

Biggar[1] has tabulated the relationships between the various matrices shown above, and this table appears in Table 8-1 in modified form. It is self-explanatory and indicates how any required matrix may be calculated from the constants or parameters of any of the other matrices.

[1] H. P. Biggar, Application of Matrices to Four Terminal Network Problems, *Electronic Eng.*, **23**: 307–309, August, 1951.

Note that the symbol | | signifies the determinant of the matrix, so that $|A| = AD - BC$, $|Z| = Z_{11}Z_{22} - Z_{12}Z_{21}$, etc. Recall also that, *for passive networks which have no internal sources of emf,* $|A| = 1$. The values of the matrix determinants in terms of parameters of the other matrices are shown in Table 8-2.

TABLE 8-1 MATRIX CONVERSION TABLE

	$[A]$	$[Z]$	$[Y]$	$[H]$
$[A]$ Transfer matrix	$\begin{bmatrix} A & B \\ C & D \end{bmatrix}$	$\begin{bmatrix} \frac{Z_{11}}{Z_{21}} & \frac{-\lvert Z\rvert}{Z_{21}} \\ \frac{1}{Z_{21}} & \frac{-Z_{22}}{Z_{21}} \end{bmatrix}$	$\begin{bmatrix} \frac{-Y_{22}}{Y_{21}} & \frac{1}{Y_{21}} \\ \frac{-\lvert Y\rvert}{Y_{21}} & \frac{Y_{11}}{Y_{21}} \end{bmatrix}$	$\begin{bmatrix} \frac{-\lvert H\rvert}{H_{21}} & \frac{H_{11}}{H_{21}} \\ \frac{-H_{22}}{H_{21}} & \frac{1}{H_{21}} \end{bmatrix}$
$[A]^{-1}$	$\begin{bmatrix} \frac{D}{\lvert A\rvert} & \frac{-B}{\lvert A\rvert} \\ \frac{-C}{\lvert A\rvert} & \frac{A}{\lvert A\rvert} \end{bmatrix}$	$\begin{bmatrix} \frac{Z_{22}}{Z_{12}} & \frac{-\lvert Z\rvert}{Z_{12}} \\ \frac{1}{Z_{12}} & \frac{-Z_{11}}{Z_{12}} \end{bmatrix}$	$\begin{bmatrix} \frac{-Y_{11}}{Y_{12}} & \frac{1}{Y_{12}} \\ \frac{-\lvert Y\rvert}{Y_{12}} & \frac{Y_{22}}{Y_{12}} \end{bmatrix}$	$\begin{bmatrix} \frac{1}{H_{12}} & \frac{-H_{11}}{H_{12}} \\ \frac{H_{22}}{H_{12}} & -\frac{\lvert H\rvert}{H_{12}} \end{bmatrix}$
$[Z]$	$\begin{bmatrix} \frac{A}{C} & \frac{-\lvert A\rvert}{C} \\ \frac{1}{C} & \frac{-D}{C} \end{bmatrix}$	$\begin{bmatrix} Z_{11} & Z_{12} \\ Z_{21} & Z_{22} \end{bmatrix}$	$\begin{bmatrix} \frac{Y_{22}}{\lvert Y\rvert} & \frac{-Y_{12}}{\lvert Y\rvert} \\ \frac{-Y_{21}}{\lvert Y\rvert} & \frac{Y_{11}}{\lvert Y\rvert} \end{bmatrix}$	$\begin{bmatrix} \frac{\lvert H\rvert}{H_{22}} & \frac{H_{12}}{H_{22}} \\ \frac{-H_{21}}{H_{22}} & \frac{1}{H_{22}} \end{bmatrix}$
$[Y]$	$\begin{bmatrix} \frac{D}{B} & \frac{-\lvert A\rvert}{B} \\ \frac{1}{B} & \frac{-A}{B} \end{bmatrix}$	$\begin{bmatrix} \frac{Z_{22}}{\lvert Z\rvert} & \frac{-Z_{12}}{\lvert Z\rvert} \\ \frac{-Z_{21}}{\lvert Z\rvert} & \frac{Z_{11}}{\lvert Z\rvert} \end{bmatrix}$	$\begin{bmatrix} Y_{11} & Y_{12} \\ Y_{21} & Y_{22} \end{bmatrix}$	$\begin{bmatrix} \frac{1}{H_{11}} & \frac{-H_{12}}{H_{11}} \\ \frac{H_{21}}{H_{11}} & \frac{\lvert H\rvert}{H_{11}} \end{bmatrix}$
$[H]$	$\begin{bmatrix} \frac{B}{D} & \frac{\lvert A\rvert}{D} \\ \frac{1}{D} & \frac{-C}{D} \end{bmatrix}$	$\begin{bmatrix} \frac{\lvert Z\rvert}{Z_{22}} & \frac{Z_{12}}{Z_{22}} \\ \frac{-Z_{21}}{Z_{22}} & \frac{1}{Z_{22}} \end{bmatrix}$	$\begin{bmatrix} \frac{1}{Y_{11}} & \frac{-Y_{12}}{Y_{11}} \\ \frac{Y_{21}}{Y_{11}} & \frac{\lvert Y\rvert}{Y_{11}} \end{bmatrix}$	$\begin{bmatrix} H_{11} & H_{12} \\ H_{21} & H_{22} \end{bmatrix}$
$[H]^{-1}$	$\begin{bmatrix} \frac{C}{A} & \frac{\lvert A\rvert}{A} \\ \frac{1}{A} & \frac{-B}{A} \end{bmatrix}$	$\begin{bmatrix} \frac{1}{Z_{11}} & \frac{-Z_{12}}{Z_{11}} \\ \frac{Z_{21}}{Z_{11}} & \frac{\lvert Z\rvert}{Z_{11}} \end{bmatrix}$	$\begin{bmatrix} \frac{\lvert Y\rvert}{Y_{22}} & \frac{Y_{12}}{Y_{22}} \\ \frac{-Y_{21}}{Y_{22}} & \frac{1}{Y_{22}} \end{bmatrix}$	$\begin{bmatrix} \frac{H_{22}}{\lvert H\rvert} & \frac{-H_{12}}{\lvert H\rvert} \\ \frac{-H_{21}}{\lvert H\rvert} & \frac{H_{11}}{\lvert H\rvert} \end{bmatrix}$

The use of the various matrix forms given in Table 8-1 is essential in the calculation of the matrix equations for interconnected four-pole networks. There are five fundamental ways in which a pair of four-terminal networks may be interconnected, as shown in Table 8-3. As indicated by the column designated Matrix equation, the combined matrix of each of these interconnected networks can be secured by multiplication or addition of the appropriate matrices. One of the most useful cases shown is that of networks in cascade. Note that *the transfer matrix of four-pole networks in cascade equals the product of the transfer matrices of the individual networks.*

TABLE 8-2 VALUES OF MATRIX DETERMINANTS IN TERMS OF PARAMETERS OF OTHER MATRICES

	A	Z	Y	H
$\|A\|$		$\frac{-Z_{12}}{Z_{21}}$	$\frac{-Y_{12}}{Y_{21}}$	$\frac{H_{12}}{H_{21}}$
$\|Z\|$	$\frac{-B}{C}$		$\|Y\|^{-1}$	$\frac{H_{11}}{H_{22}}$
$\|Y\|$	$\frac{-C}{B}$	$\|Z\|^{-1}$		$\frac{H_{22}}{H_{11}}$
$\|H\|$	$\frac{-A}{D}$	$\frac{Z_{11}}{Z_{22}}$	$\frac{Y_{22}}{Y_{11}}$	
$\|H\|^{-1}$	$\frac{-D}{A}$	$\frac{Z_{22}}{Z_{11}}$	$\frac{Y_{11}}{Y_{22}}$	

As an example of how this rule is applied, consider two networks which have the general parameters

	Network A_1	Network A_2
A	1.50	1.66
B	11.00 ohms	4.00 ohms
C	0.25 mhos	1.00 mhos
D	2.50	3.00

The transfer matrices of these networks, according to Table 8-1, are

$$[A_1] = \begin{bmatrix} 1.50 & 11.00 \\ 0.25 & 2.50 \end{bmatrix} \tag{8-12}$$

$$[A_2] = \begin{bmatrix} 1.66 & 4.00 \\ 1.00 & 3.00 \end{bmatrix} \tag{8-13}$$

According to Table 8-3, if these two networks are connected in order A_1, A_2, the transfer matrix of the combination will be the product of the original transfer matrices, or

$$[A_{1,2}] = [A_1] \cdot [A_2] \tag{8-14}$$

$$[A_{1,2}] = \begin{bmatrix} 1.50 & 11.00 \\ 0.25 & 2.50 \end{bmatrix} \cdot \begin{bmatrix} 1.66 & 4.00 \\ 1.00 & 3.00 \end{bmatrix} \tag{8-15}$$

$$[A_{1,2}] = \begin{bmatrix} (1.50)(1.66) + (11.00)(1.00) & (1.50)(4.00) + (11.00)(3.00) \\ (0.25)(1.66) + (2.50)(1.00) & (0.25)(4.00) + (2.50)(3.00) \end{bmatrix} \tag{8-16}$$

$$[A_{1,2}] = \begin{bmatrix} (2.5 + 11.00) & (6.00 + 33.00) \\ (0.416 + 2.50) & (1.00 + 7.5) \end{bmatrix} \tag{8-17}$$

$$[A_{1,2}] = \begin{bmatrix} 13.5 & 39.00 \\ 2.916 & 8.5 \end{bmatrix} \tag{8-18}$$

As an example of the parallel connection, consider the networks for which the general parameters were given in the cascade example. The parallel connection requires the use of the Y parameters, and, from Table 8-1, these are:

	Network A_1, mhos	Network A_2, mhos
$Y_{11} = \frac{D}{B}$	$\frac{2.5}{11}$	$\frac{3}{4}$
$Y_{12} = \frac{-\|A\|}{B}$	$\frac{-1}{11}$	$\frac{-1}{4}$
$Y_{21} = \frac{1}{B}$	$\frac{1}{11}$	$\frac{1}{4}$
$Y_{22} = \frac{-A}{B}$	$\frac{-1.5}{11}$	$\frac{-5}{(3)(4)}$

The admittance, or $[Y]$, matrix of these networks in parallel is therefore

$$[Y_{1,2}] = [Y_1] + [Y_2] \tag{8-19}$$

$$[Y_{1,2}] = \begin{bmatrix} \frac{2.5}{11} & \frac{-1}{11} \\ \frac{1}{11} & \frac{-1.5}{11} \end{bmatrix} + \begin{bmatrix} \frac{3}{4} & \frac{-1}{4} \\ \frac{1}{4} & \frac{-5}{(3)(4)} \end{bmatrix} \tag{8-20}$$

$$[Y_{1,2}] = \begin{bmatrix} \frac{43}{44} & \frac{-15}{44} \\ \frac{15}{44} & \frac{-73}{132} \end{bmatrix} \tag{8-21}$$

If the results obtained in Eqs. (8-21) and (8-18) are not in the desired forms, they can be readily converted to any other set of matrices by means of the relationships given in Table 8-1.

There are certain restrictions which apply to the last four cases, shown in Table 8-3. For these cases, the matrix equations shown in Table 8-3 may not always be valid. For example, the parallel case may not be applied where a balanced-H network is connected in parallel with an unbalanced-T or -π network. Guillemin[1] has summarized the methods which can be used to determine whether or not interconnected networks can be analyzed by the matrix methods just discussed.

The test for networks to be connected in parallel is shown in Fig. 8-1*a*. Terminals 1, 5 and 2, 6 are connected together and terminals 3, 4 and 7, 8 are separately shorted. If the networks are to be analyzed

TABLE 8-3 MATRIX EQUATIONS FOR INTERCONNECTED FOUR-POLE NETWORKS

Circuit configuration	Block diagram	Matrix equation
Two networks in cascade	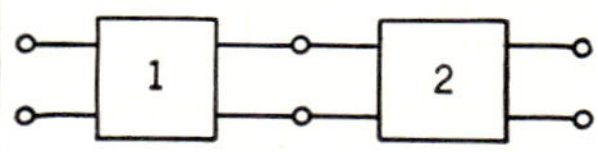	$[A_{1,2}] = [A_1] \cdot [A_2]$
Two networks with inputs and outputs in series	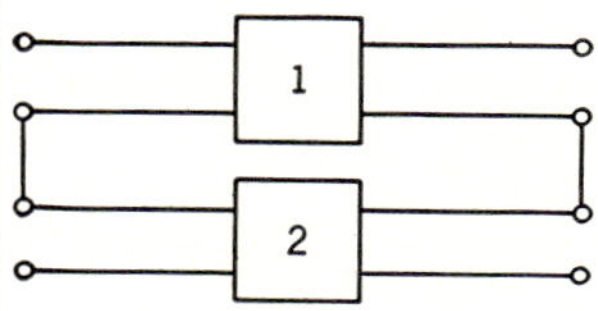	$[Z_{1,2}] = [Z_1] + [Z_2]$
Two networks with inputs and outputs in parallel	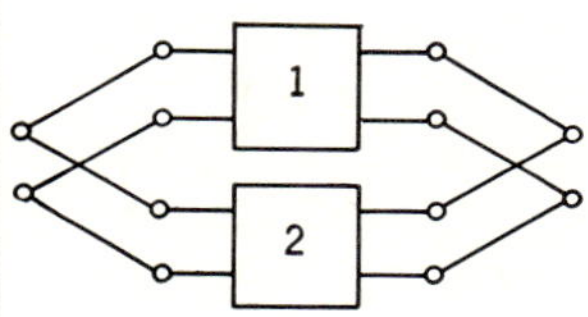	$[Y_{1,2}] = [Y_1] + [Y_2]$
Two networks with inputs in series and outputs in parallel	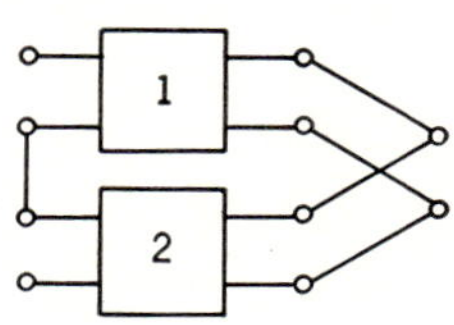	$[H_{1,2}] = [H_1] + [H_2]$
Two networks with inputs in parallel and outputs in series	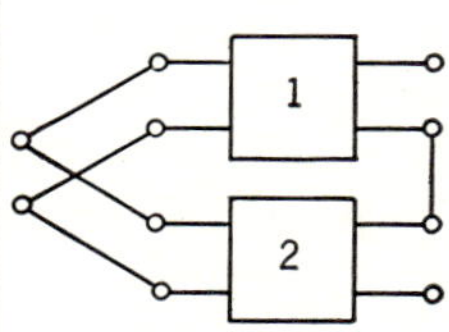	$[H_{1,2}]^{-1} = [H_1]^{-1} + [H_2]^{-1}$

by the matrix method, then, in this condition, $E_{24} = E_{68}$, or the voltage V should be zero. Also, with the networks reversed, i.e., with terminals 3, 7 and 4, 8 connected together and terminals 1, 2 and 5, 6 separately shorted, $E_{24} = E_{68}$ and once again $V = 0$. Both conditions must be satisfied for the matrix method of combining parallel networks to be valid.

[1] E. A. Guillemin, "Communication Networks," vol. II, pp. 147–151, John Wiley & Sons, Inc., New York, 1935.

In the series-connection test shown in Fig. 8-1*b*, the terminals on the left are connected in series as is normally the case, but terminals 3, 4 and 7, 8 are left unconnected or open. Then, $E_{24} = E_{57}$, or $V = 0$, if the matrix method of analysis is to be used. Also, the test must be repeated with the networks reversed, and both conditions must be satisfied.

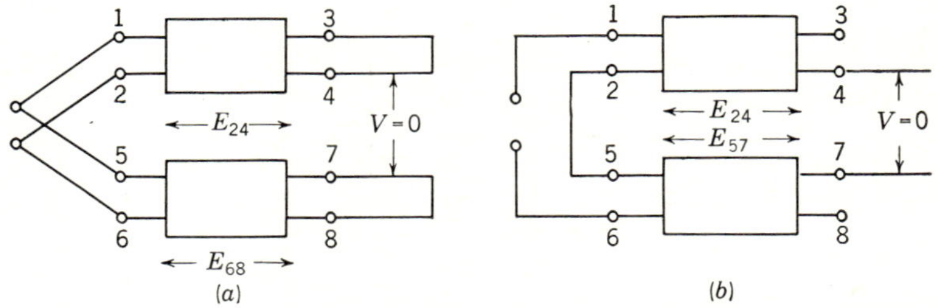

FIG. 8-1 Criterion for valid four-pole-network interconnection. (*a*) Parallel-connection test. (*b*) Series-connection test.

In the series-parallel or parallel-series-connection test, the series-connected ends of the network are arranged as shown in Fig. 8-2*a* with the opposing pairs of terminals shorted. In this case, the necessary condition is that $E_{24} = E_{67}$ so that $V = 0$. The parallel-connected ends of the network are arranged as shown in Fig. 8-2*b*, and the opposing terminals remain open. In this case, the necessary condition is that $E_{24} = E_{67}$, or $V = 0$. Both conditions must be satisfied for the matrix methods of combining networks in series-parallel or parallel-series to be valid.

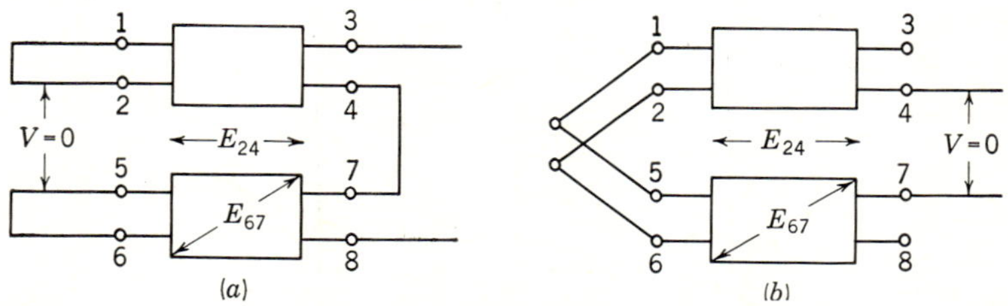

FIG. 8-2 Criterion for valid four-pole-network interconnection. (*a*) Series-parallel-connection test. (*b*) Parallel-series connection test.

In Fig. 8-3*a* and *c* are shown parallel and series cases of interconnection for which the matrix methods are not valid. It can be seen that these cases do not fulfill the tests just outlined. For example, it is obvious that, in Fig. 8-3*a*, $E_{24} \neq E_{68}$ when the right-hand terminal pairs are separately shorted, and, in Fig. 8-3*c*, $E_{24} \neq E_{57}$ when the right-hand terminal pairs are open.

Quite often the networks to be analyzed can be rearranged to satisfy the matrix analysis without alteration of their external properties. For example, in Fig. 8-3*b* is shown a rearrangement of Fig. 8-3*a* in which the series resistances in the bottom branches of the balanced-H pad (the lower network) are removed and added to its top branches. It will then have the same structure as the upper T network and may be placed in parallel with it without violating the conditions for the parallel connection. In Fig. 8-3*d*, the series arrangement in Fig. 8-3*c* has been modified

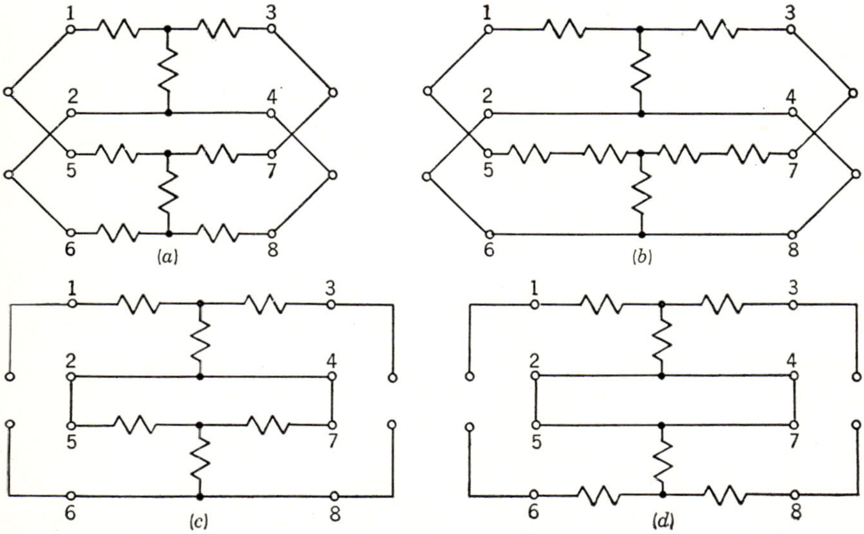

FIG. 8-3 (*a*) Parallel combination by matrices fails. (*b*) Parallel combination by matrices is successful. (*c*) Series combination by matrices fails. (*d*) Series combination by matrices is successful.

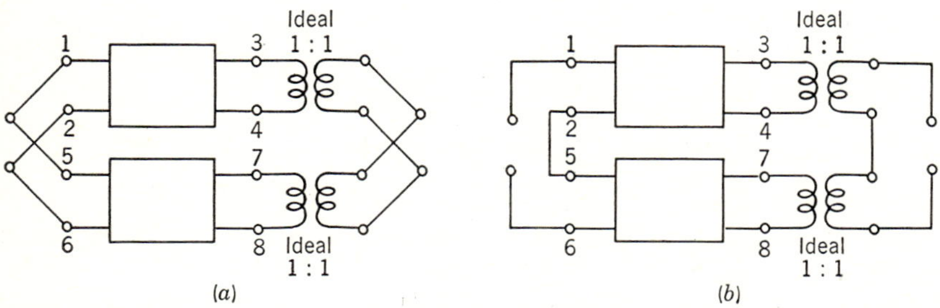

FIG. 8-4 Parallel and series interconnection by means of ideal transformers. (*a*) Parallel connection. (*b*) Series connection.

by inverting the lower T network so that the series combination is symmetrical about the horizontal center line and satisfies the tests for the series connection.

When simple measures fail, the introduction of hypothetical ideal transformers may permit the matrix analysis to be made. Examples of their use are shown in Fig. 8-4. In each case, the transformers cause the currents entering and leaving the right-hand end of each network to be equal, and dissimilar networks may thus be connected in parallel or series and analyzed by matrix methods. Quite often it is possible to eliminate these ideal transformers by suitable network manipulation, so that their use in network analysis or synthesis is not so restrictive as it might at first appear.

In conclusion, it is worthwhile to review the methods by which the network parameters can be measured. These are indicated in Table 8-4

TABLE 8-4 MATRIX PARAMETERS IN PHYSICAL TERMS

$$[A] = \begin{bmatrix} \dfrac{E_1}{E_2}\ (3,4\text{ o-c}) & \dfrac{E_1}{I_2}\ (3,4\text{ s-c}) \\ \dfrac{I_1}{E_2}\ (3,4\text{ o-c}) & \dfrac{I_1}{I_2}\ (3,4\text{ s-c}) \end{bmatrix}$$

$$[Z] = \begin{bmatrix} \dfrac{E_1}{I_1}\ (3,4\text{ o-c}) & \dfrac{-E_1}{I_2}\ (1,2\text{ o-c}) \\ \dfrac{E_2}{I_1}\ (3,4\text{ o-c}) & \dfrac{-E_2}{I_2}\ (1,2\text{ o-c}) \end{bmatrix}$$

$$[Y] = \begin{bmatrix} \dfrac{I_1}{E_1}\ (3,4\text{ s-c}) & \dfrac{-I_1}{E_2}\ (1,2\text{ s-c}) \\ \dfrac{I_2}{E_1}\ (3,4\text{ s-c}) & \dfrac{-I_2}{E_2}\ (1,2\text{ s-c}) \end{bmatrix}$$

$$[H] = \begin{bmatrix} \dfrac{E_1}{I_1}\ (3,4\text{ s-c}) & \dfrac{E_1}{E_2}\ (1,2\text{ o-c}) \\ \dfrac{I_2}{I_1}\ (3,4\text{ s-c}) & \dfrac{-I_2}{E_2}\ (1,2\text{ o-c}) \end{bmatrix}$$

NOTE: o-c = open-circuited, s-c = short-circuited. Numbers refer to terminals of four-poles.

for the $[A]$, $[Z]$, $[Y]$, and $[H]$ matrices. The values for the matrix parameters obtained by measurement will, of course, be valid only at the frequency at which the measurement is made.

8-3 Transfer Matrices of Special Four-pole Networks

Although the methods for deriving the matrices of networks connected in different ways have been developed, certain network forms appear so often that it is useful to tabulate their transfer matrices. In the cases which follow, only the transfer matrix $[A]$ of Eq. (8-7) will be evaluated, and when other matrix forms are desired ($[Z]$ or $[Y]$, etc.) they may be obtained directly from Table 8-1.

Probably the most elementary four-pole network is that shown in Fig. 8-5a, the series impedance. If the equations for the input voltage and current E_1 and I_1 are written in the same form as Eq. (8-4), then

$$\begin{aligned} E_1 &= E_2 + ZI_2 \\ I_1 &= + I_2 \end{aligned} \tag{8-22}$$

From these equations, the general network parameters are

$$\begin{aligned} A &= 1 \\ B &= Z \text{ ohms} \\ C &= 0 \\ D &= 1 \end{aligned} \tag{8-23}$$

and the corresponding transfer matrix for the series impedance is

$$[A_Z] = \begin{bmatrix} 1 & Z \\ 0 & 1 \end{bmatrix} \tag{8-24}$$

The other fundamental four-pole-network form is the shunt admit-

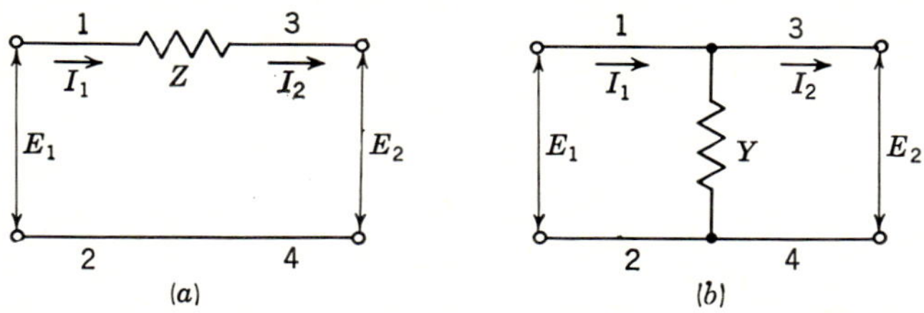

FIG. 8-5 (*a*) Fundamental network, series impedance. (*b*) Fundamental network, shunt admittance.

tance, shown in Fig. 8-5*b*. In this case,

$$\begin{aligned} E_1 &= E_2 \\ I_1 &= YE_2 + I_2 \end{aligned} \tag{8-25}$$

The general network parameters are

$$\begin{aligned} A &= 1 \\ B &= 0 \\ C &= Y \text{ mhos} \\ D &= 1 \end{aligned} \tag{8-26}$$

and the corresponding transfer matrix is

$$[A_Y] = \begin{bmatrix} 1 & 0 \\ Y & 1 \end{bmatrix} \tag{8-27}$$

Obviously, any network can be dissected into series and shunt components so that the transfer matrix of any network can be synthesized by means of Eqs. (8-24) and (8-27).

Consider first the simple L network, or voltage divider, in Fig. 8-6*a*. This is composed of a series impedance and a shunt admittance in cascade. The transfer matrix for the L network is therefore

$$[A_L] = \begin{bmatrix} 1 & Z \\ 0 & 1 \end{bmatrix} \cdot \begin{bmatrix} 1 & 0 \\ Y & 1 \end{bmatrix} = \begin{bmatrix} (1 + ZY) & Z \\ Y & 1 \end{bmatrix} \tag{8-28}$$

The reversed L network, or current divider, in Fig. 8-6*b* is composed of a shunt admittance and a series impedance in cascade, and

$$[A_{L_{rev}}] = \begin{bmatrix} 1 & 0 \\ Y & 1 \end{bmatrix} \cdot \begin{bmatrix} 1 & Z \\ 0 & 1 \end{bmatrix} = \begin{bmatrix} 1 & Z \\ Y & (1 + ZY) \end{bmatrix} \tag{8-29}$$

The T network of Fig. 8-6*c* can be analyzed in several ways. One is

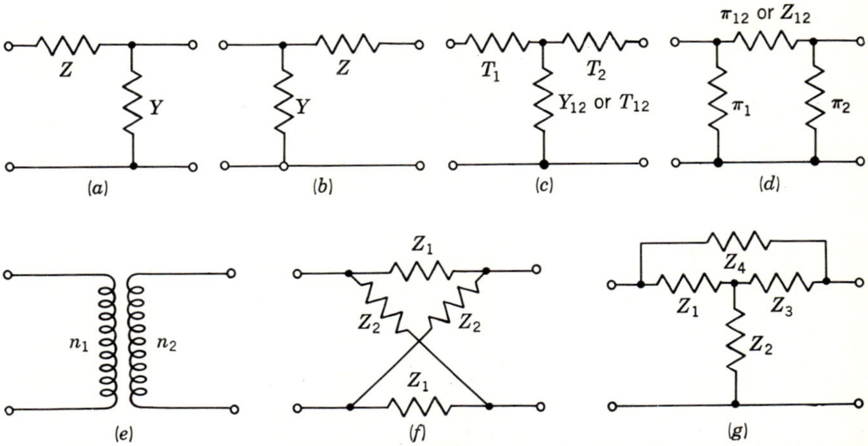

FIG. 8-6 (*a*) L network or voltage divider (*b*) L network (reversed) or current divider. (*c*) T network. (*d*) π network. (*e*) Ideal transformer. (*f*) Lattice network. (*g*) Bridged-T network.

to consider it a series impedance, and an admittance and a series impedance in cascade. Thus

$$\begin{aligned} [A_T] &= \begin{bmatrix} 1 & T_1 \\ 0 & 1 \end{bmatrix} \cdot \begin{bmatrix} 1 & 0 \\ 1/T_{12} & 1 \end{bmatrix} \cdot \begin{bmatrix} 1 & T_2 \\ 0 & 1 \end{bmatrix} \\ &= \begin{bmatrix} 1 + \dfrac{T_1}{T_{12}} & T_1 + T_2 + T_1 \cdot T_2 \cdot \dfrac{1}{T_{12}} \\ \dfrac{1}{T_{12}} & 1 + T_2 \cdot \dfrac{1}{T_{12}} \end{bmatrix} \end{aligned} \tag{8-30}$$

The π network shown in Fig. 8-6*d* can be considered as a shunt admittance, and a series impedance and a shunt admittance in cascade, or

$$\begin{aligned} [A_\pi] &= \begin{bmatrix} 1 & 0 \\ 1/\pi_1 & 1 \end{bmatrix} \cdot \begin{bmatrix} 1 & \pi_{12} \\ 0 & 1 \end{bmatrix} \cdot \begin{bmatrix} 1 & 0 \\ 1/\pi_2 & 1 \end{bmatrix} \\ &= \begin{bmatrix} 1 + \dfrac{\pi_{12}}{\pi_2} & \pi_{12} \\ \dfrac{1}{\pi_1} + \dfrac{1}{\pi_2}\left(\dfrac{1}{\pi_1} \cdot \dfrac{1}{\pi_2} \cdot \pi_{12}\right) & 1 + \dfrac{\pi_{12}}{\pi_1} \end{bmatrix} \end{aligned} \tag{8-31}$$

For the case of the ideal transformer in Fig. 8-6e the transfer matrix is

$$[A_{IT}] = \begin{bmatrix} \dfrac{n_1}{n_2} & 0 \\ 0 & \dfrac{n_2}{n_1} \end{bmatrix} \tag{8-32}$$

where n_1 and n_2 are the number of turns in primary and secondary.

It can also be shown that the $[A]$ matrices for the lattice (Fig. 8-6f) and the bridged-T network (Fig. 8-6g) are

$$[A_{LA}] = \begin{bmatrix} \dfrac{Z_1 + Z_2}{Z_2 - Z_1} & \dfrac{2Z_1Z_2}{Z_2 - Z_1} \\ \dfrac{2}{Z_2 - Z_1} & \dfrac{Z_1 + Z_2}{Z_2 - Z_1} \end{bmatrix} \tag{8-33}$$

$$[A_{BT}] = \begin{bmatrix} \dfrac{\begin{matrix} S + (Z_1 + Z_2)Z_4 & SZ_4 \\ (Z_1 + Z_3 + Z_4) & S + (Z_2 + Z_3)Z_4 \end{matrix}}{S + Z_2Z_4} \end{bmatrix} \tag{8-34}$$

In Eq. (8-34), all terms are divided by the denominator $S + Z_2Z_4$, where $S = Z_1Z_2 + Z_2Z_3 + Z_3Z_1$.

With the group of transfer matrices just developed, it is possible to secure the transfer matrix of a complex network structure by breaking it down into a series of cascaded elements or sections. The over-all transfer matrix is then the product of the matrices of the individual sections. In certain circumstances where a nonplanar network, such as the parallel-T network, is involved, it will be more convenient to analyze the network on an admittance or impedance basis, which will require a breakdown of the network into the appropriate parallel or series components rather than into cascaded sections.

It should be noted that one network can be made equivalent to another by making their transfer matrices equal. For example, a T and a π network are equivalent if their $[A]$ matrices are equal, or, in other words, if the corresponding elements in each transfer matrix can be equated. Thus, if the corresponding elements of the transfer matrices of the T and π networks in Eqs. (8-30) and (8-31) are equated, then

$$A = 1 + \frac{T_1}{T_{12}} = 1 + \frac{\pi_{12}}{\pi_2} \tag{8-35}$$

$$B = T_1 + T_2 + \frac{T_1T_2}{T_{12}} = \pi_{12} \tag{8-36}$$

$$C = \frac{1}{T_{12}} = \frac{1}{\pi_1} + \frac{1}{\pi_2} + \frac{\pi_{12}}{\pi_1\pi_2} \tag{8-37}$$

$$D = 1 + \frac{T_2}{T_{12}} = 1 + \frac{\pi_{12}}{\pi_1} \tag{8-38}$$

A simple rearrangement of these equations will yield the familiar T–π and π–T transformations.

8-4 Applications of Matrices in Passive-network Analysis

The transfer matrices of the networks obtained in the previous section specify the behavior of the network independently of the characteristics of the generator and of the load which are normally connected to the input and output terminals.

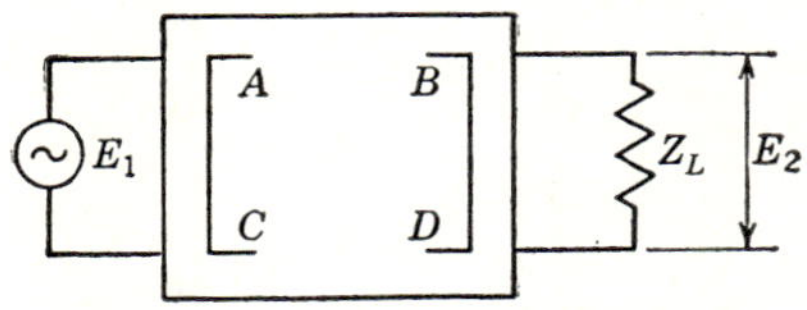

FIG. 8-7 Four-pole network with load.

Consider the simple case of a network which has the general parameters A, B, C, and D and which has a load impedance Z_L as shown in Fig. 8-7. The transfer matrix of the network and load in cascade, as shown, is

$$[A_{Z_L}] = \begin{bmatrix} A & B \\ C & D \end{bmatrix} \cdot \begin{bmatrix} 1 & 0 \\ 1/Z_L & 1 \end{bmatrix} = \begin{bmatrix} A + \dfrac{B}{Z_L} & B \\ C + \dfrac{D}{Z_L} & D \end{bmatrix} \tag{8-39}$$

The input impedance Z_{11} of the network combined with load, from Table 8-1 and Eq. (8-39), is

$$Z_{11} = \frac{A + B/Z_L}{C + D/Z_L} = \frac{AZ_L + B}{CZ_L + D} \tag{8-40}$$

If the internal impedance of the generator can be neglected (that is, E_1 is a constant-voltage source), then the ratio of the voltage across the load, E_2, to E_1 is [from Table 8-4 and Eq. (8-39)]

$$\frac{E_2}{E_1} = \frac{1}{A + B/Z_L} = \frac{1}{A + BY_L} \tag{8-41}$$

This method of network analysis has been employed by Sulzer[1] in the design of low-pass filters.

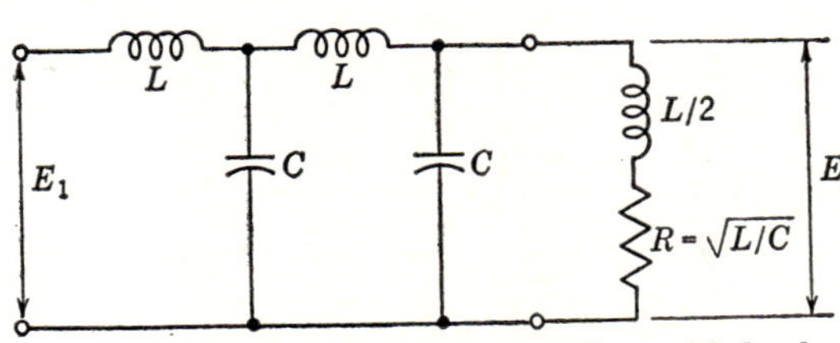

FIG. 8-8 Two-section low-pass filter with load.

[1] P. G. Sulzer, L-section Low-pass Filter Design, *Communications*, **29** (7): 22–23, July, 1949.

Consider the two-section low-pass filter shown in Fig. 8-8 which is terminated with an RL series load. The transfer matrix for this network not including the load is

$$[A_{LPF}] = \begin{bmatrix} 1 & j\omega L \\ 0 & 1 \end{bmatrix} \cdot \begin{bmatrix} 1 & 0 \\ j\omega C & 1 \end{bmatrix} \cdot \begin{bmatrix} 1 & j\omega L \\ 0 & 1 \end{bmatrix} \cdot \begin{bmatrix} 1 & 0 \\ j\omega C & 1 \end{bmatrix} \tag{8-42}$$

$$= \begin{bmatrix} (1 - \omega^2 LC) & j\omega L \\ j\omega C & 1 \end{bmatrix} \cdot \begin{bmatrix} (1 - \omega^2 LC) & j\omega L \\ j\omega C & 1 \end{bmatrix} \tag{8-43}$$

$$= \begin{bmatrix} (1 - 3\omega^2 LC + \omega^4 L^2 C^2) & (2j\omega L - j\omega^3 L^2 C) \\ (2j\omega C - j\omega^3 LC^2) & (1 - \omega^2 LC) \end{bmatrix} \tag{8-44}$$

To compute the ratio E_2/E_1 by means of Eq. (8-41), only the A and B parameters of the transfer matrix $[A_{LPF}]$ are needed. These are $A = 1 - 3\omega^2 LC + \omega^4 L^2 C^2$ and $B = 2j\omega L - j\omega^3 L^2 C$. If the load impedance is $\sqrt{L/C} + j\omega L/2$ as indicated in Fig. 8-8, E_2/E_1 is

$$\frac{E_2}{E_1} = \frac{1}{A + B/Z_L}$$

$$= \frac{1}{1 - 3\omega^2 LC + \omega^4 L^2 C^2 + (2j\omega L - j\omega^3 L^2 C)/(\sqrt{L/C} + j\omega L/2)} \tag{8-45}$$

If $L = C = 1$, the ratio becomes

$$\frac{E_2}{E_1} = \frac{1}{1 - 3\omega^2 + \omega^4 + j(2\omega - \omega^3)/(1 - j\omega/2)} \tag{8-46}$$

which is Sulzer's result. If the reader is not convinced that this solution has really been simplified as the result of using matrix methods, it is recommended that he obtain the same ratio E_2/E_1 by the conventional method of mesh analysis which was used in Chap. 6.

As a second example, consider the T-section low-pass filter with load R shown in Fig. 8-9. It is desired to calculate the ratio E_2/E_1. The transfer matrix of the network can be written directly from Eq. (8-30) and is

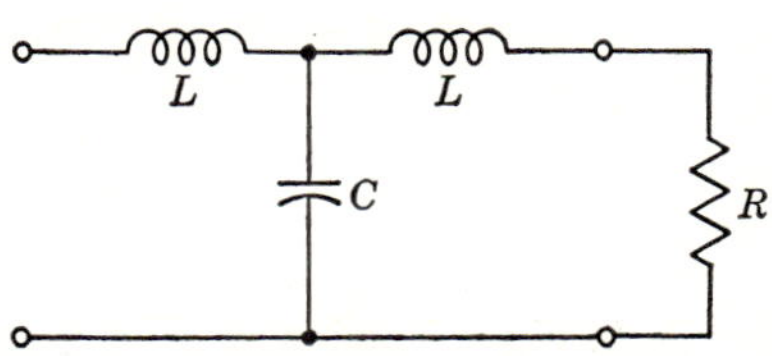

FIG. 8-9 T-section low-pass filter with load R.

$$[A_T] = \begin{bmatrix} 1 - \omega^2 LC & 2j\omega L - j\omega^3 L^2 C \\ j\omega C & 1 - \omega^2 LC \end{bmatrix} \tag{8-47}$$

From Eq. (8-41), E_2/E_1 with load R is

$$\frac{E_2}{E_1} = \frac{R}{(1 - \omega^2 LC)R + 2j\omega L - j\omega^3 L^2 C} \tag{8-48}$$

This result required only two equations!

The transfer-matrix parameters A, B, C, and D obey certain definite laws in the case of *passive* four-poles. In addition to the fact that the determinant of the transfer matrix $AD - BC$ is equal to 1, in a lossless network A and D are real and B and C are pure imaginaries. If the network is symmetrical, furthermore, $A = D$ and, for filters, it can be shown that

(*a*) In the pass band of a filter, $|(A + D)| \leq 2$.*

(*b*) At the edges of a pass band, $AD = 0$ or $BC = 0$.†

8-5 Matrices of Electron-tube Networks

Shortly after the first application of matrix algebra to passive networks, Strecker and Feldtkeller[1] and Bartlett[2] extended the matrix method to the analysis of electron-tube amplifiers. Since that time, the use of matrix algebra in studies of active networks has been less widespread than in the field of passive-network analysis, but inasmuch as the same advantages accrue through the use of the matrix notation in active-network analysis, it is to be expected that the method will gain popularity in this field also.

The various forms of matrix representation $[A]$, $[Y]$, $[Z]$, etc., previously developed for passive networks, can be used without modification in electron-tube networks. However, as was mentioned in Chap. 6, the equivalent networks will no longer be bilateral and the principle of reciprocity will be invalid. For example, $-Z_{12} \neq Z_{21}$. In the analysis of electron-tube amplifier circuits by matrix methods, tube operation is assumed to be linear (Class-A operation) and arranged so that $E_{gk} < 0$ (i.e., grid current does not occur). Under these circumstances the input-output equations of the grounded-cathode electron-tube network in Fig. 8-10*a* are

$$\begin{aligned} i_g &= 0 \\ i_p &= g_m e_g + g_p e_p \end{aligned} \tag{8-49}$$

* L. A. Pipes, The Matrix Theory of Four-terminal Networks, *Phil. Mag.*, ser. 7–33, **30**: 370–395, November, 1940.

† P. I. Richards, Applications of Matrix Algebra to Filter Theory, *Proc. I.R.E.*, **34** (3): 145–150P, March, 1946. This reference also contains information on the design and analysis of filters on an iterative basis. See also M. E. Fisher, The Matrix Approach to Filters and Transmission Lines, *Electronic Eng.*, **27** (327, 328): 198–204, 258, 263, May, June, 1955; H. L. Armstrong, Note on the Use of Tchebycheff Functions in Dealing with Iterated Networks, E. C. Ho, A General Matrix Factorization Method for Network Synthesis, and R. H. Pantell, Minimum-phase Transfer-function Synthesis, *I.R.E. Trans. Circuit Theory*, **CT-2**, June, 1955.

[1] F. Strecker and R. Feldtkeller, Theory of Low Frequency Amplifier Chains, *Arch. Elektrotech.*, **24**: 425–468, November 7, 1930.

[2] A. C. Bartlett, "Multi-stage Valve Amplifier," *Phil. Mag.*, ser. 7–10, pp. 734–738, October, 1930.

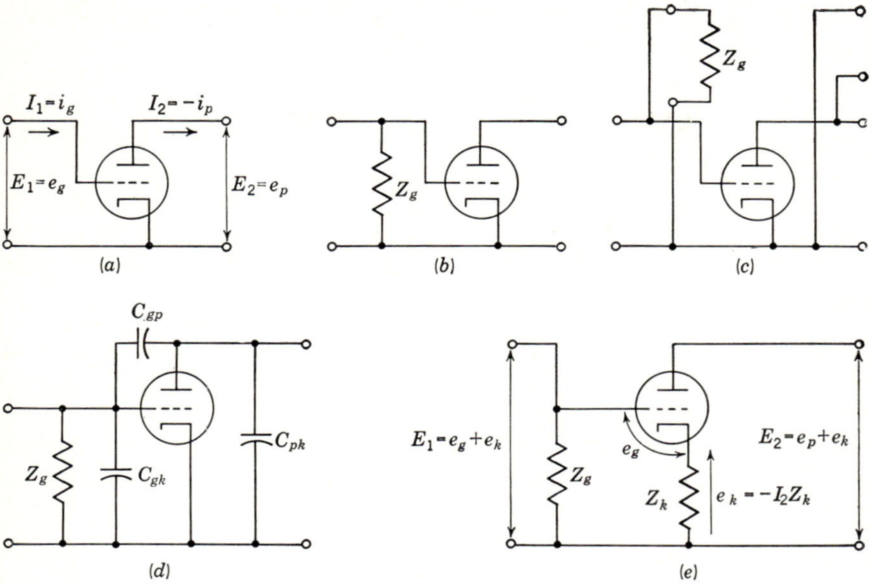

FIG. 8-10 Four-pole electron-tube equivalents. (*a*) Grounded-cathode circuit. (*b*) Grounded-cathode circuit with grid input impedance. (*c*) Four-pole equivalent of (*b*). (*d*) Grounded-cathode circuit with interelectrode capacitances added. (*e*) Grounded-cathode circuit with cathode impedance.

where g_m is the control-grid-to-plate transconductance and g_p is the plate conductance, or $1/r_p$ (where r_p is the plate resistance). Now, resorting to the current and voltage notation used previously,

$$I_1 = i_g \qquad -i_p = I_2 \qquad e_g = E_1 \qquad e_p = E_2$$

so that from Eq. (8-49),

$$\begin{aligned} I_1 &= 0 \\ I_2 &= -g_m E_1 - g_p E_2 \end{aligned} \tag{8-50}$$

In matrix form these equations become

$$\begin{bmatrix} I_1 \\ I_2 \end{bmatrix} = \begin{bmatrix} 0 & 0 \\ -g_m & -g_p \end{bmatrix} \cdot \begin{bmatrix} E_1 \\ E_2 \end{bmatrix} \tag{8-51}$$

and the admittance matrix of the grounded-cathode configuration $[Y_k]$ is therefore

$$[Y_k] = \begin{bmatrix} 0 & 0 \\ -g_m & -g_p \end{bmatrix} \tag{8-52}$$

where the subscript k indicates grounded-cathode operation.

The rank of $[Y_k]$ is less than its order because the grid-current coefficients are zero. The matrix is therefore singular, and the inverse matrix

$[Y_k]^{-1}$, or $[Z_k]$, does not exist. This is not true when there is a finite grid input resistance or impedance because in that case $I_1 = i_g \neq 0$. It will be noticed that it has been convenient to develop the admittance matrix $[Y]$ rather than the transfer matrix $[A]$. It will be found that the use of admittance matrices is to be preferred in the development of the matrix equations for any of the electron-tube configurations for the same reasons that the nodal method of analysis was preferable. Once the admittance matrix $[Y]$ has been determined, the transfer matrix can, of course, be obtained readily from the relationships in Tables 8-1 and 8-2. In this case

$$\begin{aligned} A &= \frac{-Y_{22}}{Y_{21}} = \frac{-g_p}{g_m} \\ B &= \frac{1}{Y_{21}} = \frac{-1}{g_m} \\ C &= \frac{-|Y|}{Y_{21}} = 0 \\ D &= \frac{Y_{11}}{Y_{21}} = 0 \end{aligned} \tag{8-53}$$

and the transfer matrix for the grounded-cathode amplifier is

$$A_k = \begin{bmatrix} \dfrac{-g_p}{g_m} & \dfrac{-1}{g_m} \\ 0 & 0 \end{bmatrix} \tag{8-54}$$

Since $g_p = 1/r_p$, and $\mu = g_m r_p$, Eq. (8-54) may also be written as

$$A_k = \begin{bmatrix} \dfrac{-1}{\mu} & \dfrac{-1}{g_m} \\ 0 & 0 \end{bmatrix} \tag{8-55}$$

Returning to the admittance matrix in Eq. (8-52), the effect of adding an input grid impedance of Z_g, as in Fig. 8-10*b*, can be taken into account by separating the combination into two parallel four-pole networks as in Fig. 8-10*c*. The required matrix will then be the sum of the $[Y]$ matrices for the upper and lower four-poles, or

$$[Y_k] = \begin{bmatrix} Y_G & 0 \\ 0 & 0 \end{bmatrix} + \begin{bmatrix} 0 & 0 \\ -g_m & -g_p \end{bmatrix} = \begin{bmatrix} Y_G & 0 \\ -g_m & -g_p \end{bmatrix} \tag{8-56}$$

where $Y_G = 1/Z_G$

In the same way, the interelectrode capacitances of the grounded-cathode configuration shown in Fig. 8-10*d* can be introduced by adding a third $[Y]$ matrix to Eq. (8-56). The result in this case will be

$$[Y_k] = \begin{bmatrix} Y_G & 0 \\ -g_m & -g_p \end{bmatrix} + \begin{bmatrix} j\omega(C_{gk} + C_{gp}) & -j\omega C_{gp} \\ j\omega C_{gp} & -j\omega(C_{pk} + C_{gp}) \end{bmatrix}$$

$$= \begin{bmatrix} Y_G + j\omega(C_{gk} + C_{gp}) & -j\omega C_{gp} \\ (j\omega C_{gp} - g_m) & -[g_p + j\omega(C_{pk} + C_{gp})] \end{bmatrix} \tag{8-57}$$

When an impedance Z_k is introduced in series with the cathode, degeneration occurs which will modify the $[Y_k]$ matrix. The circuit in this case is shown in Fig. 8-10*e*, from which $i_g = I_1 - Y_G E_1$, $-i_p = I_2$, $e_g = E_1 + I_2 Z_k$, $e_p = E_2 + I_2 Z_k$. When these values are substituted in Eq. (8-49),

$$\begin{aligned} I_1 &= Y_G E_1 \\ I_2 &= -g_m(E_1 + I_2 Z_k) - g_p(E_2 + I_2 Z_k) \end{aligned} \tag{8-58}$$

for which the $[Y]$ matrix is

$$[Y_k \text{ (with } Z_k)] = \begin{bmatrix} Y_g & 0 \\ \dfrac{-g_m}{1 + Z_k(g_m + g_p)} & \dfrac{-g_p}{1 + Z_k(g_m + g_p)} \end{bmatrix} \tag{8-59}$$

In the case of the grounded-plate, or cathode-follower, configuration shown in Fig. 8-11*a*, the voltages and currents are $i_g = I_1 - Y_G E_1$,

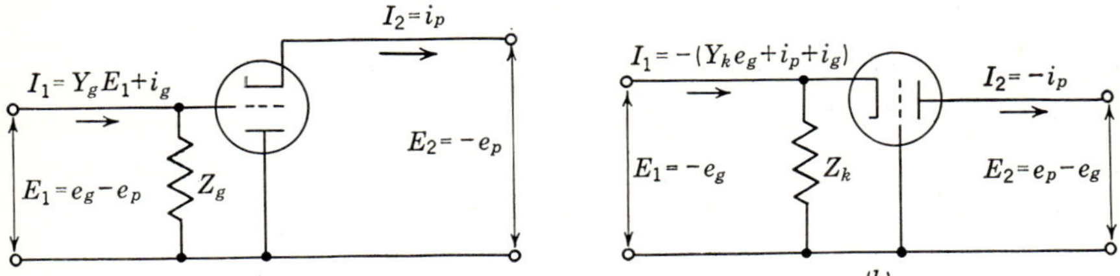

FIG. 8-11 Grounded-plate and -grid electron-tube four-pole configurations. (*a*) Grounded-plate configuration (cathode follower). (*b*) Grounded-grid configuration.

$i_p = I_2$, $e_g = E_1 - E_2$, $e_p = -E_2$. When these values are substituted in Eq. (8-49), the resulting $[Y]$ matrix is

$$[Y_p] = \begin{bmatrix} Y_g & 0 \\ g_m & -(g_m + g_p) \end{bmatrix} \tag{8-60}$$

The grounded-grid amplifier circuit is shown in Fig. 8-11*b*. The appropriate voltages and currents to be substituted in Eq. (8-49) are $i_g = Y_k E_1 + I_2 - I_1$, $-i_p = I_2$, $-e_g = E_1$, and $e_p = E_2 - E_1$; and the resulting $[Y]$ matrix is

$$[Y_G] = \begin{bmatrix} (Y_k + g_m + g_p) & -g_p \\ (g_m + g_p) & -g_p \end{bmatrix} \tag{8-61}$$

Deards[1] has tabulated the relationships which exist between the $[Y]$, $[A]$, $[Z]$, etc., matrices for the various electron-tube configurations, and this table appears in Table 8-5 in modified form.

If the gain of an amplifier is to be calculated, only the A and B parameters of the $[A]$ matrix need be used. For example, the A and B parameters of the $[A_k]$ matrix in Table 8-5 are $A = -1/\mu$, $B = -1/g_m$. From Eq. (8-41), the gain of a network with load Z_L is

$$\frac{E_2}{E_1} = \frac{1}{A + BY_L} = \frac{1}{-1/\mu - Y_L/g_m} \tag{8-62}$$

and since $g_m = \mu/r_p$,

$$\frac{E_2}{E_1} = \frac{-\mu Z_L}{Z_L + r_p} \tag{8-63}$$

which is the customary expression.

8-6 Applications of Matrices in Electron-tube-network Analysis

As an example of the application of the matrices given in Table 8-5, it is proposed to analyze the cathode-follower circuit shown in Fig. 8-12d. It will be assumed that R_B, the cathode bias resistor, is adequately bypassed by C_B, so that it can be neglected in the equivalent network. It will also be assumed that the internal impedance of the plate voltage supply is sufficiently low so that the plate is effectively grounded. Taking into account the interelectrode capacitances, the equivalent network is made up of the three four-poles shown in Fig. 8-12a, b, and c connected in parallel. The $[Y]$ matrix of the network will therefore be the sum of the $[Y]$ matrices for these four-pole networks, or

$$[Y_p] = \begin{bmatrix} 0 & 0 \\ g_m & -(g_m + g_p) \end{bmatrix} + \begin{bmatrix} Y_g & -Y_g \\ Y_g & -(Y_g + Y_k) \end{bmatrix} + \begin{bmatrix} j\omega(C_{gp} + C_{gk}) & -j\omega C_{gk} \\ j\omega C_{gk} & -j\omega(C_{pk} + C_k + C_{gk}) \end{bmatrix} \tag{8-64}$$

Note that the $[Y_p]$ matrix given in Table 8-5 was modified for use here because Z_g is connected directly to the top of Z_k rather than to ground.

The final result for $[Y_p]$ is

$$[Y_p] = \begin{bmatrix} Y_G + j\omega(C_{gp} + C_{gk}) & -(Y_G + j\omega C_{gk}) \\ (g_m + Y_G + j\omega C_{gk}) & -[g_m + g_p + Y_G + Y_k + j\omega(C_{pk} + C_k + C_{gk})] \end{bmatrix} \tag{8-65}$$

[1] S. R. Deards, Matrix Theory Applied to Thermionic Valve Circuits, *Electronic Eng.*, **24**: 264–267, June, 1952.

The voltage gain α of the network, from Eq. (8-41), with Z_k self-contained as the load so that $Z_L = \infty$, is simply $1/A$ so that, from Table 8-1,

$$\alpha = \frac{E_2}{E_1} = \frac{1}{A} = \frac{Y_{21}}{-Y_{22}} = \frac{g_m + Y_G + j\omega C_{gk}}{g_m + g_p + Y_G + Y_k + j\omega(C_{pk} + C_k + C_{gk})} \tag{8-66}$$

If the interelectrode capacitances can be ignored at low frequencies and Z_G is very large, α is reduced to the familiar expression

$$\alpha_{LF} = \frac{\mu Z_K}{r_p + Z_K(1 + \mu)} \tag{8-67}$$

The input admittance is $1/Z_{11}$, which according to Table 8-1 is $|Y|/Y_{22}$,

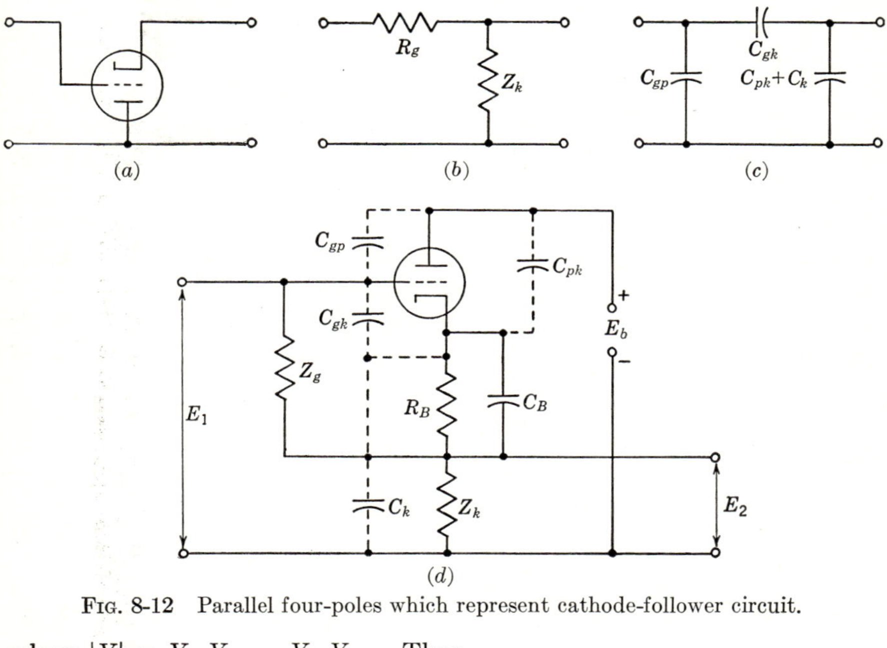

FIG. 8-12 Parallel four-poles which represent cathode-follower circuit.

where $|Y| = Y_{11}Y_{22} - Y_{12}Y_{21}$. Thus

$$Y_{\text{input}} = Y_G + j\omega(C_{gp} + C_{gk}) - \frac{(Y_G + j\omega C_{gk})(g_m + Y_G + j\omega C_{gk})}{g_m + g_p + Y_G + Y_K + j\omega(C_{pk} + C_k + C_{gk})} \tag{8-68}$$

Using the value of α obtained in Eq. (8-66), (8-68) can be reduced to

$$Y_{\text{input}} = Y_G(1 - \alpha) + j\omega C_{gp} + j\omega C_{gk}(1 - \alpha) \tag{8-69}$$

and if Y_G is due to a grid resistance R_g, then

$$Y_{\text{input}} = G_G(1 - \alpha) + j\omega[C_{gp} + C_{gk}(1 - \alpha)] \tag{8-70}$$

Very often $\alpha \cong 0.9$ and is real so that the input admittance has a real component due to the grid resistor R_g, the value of which is approximately $10R_g$. The input grid resistance is therefore multiplied by about 10. The imaginary component at the input is due to an effective capacitance of approximately $C_{gp} + 0.1C_{gk}$, so that C_{gk} has been reduced to $\frac{1}{10}$ of its normal value.

The output admittance of the circuit is, from Eq. (8-7), just $-Y_{22}$, so that

$$Y_{\text{output}} = g_m + g_p + Y_G + Y_K + j\omega(C_{pk} + C_k + C_{gk}) \tag{8-71}$$

If $1/r_p = g_p = Y_G = Y_K \ll g_m$ and the capacitances are negligible at low frequencies, then the output admittance is simply

$$Y_{\text{output}LF} \cong g_m$$

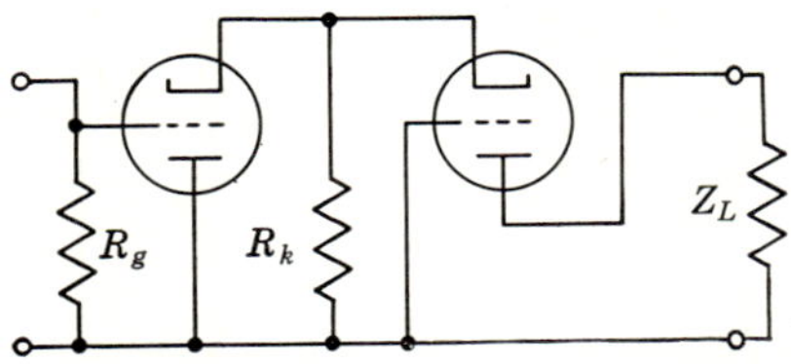

FIG. 8-13 Cathode-coupled triode amplifier as a four-terminal network with load Z_L.

As a second example, suppose that the gain of the cathode-coupled amplifier network shown in Fig. 8-13 is to be computed. Interelectrode capacitances are to be neglected, and the two triodes will be operated in such a way that the μ, r_p, and g_m of each of the triodes will be the same. The transfer matrix for the network will be the product of the transfer matrices for the grounded-plate amplifier and the grounded-grid amplifier as given in Table 8-5, or

$$[A_{P,G}] = \begin{bmatrix} \dfrac{1+\mu}{\mu} & \dfrac{1}{g_m} \\ \dfrac{1+\mu}{\mu Z_G} & \dfrac{1}{g_m Z_G} \end{bmatrix} \cdot \begin{bmatrix} \dfrac{1}{1+\mu} & \dfrac{r_p}{1+\mu} \\ \dfrac{1}{Z_K(1+\mu)} & \dfrac{r_p + Z_K(1+\mu)}{Z_K(1+\mu)} \end{bmatrix} \tag{8-72}$$

$$[A_{P,G}] = \begin{bmatrix} \dfrac{(1+\mu)Z_K + r_p}{\mu(1+\mu)Z_K} & \dfrac{2r_p(1+\mu)Z_K + r_p^2}{\mu(1+\mu)Z_K} \\ \dfrac{(1+\mu)Z_K + r_p}{\mu(1+\mu)Z_K Z_G} & \dfrac{2r_p(1+\mu)Z_K + r_p^2}{\mu(1+\mu)Z_K Z_G} \end{bmatrix} \tag{8-73}$$

The gain α of the network with load Z_L is then, from Eq. (8-41),

$$\alpha = \frac{Z_L}{AZ_L + B} \tag{8-74}$$

where A and B are the parameters from the transfer matrix $[A_{P,G}]$ given in Eq. (8-73). Thus, the gain after simplification is

$$\alpha = \frac{Z_L[\mu(1+\mu)Z_K]}{Z_L[(1+\mu)Z_K + r_p] + 2r_p(1+\mu)Z_K + r_p^2} \tag{8-75}$$

Table 8-5 Matrices of Various Electron-tube Configurations

	Circuit configuration			
	Grounded cathode	Grounded cathode with cathode impedance	Grounded plate	Grounded grid
$[A]$	$\begin{bmatrix} -\frac{1}{\mu} & -\frac{1}{g_m} \\ -\frac{1}{\mu Z_g} & -\frac{1}{g_m Z_g} \end{bmatrix}$	$\begin{bmatrix} -\frac{1}{\mu} & -\frac{1 + Z_K(g_m + g_p)}{g_m} \\ -\frac{1}{\mu Z_g} & -\frac{1 + Z_K(g_m + g_p)}{g_m Z_g} \end{bmatrix}$	$\begin{bmatrix} \frac{1+\mu}{\mu} & \frac{1}{g_m} \\ \frac{1+\mu}{\mu Z_g} & \frac{1}{g_m Z_g} \end{bmatrix}$	$\begin{bmatrix} \frac{1}{1+\mu} & \frac{r_p}{1+\mu} \\ \frac{1}{Z_K(1+\mu)} & \frac{r_p + Z_K(1+\mu)}{Z_K(1+\mu)} \end{bmatrix}$
$[Y]$	$\begin{bmatrix} Y_g & 0 \\ -g_m & -g_p \end{bmatrix}$	$\begin{bmatrix} Y_g & 0 \\ -\frac{g_m}{1 + Z_K(g_m + g_p)} & -\frac{g_p}{1 + Z_K(g_m + g_p)} \end{bmatrix}$	$\begin{bmatrix} Y_g & 0 \\ g_m & -(g_m + g_p) \end{bmatrix}$	$\begin{bmatrix} Y_K + g_m + g_p & -g_p \\ g_m + g_p & -g_p \end{bmatrix}$
$[Z]$	$\begin{bmatrix} Z_g & 0 \\ -\mu Z_g & -r_p \end{bmatrix}$	$\begin{bmatrix} Z_g & 0 \\ -\mu Z_g & -[r_p + Z_K(1+\mu)] \end{bmatrix}$	$\begin{bmatrix} Z_g & 0 \\ \frac{\mu Z_g}{1+\mu} & \frac{-r_p}{1+\mu} \end{bmatrix}$	$\begin{bmatrix} Z_K & -Z_K \\ Z_K(1+\mu) & -[r_p + Z_K(1+\mu)] \end{bmatrix}$
$[H]$	$\begin{bmatrix} Z_g & 0 \\ -g_m Z_g & -g_p \end{bmatrix}$	$\begin{bmatrix} Z_g & 0 \\ \frac{-g_m Z_g}{1 + Z_K(g_m + g_p)} & \frac{-g_p}{1 + Z_K(g_m + g_p)} \end{bmatrix}$	$\begin{bmatrix} Z_g & 0 \\ g_m Z_g & -(g_m + g_p) \end{bmatrix}$	$\begin{bmatrix} \frac{r_p Z_K}{r_p + Z_K(1+\mu)} & \frac{Z_K}{r_p + Z_K(1+\mu)} \\ \frac{Z_K(1+\mu)}{r_p + Z_K(1+\mu)} & \frac{-1}{r_p + Z_K(1+\mu)} \end{bmatrix}$
	See circuit of Fig. 8-10*b*	See circuit of Fig. 8-10*e*	See circuit of Fig. 8-11*a*	See circuit of Fig. 8-11*b*

The same result obtained by the conventional mesh analysis would have involved a considerably more complicated set of equations (see Korman[1]).

When a large number of identical amplifier stages are cascaded, a series of transfer-matrix products equivalent to

$$[A]^m = \begin{bmatrix} A & B \\ C & D \end{bmatrix}^m \tag{8-76}$$

must be evaluated, where m is the number of stages. Brown and Bennett[2] give a result for $[A]^m$ which is

$$[A]^m = \frac{1}{S_2 - S_1} \begin{bmatrix} -S_1 & S_2 \\ -1 & 1 \end{bmatrix} \cdot \begin{bmatrix} \lambda_1{}^m & 0 \\ 0 & \lambda_2{}^m \end{bmatrix} \cdot \begin{bmatrix} 1 & -S_2 \\ 1 & -S_1 \end{bmatrix} \tag{8-77}$$

$$\text{where } S_1 = \frac{(A - D) + \sqrt{(A + D)^2 - 4\eta}}{2C}$$

$$S_2 = \frac{(A - D) - \sqrt{(A + D)^2 - 4\eta}}{2C}$$

$$\eta = AD - BC = \lambda_1\lambda_2$$

$$S_1 + S_2 = \frac{A - D}{C}$$

$$S_1S_2 = \frac{-B}{C}$$

$$S_1 - S_2 = \frac{\sqrt{(A + D)^2 - 4\eta}}{C}$$

$$\lambda_1 = \frac{(A + D) + \sqrt{(A + D)^2 - 4\eta}}{2}$$

$$\lambda_2 = \frac{(A + D) - \sqrt{(A + D)^2 - 4\eta}}{2}$$

$$\lambda_1 + \lambda_2 = A + D$$

The expanded form of Eq. (8-77) is

$$[A]^m = \frac{1}{S_2 - S_1} \begin{bmatrix} (S_2\lambda_2{}^m - S_1\lambda_1{}^m) & S_1S_2(\lambda_1{}^m - \lambda_2{}^m) \\ (\lambda_2{}^m - \lambda_1{}^m) & (S_2\lambda_1{}^m - S_1\lambda_2{}^m) \end{bmatrix} \tag{8-78}$$

[1] N. Korman, Cathode-coupled Triode Amplifiers, *Proc. I.R.E.*, **35** (1): 48, January, 1947.

[2] J. S. Brown and F. D. Bennett, The Application of Matrices to Vacuum-tube Circuits, *Proc. I.R.E.*, **36** (7): 848–849, July, 1948. See also H. L. Armstrong, An Approximate Treatment of Cascaded Four-terminal Networks, *Electronic Eng.*, **27** (325): 130–131, March, 1955.

Quite often it is necessary to determine the stability of a network which consists of one or more electron-tube amplifier stages equipped with a closed feedback loop. If the feedback loop is opened at an arbitrary point, assume that $[A_0]$ represents the over-all transfer matrix around the whole severed loop. Then, the input and output voltages will be related by the equation

$$\begin{bmatrix} E_1 \\ I_1 \end{bmatrix} = [A_0] \cdot \begin{bmatrix} E_2 \\ I_2 \end{bmatrix} \tag{8-79}$$

Now it is clear that the network will be on the threshold of instability after the loop is closed if

$$\begin{bmatrix} E_1 \\ I_1 \end{bmatrix} = \begin{bmatrix} E_2 \\ I_2 \end{bmatrix} \tag{8-80}$$

that is, if the complete signal is transmitted through the entire loop with zero gain or loss, and zero phase shift. Under these conditions, by substituting Eq. (8-80) into (8-79), it will be found that

$$A_0 + D_0 - |A_0| - 1 = 0 \tag{8-81}$$

where A_0 and D_0 are the parameters of the matrix $[A_0]$ and $|A_0|$ is the determinant of the matrix $[A_0]$.

Equation (8-81) is the generalized transfer-matrix stability criterion and represents the sum of the terms on the principal diagonal of the over-all open-loop transfer matrix $[A_0]$, minus the determinant $|A_0|$ of the same matrix, minus 1. The network will oscillate spontaneously upon closing the loop for those complex frequencies which are roots of Eq. (8-81) and have positive real parts.

If the loop is cut at the effectively zero-admittance grid input of one of the grounded-cathode electron-tube amplifier stages it contains, the criterion given by Eq. (8-81) reduces to

$$A_0 - 1 = 0 \tag{8-82}$$

because the amplifier stage is unilateral. This is equivalent to the Nyquist criterion

$$\mu\beta - 1 = 0 \tag{8-83}$$

However, the generalized criterion given by Eq. (8-81) applies to networks with unilateral *or* bilateral transmission properties, and the loop for which the matrix coefficients are computed may be cut at any desired point.[1]

[1] P. M. Honnell, Transmission Matrix Stability Criterion, *Elec. Eng.*, **70** (7): 580, July, 1951. See also C. C. Cheng, Neutralization and Unilateralization, *I.R.E. Trans. Circuit Theory*, **CT-2,** 138–145, June, 1955.

The matrix methods discussed in this section can obviously be applied to any of a wide variety of electron-tube-network problems which are amenable to analysis by means of the equivalent plate-circuit theorem. Interesting applications of matrix methods to such problems, include the analysis of the resistance-coupled amplifier by Pinciroli and Taraboletti,[1] phase-shift oscillators by Hinton,[2] the class B modulator by Guertler,[3] band-pass amplifiers by Jelonek and Sidorowicz,[4] the cascode amplifier by Wallman, Macnee, and Gadsden,[5] and the distributed amplifier by Payne.[6] Also, Shekel[7] has developed a method of writing an "indefinite" admittance matrix for electron-tube networks which does not require the use of a reference node at ground potential. This approach is particularly helpful in analyzing circuits in which no tube electrode is grounded, such as in series-tube electronically regulated power supplies, or which are multiterminal transducers such as distributed amplifier stages. (See also the following section.)

8-7 Matrices of Transistor Networks

The transistor network can be represented as an active four-pole network in a manner similar to that employed for the electron-tube network. Giacoletto has tabulated a complete set of four-pole parameters for such networks.[8] Shekel[9] has presented a system of matrix analysis for transistor networks which will be described here.

The characteristics of a point-contact transistor can be specified in terms of its terminal impedances, i.e., input and transfer impedance or admit-

[1] A. Pinciroli and A. Taraboletti, Matrix Analysis of Linear Networks including Active Quadripoles, *Alta Frequenza,* **18**: 73–82, April, 1949.

[2] W. R. Hinton, The Design of *RC* Oscillator Phase Shifting Networks, *Electronic Eng.,* **22**: 13–17, January, 1950.

[3] R. Guertler, An Introduction to the Matrix Method of Solving Electrical Network Problems, *J. Inst. Engrs. Australia,* **22**: 46–52, March, 1950.

[4] Z. Jelonek and R. Sidorowicz, Bandpass Amplifiers, Investigation of Design and Stability, *Wireless Engr.,* **31** (4): 84–95, April, 1954.

[5] H. Wallman, A. B. Macnee, and C. P. Gadsden, A Low-noise Amplifier, *Proc. I.R.E.,* **36** (6), June, 1948.

[6] D. V. Payne, Distributed Amplifier Theory, *Proc. I.R.E.,* **41** (6): 759–762, June, 1953.

[7] J. Shekel, Voltage-reference Node, Its Transformations in Nodal Analysis, *Wireless Engr.,* **31** (1): 6–10, January, 1954; Matrix Analysis of Multi-terminal Transducers, *Proc. I.R.E.,* **42** (5): 840–847, May, 1954.

[8] L. J. Giacoletto, Terminology and Equations for Linear Active Four-terminal Networks Including Transistors, *RCA Rev.,* **14** (1), March, 1953. See also R. L. Pritchard, Small-signal Parameters for Transistors, *Elec. Eng.,* **73** (10): 902–905, October, 1954.

[9] J. Shekel, Matrix Representation of Transistor Circuits, *Proc. I.R.E.,* **40** (11): 1493–1497, November, 1952.

tance. For example, in a given point-contact transistor,[1] the open-circuit impedances measured at low frequencies (5 kc) under a specified set of operating conditions are

$$Z_{11} = 530 \text{ ohms} \qquad Z_{12} = 290 \text{ ohms}$$
$$Z_{21} = 34{,}000 \text{ ohms} \qquad Z_{22} = 19{,}000 \text{ ohms}$$

where 1 denotes the emitter terminal, 2 is the collector terminal, and the base terminal 3 is grounded as in Fig. 8-14*a*.

The corresponding impedance matrix $[Z_b]$ is

$$[Z_b] = \begin{bmatrix} 530 & 290 \\ 34{,}000 & 19{,}000 \end{bmatrix} \tag{8-84}$$

where the subscript b indicates grounded-base operation. The admittance matrix $[Y_b]$ is $[Z_b]^{-1}$, or Eq. (8-84) inverted:

$$[Y_b] = \begin{bmatrix} 90.48 & -1.38 \\ -161.90 & 2.52 \end{bmatrix} \times 10^{-3} \tag{8-85}$$

A third row and column are now added in such a way as to make the sum of every row and column equal to zero. This gives the *indefinite* admittance matrix of the transistor,

$$[Y_b] = \begin{bmatrix} 90.48 & -1.38 & -89.10 \\ -161.90 & 2.52 & 159.38 \\ 71.42 & -1.14 & -70.28 \end{bmatrix} \times 10^{-3} \tag{8-86}$$

The use of the indefinite nonsymmetric matrix form in Eq. (8-86) permits complete flexibility in changing the grounded terminal. Once this admittance matrix is known, it is immaterial which terminal, if any, is grounded. In transistor circuits, this permits rapid calculation of the operating performance of grounded-base, grounded-emitter, or grounded-collector amplifier stages, or of stages where all three terminals are ungrounded. When any terminal is grounded, the indefinite admittance matrix is modified by crossing out the appropriate row and column. The shorting of two terminals together is represented by adding the corresponding rows and columns to form one row and column. Note that, because the sum of any row or column is zero, the matrix in Eq. (8-86) is a singular matrix. It is described as indefinite because the reference or grounded terminal is undefined.

The manner in which rows and columns are eliminated as various terminals in a network are grounded can be illustrated as follows. Con-

[1] See R. M. Ryder and R. J. Kircher, Some Circuit Aspects of the Transistor, *Bell System Tech. J.*, **28** (3): 373, July, 1949.

sider the general indefinite admittance matrix

$$[Y] = \begin{bmatrix} Y_{11} & Y_{12} & Y_{13} & Y_{14} \\ Y_{21} & Y_{22} & Y_{23} & Y_{24} \\ Y_{31} & Y_{32} & Y_{33} & Y_{34} \\ Y_{41} & Y_{42} & Y_{43} & Y_{44} \end{bmatrix} \tag{8-87}$$

in which

$$\begin{aligned} Y_{14} &= -[Y_{11} + Y_{12} + Y_{13}] \\ Y_{41} &= -[Y_{11} + Y_{21} + Y_{31}] \\ Y_{44} &= [Y_{11} + Y_{12} + Y_{13}] + [Y_{21} + Y_{22} + Y_{23}] \\ &\qquad + [Y_{31} + Y_{32} + Y_{33}] \end{aligned}$$

so that the sum of every row and column is zero.

If terminal 2 is grounded, Eq. (8-87) is altered to

$$[Y_2] = \begin{bmatrix} Y_{11} & Y_{13} & Y_{14} \\ Y_{31} & Y_{33} & Y_{34} \\ Y_{41} & Y_{43} & Y_{44} \end{bmatrix} \tag{8-88}$$

If a conductance G is connected between terminals 1 and 3, the admittance of the combination will be

$$\begin{bmatrix} Y_{11} + G & Y_{13} - G & Y_{14} \\ Y_{31} - G & Y_{33} + G & Y_{34} \\ Y_{41} & Y_{43} & Y_{44} \end{bmatrix}$$

Also, an element connected between any terminal and ground will appear only once on the main diagonal of the matrix. Thus G connected between terminal 1 and ground alters Y_{11} to $(Y_{11} + G)$.

Considering the indefinite admittance matrix of Eq. (8-86), the first row and column are associated with the emitter, the second with the collector, and the third with the base. If the emitter and collector are shorted together and grounded, the first and second rows and columns will be eliminated and the matrix degenerates into a single negative conductance of -70.28 millimhos. The negative sign here is indicative of a short circuit instability.

When the indefinite admittance matrix of a transistor in combination with the circuit elements including load impedance has been obtained the voltage, current, and power gain can be calculated as follows. The forward voltage gain of a network with the load and generator impedance or admittances incorporated in the matrix is simply $1/A$, so that, from Table 8-1,

$$\text{Forward voltage gain } \mu_f = -\frac{Y_{21}}{Y_{22}} = \frac{Z_{21}}{Z_{11}} = -\frac{H_{21}}{|H|} \tag{8-89}$$

The forward current gain with the load and generator impedances or admittances incorporated in the matrix is $1/D$, or

$$\text{Forward current gain } \alpha_f = \frac{Y_{21}}{Y_{11}} = -\frac{Z_{21}}{Z_{22}} = H_{21} \tag{8-90}$$

where the direction of the current I_2 in the four-pole equivalent network is opposite to that assumed here, the forward current gain is negative and, for this reason, is sometimes described in manufacturer's data sheets as the negative forward current amplification ratio. However, the value of this ratio is usually specified as a positive quantity when so identified.

The forward power gain PG_f is defined as the ratio of the power delivered to the load Z_L, to the maximum power available from the generator. The power in the load is $V_2^2G_L$, where G_L is the conductance of the load and the power available from the generator is $(I_1/2)^2/G_G$, where G_G is the conductance of the generator. Since $V_2/I_1 = Z_{21}$, the forward power gain with the load and generator impedances or admittances incorporated in the matrix is

$$\begin{aligned} \text{PG}_f &= \frac{V_2^2G_L}{I_1^2/4G_G} = 4\left(\frac{V_2}{I_1}\right)^2 G_GG_L \\ &= 4G_GG_LZ_{21}^2 = 4G_GG_L\frac{Y_{21}^2}{|Y|^2} = 4G_GG_L\frac{H_{21}^2}{H_{22}^2} \end{aligned} \tag{8-91}$$

Consider the grounded-base amplifier circuit shown in Fig. 8-14*a*. Assume that $R_G = 500$ ohms and $R_L = 20{,}000$ ohms so that $G_G = 2$ millimhos and $G_L = 0.05$ millimho. Under these conditions R_G and R_L will approximately match the values of Z_{11} and Z_{22}.

Since this is the grounded-base case, the third row and column are omitted from Eq. (8-86). G_G and G_L can now be added directly to Y_{11} and Y_{22} of the modified matrix so that

$$[Y_b] = \begin{bmatrix} 92.48 & -1.38 \\ -161.90 & 2.57 \end{bmatrix} \times 10^{-3} \tag{8-92}$$

The voltage gain in the forward direction, from Eq. (8-89), is

$$\mu_f = \frac{-Y_{21}}{Y_{22}} = \frac{161.9}{2.57} = 63 \tag{8-93}$$

The gain is positive, so there is no phase reversal.

The forward-current amplification factor α_f, from Eq. (8-90), is

$$\alpha_f = \frac{Y_{21}}{Y_{11}} = \frac{-161.90}{92.48} = -1.75 \tag{8-94}$$

and from Eq. (8-91),

$$\text{PG (forward)} = \frac{4Y_{21}^2}{R_LR_G|Y|^2} \tag{8-95}$$

and in this case

PG (forward)

$$= \frac{(4)(161.9 \times 10^{-3})^2}{(20{,}000)(500)[(92.48)(2.57)(10^{-6}) - (161.90)(1.38)(10^{-6})]^2}$$

$$= 51 = 17.07 \text{ db} \qquad (8\text{-}96)$$

If the same transistor is to be analyzed in grounded-collector operation as shown in Fig. 8-14*b*, the indefinite matrix in Eq. (8-86) must be

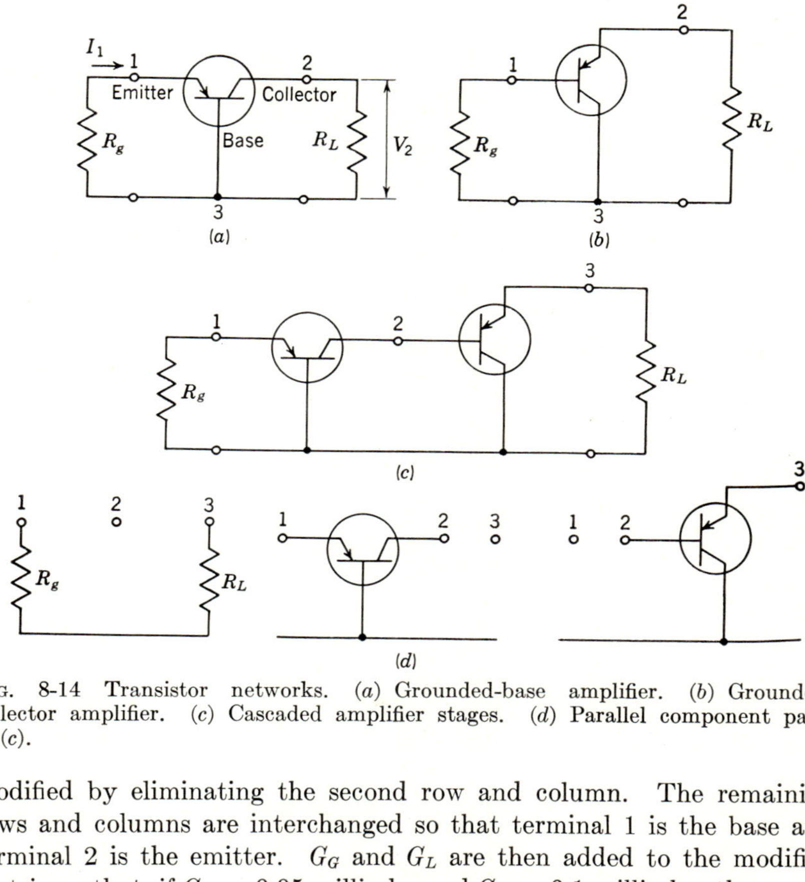

FIG. 8-14 Transistor networks. (*a*) Grounded-base amplifier. (*b*) Grounded-collector amplifier. (*c*) Cascaded amplifier stages. (*d*) Parallel component parts of (*c*).

modified by eliminating the second row and column. The remaining rows and columns are interchanged so that terminal 1 is the base and terminal 2 is the emitter. G_G and G_L are then added to the modified matrix so that, if $G_G = 0.05$ millimho and $G_L = 0.1$ millimho, then

$$[Y_c] = \begin{bmatrix} -70.33 & 71.42 \\ -89.10 & 90.58 \end{bmatrix} \times 10^{-3} \qquad (8\text{-}97)$$

By the same methods used previously,

$$\mu_f = \frac{89.10}{90.58} = 0.984 \qquad (8\text{-}98)$$

The admittance matrix of two transistors cascaded as shown in Fig. 8-14*c* consists of the sum of the admittance matrices of the parallel components shown in Fig. 8-14*d*.

The matrix for the component consisting of G_G and G_L, taking the terminals as shown, is

$$[Y_1] = \begin{bmatrix} G_G & 0 & 0 \\ 0 & 0 & 0 \\ 0 & 0 & G_L \end{bmatrix} \tag{8-99}$$

The matrix for the first transistor is that for the usual grounded-base condition, with a third row and column of zeroes corresponding to the isolated third terminal, or

$$[Y_2] = \begin{bmatrix} 90.48 & -1.38 & 0 \\ -161.90 & 2.52 & 0 \\ 0 & 0 & 0 \end{bmatrix} \times 10^{-3} \tag{8-100}$$

The matrix for the second transistor is the indefinite admittance matrix with the second row and column omitted and the rows and columns interchanged to conform with the terminal numbering shown:

$$[Y_3] = \begin{bmatrix} 0 & 0 & 0 \\ 0 & -70.28 & 71.42 \\ 0 & -89.10 & 90.48 \end{bmatrix} \times 10^{-3} \tag{8-101}$$

The sum of $[Y_1]$, $[Y_2]$, and $[Y_3]$, with $G_G = 2$ millimhos and $G_L = 0.1$ millimho, is

$$[Y_{cas}] = \begin{bmatrix} 92.48 & -1.38 & 0 \\ -161.90 & -67.76 & 71.42 \\ 0 & -89.10 & 90.58 \end{bmatrix} \times 10^{-3} \tag{8-102}$$

The forward voltage gain, since Z_{ik} is the cofactor of Y_{ki} divided by $|Y|$, is

$$\mu_f = \frac{Z_{31}}{Z_{11}} = \frac{\begin{vmatrix} -161.90 & -67.76 \\ 0 & -89.10 \end{vmatrix}}{\begin{vmatrix} -67.76 & 71.24 \\ -89.10 & 90.58 \end{vmatrix}} = 64 \tag{8-103}$$

In the case of the *n-p-n*, *p-n-p*, or other types of junction transistor, the operating characteristics are generally furnished by the manufacturer in terms of the [*H*] parameters because these are the most convenient to measure with accuracy. For all transistors, however, any one set of parameters may be transformed to another using the relationships given in Table 8-1. Where only the emitter, collector, and base resistances and the forward current gain (amplification factor) are given for the grounded base condition, the $[Z]$ parameters are

$$\begin{aligned} Z_{11} &= (r_e + r_b) \\ Z_{12} &= r_b \\ Z_{21} &= (r_b + \alpha r_c) \\ Z_{22} &= (r_b + r_c) \end{aligned}$$

where r_e is the emitter resistance, r_b is the base resistance, r_c is the collector resistance, and α is the forward current gain. Care must be taken to correct the parameter values given in the manufacturer's specification sheet in the event an operating point other than that stipulated by manufacturer for these values is chosen.

As an example of computations for the junction transistor, consider the type 2N77 *p-n-p* junction transistor. For a d-c collector voltage of -4 volts and a d-c collector current of -0.7 ma, the $[H]$ parameters for the grounded or common emitter connection are $H_{11} = 2{,}720$ ohms, $H_{12} = 323 \times 10^{-6}$, $H_{21} = 55$, $H_{22} = 14 \times 10^{-6}$ mho, and

$$|H| = 20{,}315 \times 10^{-6}$$

The corresponding $[Y]$ values are

$$\begin{aligned} Y_{11} &= \frac{1}{H_{11}} = \frac{1}{2{,}720} = 0.368 \times 10^{-3} \qquad (8\text{-}104) \\ Y_{12} &= \frac{-H_{12}}{H_{11}} = \frac{-323 \times 10^{-6}}{2{,}720} = -0.119 \times 10^{-6} \\ Y_{21} &= \frac{H_{21}}{H_{11}} = \frac{55}{2{,}720} = 20.2 \times 10^{-3} \\ Y_{22} &= \frac{|H|}{H_{11}} = \frac{20{,}315 \times 10^{-6}}{2{,}720} = 7.46 \times 10^{-6} \end{aligned}$$

The $[Y_e]$ matrix without generator or load admittances is

$$[Y_e] = \begin{bmatrix} 0.368 & -0.000119 \\ 20.2 & 0.00746 \end{bmatrix} \times 10^{-3} \qquad (8\text{-}105)$$

With a load resistance of 100,000 ohms and an input resistance of 1,980 ohms, $G_L = 0.01$ millimho and $G_G = 0.505$ millimho. Incorporating these in Eq. (8-105),

$$[Y_e] = \begin{bmatrix} 0.873 & -0.000119 \\ 20.2 & 0.01746 \end{bmatrix} \times 10^{-3} \qquad (8\text{-}106)$$

Using the values of Y_{21} and $|Y_e|$ derived from Eq. (8-106), the forward power gain for the grounded emitter connection is approximately 26,300, or 44.2 db. The gains for the grounded collector and base connections in single stages or in cascade can be computed by means of the indefinite matrix method described above.

It should be noted that transistors are nonunilateral, or, in other words, that they exhibit bidirectional transmission between input and output terminals. In general, the backward transmission is considerably lower than the forward transmission. However, its effects on the circuit properties of the transistor cannot be neglected. The backward transmission has the effect of making the input impedance of several cascaded transistor stages a function of the characteristics of each of the stages interacting with the others. Also, the internal feedback may cause oscillations even in the absence of an external feedback loop.

These undesirable effects can occur only when the parameters Y_{12} or $H_{12} \neq 0$. Methods for neutralizing this condition are described by Stern, Aldridge, and Chow.[1]

8-8 Other Applications of Matrix Methods

Matrix methods have come into wide use in fields other than elementary network analysis. More advanced applications include the matrix analysis of transmission lines, wave-guide systems, magnetic amplifiers, piezoelectric crystals, and networks more complex than the four-pole. References of interest include the following:

Applebaum, S.: Gamma Correction in Constant Luminance Color Television Systems, *Proc. I.R.E.*, **40** (10): 1185–1195, October, 1952.

Belevitch, V.: Synthesis of 2*n*-terminal Networks, *Wireless Engr.*, **28** (4): 128–129, April, 1951.

Bond, W. S.: The Mathematics of the Physical Properties of Crystals, *Bell System Tech. J.*, **22** (1): 1–72, January, 1943.

Brogle, A. P.: The Design of Reactive Equalizers, *Bell System Tech. J.*, vol. 28, no. 4, October, 1949.

Clogston, A. M.: Reduction of Skin Effect Losses by the Use of Laminated Conductors, *Proc. I.R.E.*, **39** (7): 767–782, July, 1951.

Collins, G. B.: "Microwave Magnetrons," Radiation Laboratory Series, vol. 6, McGraw-Hill Book Co., Inc., New York, 1948.

Estrin, G.: The Effects of Anisotropy in a Three-dimensional Array of Conducting Disks, *Proc. I.R.E.*, **39** (7): 821–826, July, 1951.

Fruhauf, H.: On Matrix Algebra and the Application of Normalized Magnitudes in Electrical Engineering, *Nachr. Tech.*, **3**: 263–268, June, 1953.

Haus, H. A., and F. N. H. Robinson: The Minimum Noise Figure of Microwave Beam Amplifiers, *Proc. I.R.E.*, **43** (8): 981–991, August, 1955.

Ho, E., and D. L. Trautman: Synthesis of Resistively-terminated *RLC* Ladder Networks, *Convention Rec. I.R.E.*, Pt. 2, 14–15, 1954.

[1] A. P. Stern, C. A. Aldridge, and W. F. Chow, Internal Feedback and Neutralization of Transistor Amplifiers, *Proc. I.R.E.*, **43** (7): 838–847, July, 1955. See also C. C. Cheng, loc. cit.

Kaufman, A.: Conditions of Validity of Matrix Analysis for Quadripole Assemblies: Applications to Feedback Networks, *Onde élect.*, **31**: 396–405 and 446–452, October and November, 1951.

Kirschner, V.: Generalized Representation of a Valve Circuit by an Iterative Matrix, *Arch. elekt. Ubertragung,* **5**(4): 190–196, April, 1951.

Knausenberger, G. E.: A Note on the Scattering Matrix of an Active Linear Two-terminal Pair Network; J. K. Percus: Matrix Analysis of Oriented Graphs with Irreducible Feedback Loops; H. A. Schulke: Matrix Factorization; *I.R.E. Trans. Circuit Theory*, **CT-2,** 112–116, 117–127, 128–132, June, 1955.

Makimoto, T.: Characteristic Matrix and Impedance Matrix in Microwave Passive Network, *Mem. Inst. Sci. Ind. Research Osaka Univ.*, **7**: 42–56, 1950.

Montgomery, C. G.: "Technique of Microwave Measurements," Radiation Laboratory Series, vol. 11, McGraw-Hill Book Co., Inc., New York, 1947.

Montgomery, C. G., R. H. Dicke, and E. M. Purcell: "Principles of Microwave Circuits," Radiation Laboratory Series, vol. 8, McGraw-Hill Book Co., Inc., 1948.

Percival, W. S.: Improved Matrix and Determinant Methods for Solving Networks, *Proc. Inst. Elec. Engrs.* (*London*), Monograph No. 96, April, 1954.

Riblet, H. J.: A Mathematical Theory of Directional Couplers, *Proc. I.R.E.*, **35** (11): 1307–1313, November, 1947.

Riblet, H. J.: The Short-slot Hybrid Junction, *Proc. I.R.E.*, **40** (2): 180–184, February, 1952.

Rice, S. O.: Reflections from Circular Bends in Rectangular Wave Guides—Matrix Theory, *Bell System Tech. J.*, vol. 27, no. 2, April, 1948.

Richards, P. I.: Universal Optimum-response Curves for Arbitrarily Coupled Resonators, *Proc. I.R.E.*, **34** (9): 624–629, September, 1946.

Rowlands, R. O.: Impedance Transformations in Four-element Band-pass Filters, *Proc. I.R.E.*, vol. 37, no. 11, November, 1949.

Sandeman, E. K.: Transmission Line Filters, *Wireless Engr.*, **26**: 11–25, January, 1949.

Storch, L.: The Multisection *RC* Filter Network Problem, *Proc. I.R.E.*, **39** (11): 1456–1458, November, 1951.

Tinkham, M., and M. W. P. Strandberg: The Excitation of Circular Polarization in Microwave Cavities, *Proc. I.R.E.*, **43** (6): 734–738, June, 1955.

Tournier, M.: Physical Properties of Crystals and Their Symmetry, *Elec. Commun.*, vol. 24, no. 4, December, 1947.

VerPlanck, D. W., and M. Fishman: An Analysis of Interlinked Electric and Magnetic Networks with Applications to Magnetic Amplifiers, *Proc. I.R.E.*, **37** (9): 1021–1027, September, 1949.

Watson, W. H.: Matrix Methods in Transmission Line and Impedance Calculations, *J. Inst. Elec. Engrs.* (*London*), pt. III-A, **93** (4): 737–746, 1946.

Problems

Prob. 8-1 Show that the transfer matrix $[A]$ for the coupled network shown in Fig. 8-15 is

$$\begin{bmatrix} A & B \\ C & D \end{bmatrix} = \begin{bmatrix} \dfrac{L_1}{M} & \dfrac{j\omega(L_1L_2 - M^2)}{M} \\ \dfrac{-j}{M\omega} & \dfrac{L_2}{M} \end{bmatrix}$$

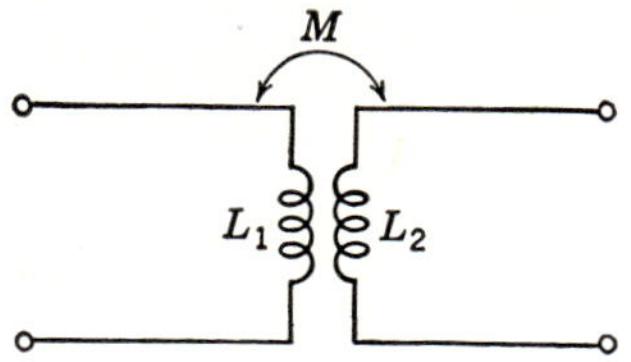

FIG. 8-15 Coupled network.

Prob. 8-2 Shown in Fig. 8-16 is the equivalent network of a class-B modulator stage in which all secondary quantities have been referred to the primary side of the modulation transformer. Find the ratio e_2/e_1

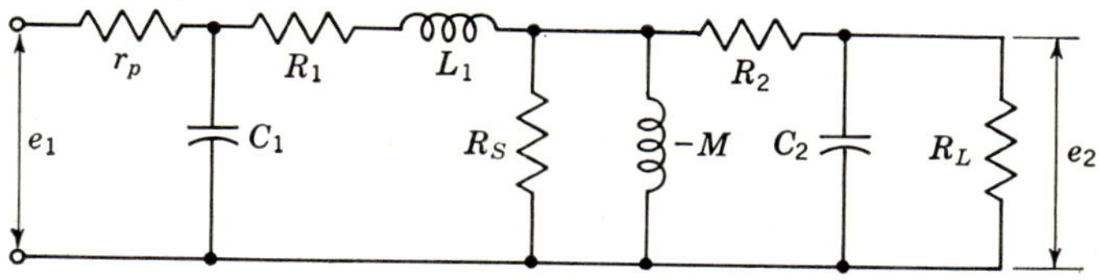

FIG. 8-16 Class-B modulator equivalent.

by means of a cascade connection of a T and π sections made up of the elements shown.

Prob. 8-3 Prepare a table which gives the arms of the T and π four-pole networks in terms of the matrix parameters for the matrices $[A]$, $[Z]$, $[Y]$, and $[H]$.

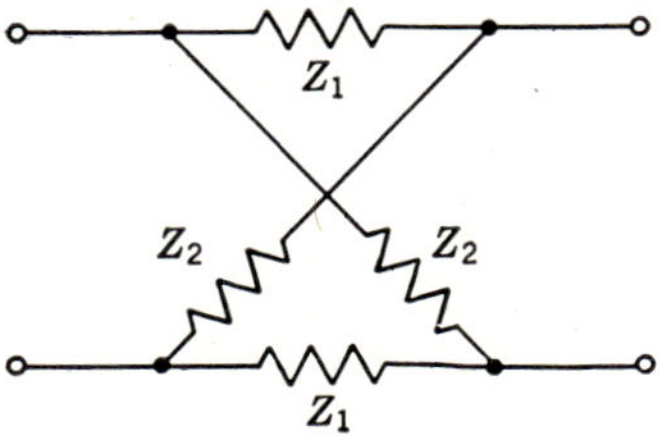

FIG. 8-17

Prob. 8-4 (*a*) Show by matrix methods that for the lattice shown in Fig. 8-17, with a load $Z_L = \sqrt{Z_1Z_2}$ (see pages 127 and 128)

$$\frac{E_1}{E_2} = \frac{\sqrt{Z_2/Z_1} + 1}{\sqrt{Z_2/Z_1} - 1}$$

(*b*) Show that the impedance matrix of the lattice is

$$[Z] = \begin{bmatrix} \dfrac{Z_1 + Z_2}{2} & \dfrac{Z_2 - Z_1}{2} \\ \dfrac{Z_2 - Z_1}{2} & \dfrac{Z_1 + Z_2}{2} \end{bmatrix}$$

Prob. 8-5* If the networks shown in Fig. 8-18*a* and *b* are connected in series, the network shown in Fig. 8-18*c* results. Show by means of the

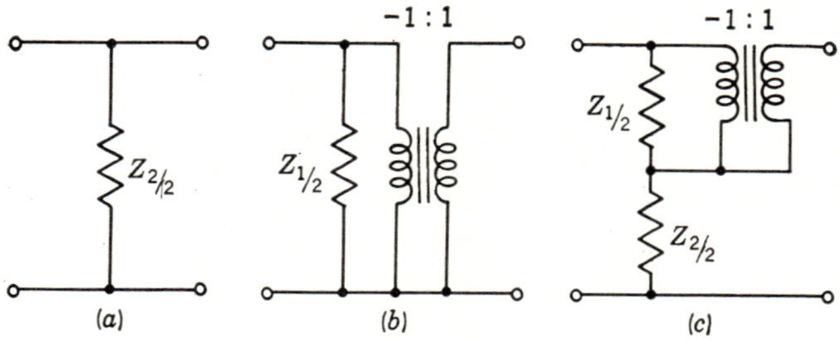

FIG. 8-18

impedance matrix of the lattice that network *c* is a three-terminal equivalent of the lattice.

Prob. 8-6 (*a*) Find the combined impedance matrix [*z*] of two four-pole networks connected in tandem.

(*b*) Two symmetrical lattices connected in tandem may be replaced by a single equivalent symmetrical lattice. Using the results of (*a*), determine the necessary criterion.

Prob. 8-7 Shown in Fig. 8-19 is a π network used as a harmonic filter in a radio transmitter.

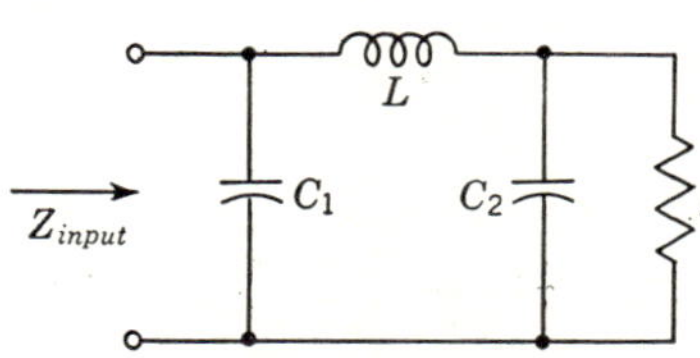

FIG. 8-19 Harmonic filter.

(*a*) Show that the input impedance *Z* is equal to *R* when

$$\omega L = \frac{1}{\omega C_1} = \frac{1}{\omega C_2} = R$$

(*b*) Show that under these conditions the phase shift of the filter is 90°, so that the network acts as a quarter-wave network.

(*c*) Determine the Z_{input} at frequencies ± 10 per cent from the operating frequencies of the filter at which the conditions in (*a*) prevail.

(*d*) Show that this filter is the equivalent of a mid-shunt section of a

* C. Willoner and F. Thelka, A Phase-shift Oscillator with Wide-range Tuning, *Proc. I.R.E.*, **36** (9): 1096–1100, September, 1948.

prototype low-pass filter operated at $1/\sqrt{2}$ times the cutoff frequency of the filter, with a termination of $\sqrt{2}$ times its characteristic impedance.

Prob. 8-8* The voltage at the input terminals of a network which has the general parameters A, B, C, and D is held constant.

(*a*) Show that, if the network load current is independent of the load impedance, $A = 0$.

(*b*) Show that for the T, π, or L network, this requirement may be met if the reactance of the series impedance is equal to the negative of the

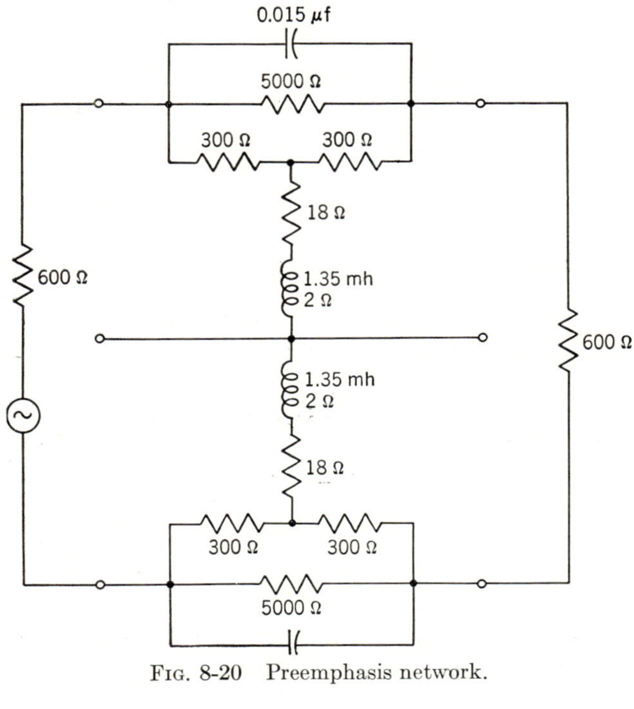

FIG. 8-20 Preemphasis network.

reactance of the shunt impedance. (This is the condition for Boucherot's constant-current circuit.)

(*c*) Under the same conditions, show that the voltage across the load will be independent of load impedance if $B = 0$.

(*d*) Show that in a symmetrical-T network, if the series impedances are each $-j2x$, the shunt arm must be jx to satisfy the condition in (*c*).

Prob. 8-9 In Fig. 8-20 is shown the circuit diagram of a 75-μsec preemphasis network used in the aural section of a television transmitter. By matrix analysis, confirm that this network has the response curve

* R. Guertler, An Introduction to the Matrix Method of Solving Electrical Network Problems, *J. Inst. Engrs. Australia,* **22:** 46–52, March, 1950.

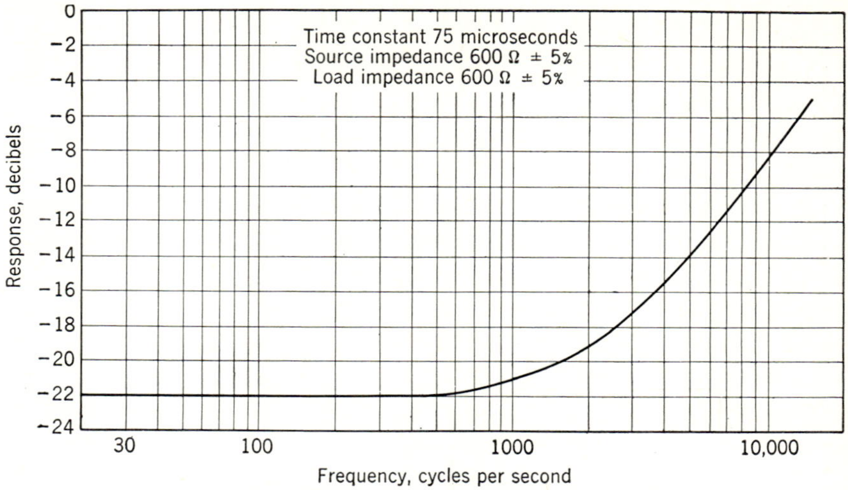

FIG. 8-21 Preemphasis-network characteristic.

shown in Fig. 8-21 when fed from a 600-ohm source and terminated in 600 ohms as indicated.

Prob. 8-10* In Fig. 8-22 is shown the circuit diagram of a loudness-weighting equalizer used in determining the loudness-efficiency ratings

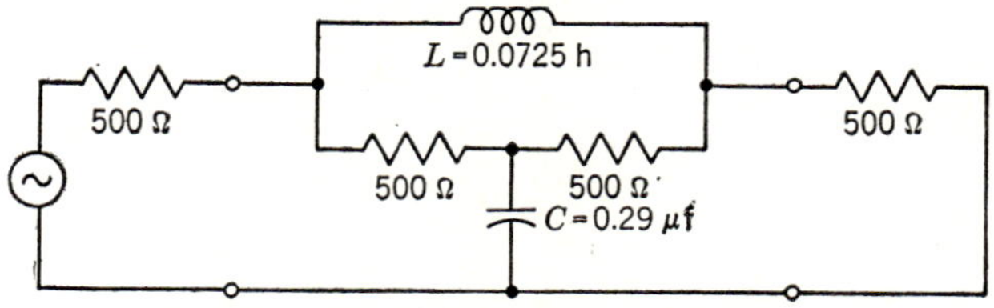

FIG. 8-22

for loudspeakers. Calculate by means of matrices and plot the insertion loss of this network in decibels as a function of frequency between 300 and 4,000 cps.

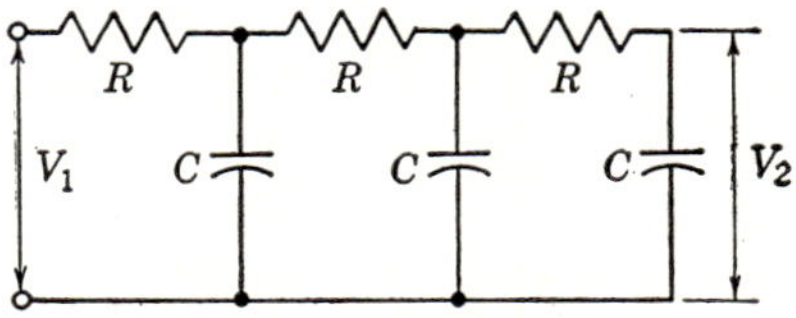

FIG. 8-23 Vertical integrator network.

Prob. 8-11 (*a*) In Fig. 8-23 is shown a three-stage integration network used in the vertical sync circuit of a television receiver. Using

* H. Hopkins and N. Stryker, A Proposed Loudness-efficiency Rating for Loudspeakers and the Determination of System Power Requirements for Enclosures, *Proc. I.R.E.*, **36** (3), March, 1948.

matrices, determine the voltage ratio V_2/V_1, at any frequency f, in terms of f and f_0, where $f_0 = 1/(2\pi RC)$.

(b) Show that the absolute value of the input-output voltage ratio is

$$\left|\frac{V_2}{V_1}\right| = \frac{1}{\sqrt{1 + 26f^2/f_0^2 + 13f^4/f_0^4 + f^6/f_0^6}}$$

(c) If vertical sync pulses are to pass through the network with little alteration, the network should have relatively little attenuation at the twentieth harmonic of the vertical sync frequency (60 cps). If, at the same time, the ratio $|V_2/V_1|$ for the horizontal sync frequency (15,750 cps) is to be 0.01, what will the attenuation of the network be for the harmonic above?

Prob. 8-12 Show that in a grounded-grid electron-tube amplifier (a) the input impedance, neglecting interelectrode capacitances, is $1/g_m$ if $R_L \ll r_p$ and $\mu \gg 1$; (b) the stage gain is $g_m R_L$, where R_L is the load, under the conditions given in (a).

Prob. 8-13 Find the admittance matrix of a grounded-cathode amplifier circuit which has the feedback network shown in Fig. 8-24 con-

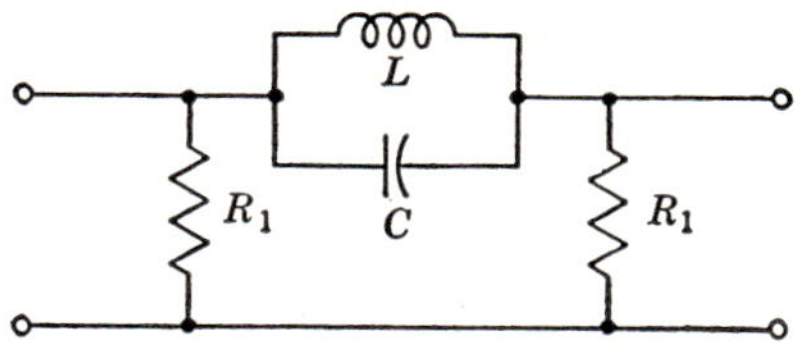

FIG. 8-24

nected between the output and input terminals. Neglecting the interelectrode capacitances, determine the stability of the network in terms of μ and r_p.

Prob. 8-14 The impedance matrix of a junction transistor is given as

$$[Z_b] = \begin{bmatrix} 199 & 182 \\ 0.502 \times 10^6 & 0.5145 \times 10^6 \end{bmatrix}$$

where the values are in ohms. Calculate α_f, μ_f, and the power gain. In matched operation use $R_6 \cong Z_{11}$ and $R_L \cong Z_{22}$ (see page 199) for the grounded-base, grounded-emitter, and grounded-collector configurations. Confirm the following characteristics of the junction transistor:

(a) Grounded-emitter operation: large current amplification, input impedance low, and phase reversal.

(b) Grounded-collector operation: large current amplification, input impedance high, no phase reversal.

(*c*) Grounded-base operation: less than unity current amplification, very low input impedance, no phase reversal.

Prob. 8-15 In Fig. 8-25 are shown the equivalent networks and α_f values for a point-contact and a junction transistor. Determine the $[Z_b]$ and $[Y_b]$ matrices for each of these units.

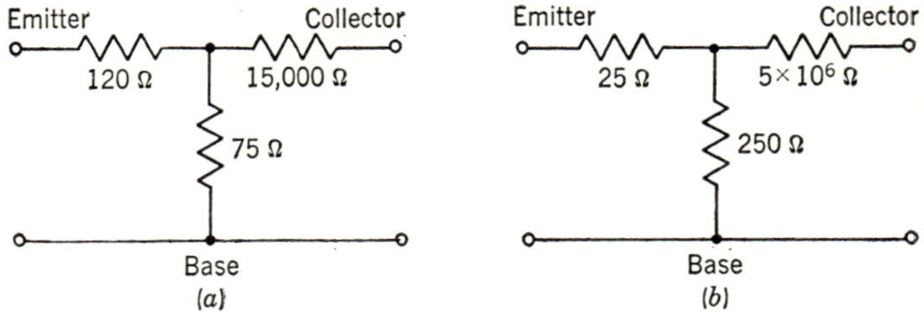

FIG. 8-25 Transistor equivalents. (*a*) Grounded-base point-contact transistor $\alpha_f = 2.5$. (*b*) Grounded-base junction transistor $\alpha_f = 0.95$.

Prob. 8-16* In a color-television system, if

$$\begin{bmatrix} R \\ G \\ B \end{bmatrix} = \begin{bmatrix} 1.910 & -0.532 & -0.288 \\ -0.985 & 1.999 & -0.028 \\ 0.058 & -0.118 & 0.898 \end{bmatrix} \cdot \begin{bmatrix} X \\ Y \\ Z \end{bmatrix}$$

show that

$$\begin{bmatrix} X \\ Y \\ Z \end{bmatrix} = \begin{bmatrix} 0.607 & -0.174 & 0.200 \\ -0.299 & 0.587 & -0.114 \\ 0.000 & -0.066 & 1.116 \end{bmatrix} \cdot \begin{bmatrix} R \\ G \\ B \end{bmatrix}$$

Prob. 8-17 Using the data given in Sec. 8-7 for the type 2N77 junction transistor, compute the forward power gain in decibels (*a*) in the grounded-base connection with $R_L = 500{,}000$ ohms and $R_g = 215$ ohms; (*b*) in the grounded-collector connection with $R_L = 10{,}000$ ohms and $R_g = 0.5$ megohm.

Prob. 8-18 The type 2N105 *p-n-p* junction transistor has the following grounded or common emitter characteristics:

	A	*B*
Collector voltage =	−4	−1.3 volts
Collector current =	−0.7	−0.3 ma
H_{11} =	2,880	4,800 ohms
H_{12} =	5.5×10^{-4}	9.1×10^{-4}
H_{21} =	55	45
H_{22} =	16.3×10^{-6}	12.4×10^{-6} mhos

* S. Applebaum, Gamma Correction in Constant Luminance Color Television Systems, *Proc. I.R.E.*, **40** (10): 1195, October, 1952.

Compute the forward power gains in decibels for the following values of R_L and R_g and conditions A and B above:

	R_L (ohms)	R_g (ohms)
Grounded emitter: A =	20,000	2,300
B =	4,700	4,700
Grounded base: A =	500,000	180
Grounded collector: B =	13,000	500,000

Installation of the "doughnut" of the 180-million-volt synchrotron. (*National Bureau of Standards.*)

PART 4

Theory and Applications of Series

Chapter 9

BASIC PROPERTIES AND APPLICATIONS

9-1 Definitions

Since *sequences* and *series* are in one-to-one correspondence, the properties of both will be presented in this section.

A sequence is defined as a succession of terms which is created in accord with some rule, while a series is merely the indicated sum of a sequence.

$$1, 8, 27, 64 \tag{9-1}$$

represents a sequence, while

$$1 + 8 + 27 + 64 \tag{9-2}$$

is an example of a series.

The number of terms which make up a sequence or series is an important consideration. If the number of terms is limited to some finite number n, then the series or sequence is a finite series or finite sequence. Expressions (9.1) and (9.2) exemplify finite sequences and finite series, respectively.

$$1, 8, 27, 64, 125, \ldots \tag{9-3}$$

and

$$1 + 8 + 27 + 64 + 125 + \cdots \tag{9-4}$$

may be used to show infinite sequences and infinite series, respectively.

The formula for the *general term* of a sequence or series is necessary for the operations required in calculations with series. The general term u_n is illustrated by the following:

EXAMPLE 1. Find u_n for the series

$$1 + \frac{x}{1!} + \frac{x^2}{2!} + \frac{x^3}{3!} + \frac{x^4}{4!} + \cdots \tag{9-5}$$

The general term u_n is given by

$$u_n = \frac{x^{n-1}}{(n-1)!} \tag{9-6}$$

where n is the number of the term. The notation $(n-1)!$, called factorial $(n-1)$, is defined by

$$(n-1)! = 1 \cdot 2 \cdot 3 \cdot 4 \cdot \ . \ . \ . \ \cdot (n-2)(n-1) \tag{9-7}$$

0! is defined as 1, or

$$0! = 1 \tag{9-8}$$

EXAMPLE 2. Find the general term of

$$1 - \tfrac{1}{2} + \tfrac{1}{3} - \tfrac{1}{4} + \tfrac{1}{5} - \cdots \tag{9-9}$$

The general term u_n is given by

$$u_n = (-1)^{n+1}\frac{1}{n} \tag{9-10}$$

EXAMPLE 3. Find the general term of

$$\frac{x^2}{2} + \frac{x^4}{2 \cdot 4} + \frac{x^6}{2 \cdot 4 \cdot 6} + \cdots \tag{9-11}$$

The general term may be represented by

$$u_n = \frac{x^{2n}}{2^n(n!)} \tag{9-12}$$

EXAMPLE 4

$$\frac{2x}{1} - \frac{4x^2}{1 \cdot 3} + \frac{8x^3}{1 \cdot 3 \cdot 5} - \frac{16x^4}{1 \cdot 3 \cdot 5 \cdot 7} + \cdots \tag{9-13}$$

The general term is given by

$$u_n = \frac{(-1)^{n+1}(2x)^n}{\dfrac{(2n-1)!}{2^{n-1}[(n-1)!]}} \tag{9-14}$$

or

$$u_n = \frac{(-1)^{n+1}2^{n-1}[(n-1)!](2x)^n}{(2n-1)!} \tag{9-15}$$

$$u_n = \frac{(-1)^{n+1}[(n-1)!]2^{2n-1}x^n}{(2n-1)!} \tag{9-16}$$

When the number of terms of a series is considered as a variable, then the sum of the series is a function of the number of terms, e.g.,

$$S_n = u_1 + u_2 + u_3 + \cdots + u_n \tag{9-17}$$

Suppose n is finite; then it follows immediately that S_n is a finite quantity. If, however, the number of terms n increases without limit, or $n \to \infty$, then, if S_n as $n \to \infty$ is a unique finite quantity, i.e., the limit $S_n = S$, which is finite, then

$$S = \sum_{n=1}^{\infty} u_n \tag{9-18}$$

is called a *convergent* series. If S does not approach a unique finite quantity as $n \to \infty$, then the series

$$\sum_{n-1}^{\infty} u_n \tag{9-19}$$

is divergent.

Consider the geometric series

$$S_n = a + ar + ar^2 + ar^3 + \cdots + ar^{n-2} + ar^{n-1} \tag{9-20}$$

If (9-20) is multiplied by r, then

$$rS_n = ar + ar^2 + ar^3 + \cdots + ar^n \tag{9-21}$$

Subtracting (9-21) from (9-20) gives

$$(r - 1)S_n = ar^n - a \tag{9-22}$$

or

$$S_n = \frac{a - ar^n}{1 - r} \tag{9-23}$$

This formula for S_n is now studied with various values of r as n increases without limit. Equation (9-23) may be written in the form

$$S_n = \frac{a}{1 - r} - \frac{ar^n}{1 - r} \tag{9-24}$$

If $|r| < 1$, then both terms have finite denominators, but as $n \to \infty$ then, in the limit, the second term in (9.24) will approach zero, and

$$\lim_{n \to \infty} S_n = \frac{a}{1 - r} \tag{9-25}$$

Hence S_n converges to S as $n \to \infty$, where

$$S = \frac{a}{1 - r} \tag{9-26}$$

If $|r| > 1$, then by a consideration similar to the above it is found that S_n will become infinite, or the geometric series is *divergent* for $|r| > 1$.

When $r = 1$, then S_n becomes

$$S_n = a + a + a + \cdots + a \tag{9-27}$$

It is obvious that the sum of this series increases without limit as $n \to \infty$; hence for $r = 1$ the series is divergent. For the case where $r = -1$, the series takes on the form

$$S_n = a - a + a - \cdots + (-a)^{n+1} \tag{9-28}$$

For n an even integer, $S_n = 0$, and for n an odd integer, $S_n = 1$. Since this series has no unique sum, but behaves in the manner described, it is called an *oscillatory series* and is not convergent as $n \to \infty$.

EXAMPLE 1. Consider the geometric series

$$S = 1 + \frac{1}{3} + \frac{1}{9} + \cdots + \frac{1}{3^{n-1}} + \cdots \tag{9-29}$$

In this geometric series, $r = 1/3$ and $a = 1$. Then

$$S = \frac{a}{1 - r} \tag{9-30}$$

will give

$$S = \frac{1}{1 - 1/3} \tag{9-31}$$

$$S = 3/2 \tag{9-32}$$

EXAMPLE 2. Express the repeating decimal fraction $0.13\ddot{1}\ddot{3}$* as a common fraction.

Since

$$0.13\ddot{1}\ddot{3} = \frac{13}{10^2} + \frac{13}{10^4} + \frac{13}{10^6} + \cdots \tag{9-33}$$

it follows that (9-33) is an infinite geometric series whose first term a is $13/10^2$, and $r = 1/10^2$. Hence

$$S = \frac{a}{1 - r} \tag{9-34}$$

$$S = \frac{13/10^2}{1 - 1/10^2} \tag{9-35}$$

or

$$S = 13/99 \tag{9-36}$$

9-2 Tests for the Convergence of Infinite Series

In Fig. 9-1a, if S_n is a variable that represents the sum of n terms of the series, and hence is a function of n, and if S_n is never greater than a fixed

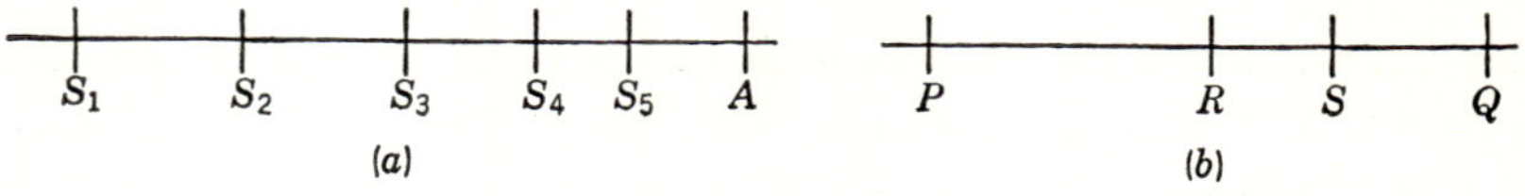

FIG. 9-1

constant A, then as $n \to \infty$, the sum S_n will approach a limit S,

$$\lim_{n \to \infty} S_n = S \tag{9-37}$$

which is of such a magnitude that $S \leq A$.

EXAMPLE. Consider the line segment PQ (Fig. 9-1b) of length 1. Divide this segment into two equal parts PR and RQ; then divide RQ

* The double dot on 13 indicates that the 13 is repeated over and over again ad infinitum to form the repeating decimal fraction.

into two equal parts RS and SQ; and repeat the process n times. It follows that

$$\begin{aligned} S_1 &= \tfrac{1}{2} \\ S_2 &= \tfrac{1}{2} + \tfrac{1}{4} \\ S_3 &= \tfrac{1}{2} + \tfrac{1}{4} + \tfrac{1}{8} \\ &\cdot \\ &\cdot \\ &\cdot \\ S_n &= \tfrac{1}{2} + \tfrac{1}{4} + \tfrac{1}{8} + \cdots + \frac{1}{2^n} \end{aligned} \tag{9-38}$$

Now

$$\lim_{n \to \infty} S_n = S \tag{9-39}$$

or

$$S = 1 \tag{9-40}$$

Further information on the behavior of convergency in series is given by the statement that if S_n is a variable which decreases as n increases, but which is either equal to or smaller than some quantity C, then as n increases without limit S_n will approach S in the limit, or $S_n \to S$, which is of such a magnitude that S is not less than C.

Suppose the series

$$S = u_1 + u_2 + u_3 + u_4 + \cdots + u_n + \cdots \tag{9-41}$$

to be convergent. Denote the sums $S_1, S_2, \ldots$ by the following:

$$\begin{aligned} &S_1 = u_1 \qquad S_2 = u_1 + u_2 \qquad S_3 = u_1 + u_2 + u_3 \\ &S_4 = u_1 + u_2 + u_3 + u_4 \qquad \cdots \end{aligned} \tag{9-42}$$

It follows that the values $S_1, S_2, \ldots, S_n, \ldots$ can be represented on a directed line. The points thus plotted will approach the point that is defined by S as n increases. It therefore follows that

$$\lim_{n \to \infty} u_n = 0 \tag{9-43}$$

in a convergent series, but if the condition of (9-43) is satisfied, the series still may be either divergent or convergent. In mathematical language, this may be stated in the following manner. For a series to converge, it is necessary, but not sufficient, that term u_n approach zero as $n \to \infty$.

The harmonic series with the general term

$$u_n = \frac{1}{n} \tag{9-44}$$

is of such a nature that

$$\lim_{n \to \infty} u_n = \lim_{n \to \infty} \frac{1}{n} \to 0 \tag{9-45}$$

but it will be shown that the harmonic series, although it has the limiting value for its general term equal to zero as $n \to \infty$, is divergent.

This consideration of a necessary but not sufficient condition for convergence is of great utility in the investigation of numerous series, since if the limiting value of u_n does not approach zero as $n \to \infty$, it is immediately concluded that the series undergoing investigation is divergent.

Cauchy's Integral Test. If a series of positive terms is represented by

$$S = u_1 + u_2 + u_3 + \cdots + u_n + \cdots \tag{9-46}$$

where the terms are of such magnitude that

$$u_n > u_{n+1} \qquad \text{for all } n \geq 1 \tag{9-47}$$

then if a positive decreasing function $f(n)$, for $n > 1$, is such that

$$f(n) = u_n$$

the series (9-46) converges if

$$I = \int_1^\infty f(n)\,dn \tag{9-48}$$

exists, i.e., is a unique finite quantity; if the series is divergent, this integral becomes infinite upon evaluation. (In place of unity as a lower limit, any finite quantity may be used.)

EXAMPLE. The K series is defined by the relation

$$\frac{1}{1^K} + \frac{1}{2^K} + \frac{1}{3^K} + \cdots + \frac{1}{n^K} + \cdots \tag{9-49}$$

The function to be considered here is $1/n^K = f(n)$. The integral involved in the *Cauchy integral test* for this series is given by

$$I = \int_1^\infty \frac{dn}{n^K}$$

This integrates into

$$\left[\left(\frac{1}{1-K}\right) n^{1-K}\right]_1^\infty \qquad \text{for } K \neq 1 \tag{9-50}$$

or

$$\left[\ln n\right]_1^\infty \qquad \text{for } K = 1 \tag{9-51}$$

Hence for $K = 1$, the integral does not have a finite value. In the case where $K > 1$,

$$\left[\left(\frac{1}{1-K}\right) n^{1-K}\right]_1^\infty \tag{9-52}$$

is finite for $n \to \infty$; in the case where $K < 1$, (9-52) represents an expression which increases without limit.

It is concluded that the K series is of such a nature that for $K \leq 1$ the series is divergent, and for $K > 1$ the series is convergent.

The Comparison Test. If for a given series of positive terms

$$\sum_{n=1}^{\infty} u_n = u_1 + u_2 + u_3 + \cdots + u_n + \cdots \tag{9-53}$$

there is a series of positive terms

$$\sum_{n=1}^{\infty} v_n = v_1 + v_2 + v_3 + \cdots + v_n + \cdots$$

which is known to be convergent, and if $u_n \leq v_n$ for corresponding n, then (9-53) is also convergent, since its sum Σu_n does not exceed the finite sum Σv_n of a series known to be convergent.

If, however, for the series of (9-53) there is a series

$$\sum_{n=1}^{\infty} w_n = w_1 + w_2 + w_3 + \cdots + w_n + \cdots$$

which is known to be divergent, and if $u_n \geq w_n$ for corresponding n, then the given series (9-53) is also divergent, since its sum exceeds that of a series known to be divergent. A multiple (or submultiple) of a divergent series is divergent; likewise, a multiple of a convergent series is convergent.

EXAMPLE 1. Show that the series

$$1 + \frac{1}{2^2} + \frac{1}{3^3} + \frac{1}{4^4} + \cdots$$

is convergent.

The general term of the above series may be expressed by

$$u_n = \frac{1}{(n-1)^{n-1}}$$

assuming terms from $n = 2, \ldots,$ are given. For comparison purposes, consider the convergent geometric series of ratio $\frac{1}{2}$.

$$1 + \frac{1}{2} + \frac{1}{2^2} + \frac{1}{2^3} + \cdots$$

where the general term is given by

$$v_n = \frac{1}{2^{n-1}}$$

making a similar assumption for this series. The first terms of these respective series are equal, but for all terms $n \geq 2$, the value of $u_n < v_n$. Therefore, the given series is convergent.

EXAMPLE 2. Show that the series

$$\frac{1}{1+100}+\frac{1}{2+100}+\frac{1}{3+100}+\cdots+\frac{1}{n+100}+\cdots$$

is divergent.

Consider the series

$$\frac{1}{100+100}+\frac{1}{200+100}+\frac{1}{300+100}+\cdots+\frac{1}{100(n+1)}+\cdots$$

It follows, for

$$u_n=\frac{1}{n+100} \qquad w_n=\frac{1}{100(n+1)}$$

that

$$u_n > w_n$$

for all n. Since

$$\sum_{n=1}^{\infty}\frac{1}{n+1}$$

is the harmonic series which is divergent, it follows that

$$\sum_{n=1}^{\infty} w_n$$

which is a submultiple of this series, is divergent. Therefore, the given series is divergent.

The Ratio Test. This test may be stated in the following form:

(*a*) Write a formula for the nth and one for the $(n+1)$th term of the given series, u_n and u_{n+1}, respectively.

(*b*) Set up the absolute value of the ratio of these terms, namely,

$$\left|\frac{u_{n+1}}{u_n}\right|$$

(*c*) Find the limiting value of the ratio (*b*) as the number of terms n is permitted to increase without limit, that is, $n \to \infty$, and denote the absolute value obtained by

$$\rho=\lim_{n\to\infty}\left|\frac{u_{n+1}}{u_n}\right|$$

(*d*) If ρ exceeds unity in magnitude, that is, $|\rho| > 1$, then the given series is divergent; if $|\rho|$ is less than unity in magnitude, that is, $|\rho| < 1$, the series is convergent, while if $|\rho| = 1$, then the test fails and some other method must be employed to determine the nature of the convergency or divergency of the series.

EXAMPLE 1. Examine for convergence by the test-ratio test the series

$$\frac{2}{2^2} + \frac{2^2}{3^2} + \frac{2^3}{4^2} + \frac{2^4}{5^2} + \cdots$$

The general, or nth, term, is expressed as

$$u_n = \frac{2^n}{(n+1)^2}$$

and the $(n + 1)$th term is given by

$$u_{n+1} = \frac{2^{n+1}}{(n+2)^2}$$

Then

$$\frac{u_{n+1}}{u_n} = \frac{2^{n+1}}{(n+2)^2}\,\frac{(n+1)^2}{2^n} = 2\left(\frac{n+1}{n+2}\right)^2$$

$$\lim_{n\to\infty}\left|\frac{u_{n+1}}{u_n}\right| = \lim_{n\to\infty}\left|2\left(\frac{n+1}{n+2}\right)^2\right| = \lim_{n\to\infty}\left|2\left(\frac{1+1/n}{1+2/n}\right)^2\right| = 2 > 1$$

Hence the given series is divergent.

If a series is of such a character that its terms undergo an alternation in algebraic sign, it is called an *alternating series.*

Consider the alternating series

$$U_1 - U_2 + U_3 - U_4 + \cdots + (-1)^{n+1}U_n + \cdots \tag{9-54}$$

where

$$|U_{n+1}| < |U_n|$$

for all n.

Suppose that n is an even integer; then the sum of the first n terms of the series may be written

$$S_n = (U_1 - U_2) + (U_3 - U_4) + \cdots + (U_{n-1} - U_n) \tag{9-55}$$

or

$$S_n = U_1 - (U_2 - U_3) - (U_4 - U_5) - \cdots - (U_{n-2} - U_{n-1}) - U_n \tag{9-56}$$

In (9-55) and (9-56), the quantity in each parentheses is positive in magnitude. Hence, as S_n increases with n, S_n is less than U_1. It follows that S_n approaches some limit. But S_{n+1} will also approach the same limit, for from fundamental concept,

$$S_{n+1} = S_n + U_{n+1} \tag{9-57}$$

For this expression to be valid, it follows that

$$\lim_{n\to\infty} U_{n+1} = 0 \tag{9-58}$$

This leads to the conclusion that for n either an even or odd integer, as $n \to \infty$, S_n will approach a finite limit. This fact is necessary and sufficient to establish convergence of alternating series.

Note that the error that results in the evaluation of an alternating series after n terms is always less in magnitude than the $(n + 1)$th term of the series under consideration.

EXAMPLE 2. Consider for possible convergence the alternating series

$$\tfrac{1}{2} - \tfrac{1}{4} + \tfrac{1}{6} - \tfrac{1}{8} + \cdots$$

The general term is given by

$$U_n = (-1)^{n+1} \frac{1}{2n}$$

and

$$\lim_{n \to \infty} U_n \to 0$$

This alternating series is consequently convergent.

An alternating infinite series which is convergent as given but which becomes divergent when all its negative quantities are replaced by positive ones of equal magnitude is called *conditionally* convergent; whereas, if it remains convergent under such a transformation, then it is *absolutely convergent.*

The ratio test is useful only in the establishment of absolute convergence; it cannot be used for reaching conclusions regarding conditional convergence.

EXAMPLE 3. Consider the following series for possible absolute and conditional convergence:

$$\tfrac{1}{2} - \tfrac{2}{5} + \tfrac{3}{10} - \tfrac{4}{17} + \cdots$$

The general term is given by

$$U_n = (-1)^{n+1} \frac{n}{n^2 + 1}$$

and

$$\lim_{n \to \infty} (-1)^{n+1} \frac{n}{n^2 + 1} \to 0$$

Hence the given series is convergent, since it is a necessary and sufficient condition for convergence of an alternating series that its nth term approach zero as n tends to become infinite.

If all of the negative signs are replaced by positive ones in the given series, the Cauchy integral test may be applied. The integral

$$\int_1^\infty \frac{n\,dn}{n^2 + 1}$$

is found upon evaluation to approach infinity in value.

The conclusion is then reached that the series given in this example is *conditionally convergent* but does not have the property of absolute convergence.

EXAMPLE 4. It is desired to investigate for convergence the alternating series

$$4/3 - 5/5 + 6/7 - 7/9 + \cdots$$

Since the general term is representable by the fraction

$$U_n = (-1)^{n+1} \frac{n+3}{2n+1}$$

it follows immediately that this alternating series is divergent as the limiting value of U_n as $n \to \infty$ is $1/2$.

The series with positive signs replacing all negative ones in the original alternating series is also divergent, since series of positive terms has as a necessary but not sufficient condition that the general term of the series approach zero as n increases without limit.

Hence, it is seen that the series given in this example is neither conditionally nor absolutely convergent.

9-3 Power Series

A series which assumes the form

$$S = c_0 + c_1x + c_2x^2 + \cdots + c_nx^n + \cdots \tag{9-59}$$

where $c_0, c_1, \ldots, c_n, \ldots$ are constant terms and x is a variable, is called a power series in x. To investigate the extent of possible values of x which make (9-59) a convergent series, the ratio-test method is used, and

$$\lim_{n\to\infty} \left| \frac{c_{n+1}}{c_n} \frac{x^{n+1}}{x^n} \right| \tag{9-60}$$

would give a convergent series for those values which make this quantity less than unity.

If

$$\lim_{n\to\infty} \left| \frac{c_{n+1}}{c_n} \right| < \beta \tag{9-61}$$

then

$$\beta|x| < 1 \tag{9-62}$$

or

$$|x| < \frac{1}{\beta} \tag{9-63}$$

satisfies the inequality

$$\lim_{n\to\infty} \left(\frac{c_{n+1}}{c_n} \right) |x| < 1 \tag{9-64}$$

The interval $-1/\beta < x < 1/\beta$, which defines those values of x in (9-64) for which (9-59) is convergent, is called the *interval of convergence*. The values at the end of the interval must be examined separately before inclusion or exclusion of them in the interval of convergence.

EXAMPLE. Ascertain the range of values for the variable x which makes the series below convergent:

$$1 + \frac{x}{3} + \frac{x^2}{5} + \frac{x^3}{7} + \cdots + \frac{x^{n-1}}{2n-1} + \cdots$$

The ratio for this series is given by

$$\left| \frac{U_n + 1}{U_n} \right| = \frac{2n-1}{2n+1} |x|$$

which has for its limiting value as n increases without limit, the value $|x|$. Therefore the series converges for all values of x for which $|x| < 1$ or for $-1 < x < 1$. It now remains necessary to investigate whether the end-points of this interval are values which are included in the range of convergency. If $x = -1$ is substituted in the given series, there is obtained the series of constant terms

$$1 - \tfrac{1}{3} + \tfrac{1}{5} - \tfrac{1}{7} + \cdots$$

which is an alternating series with its general term approaching zero as n becomes infinitely large. This series is then convergent for $x = -1$. If $x = 1$, then the series which is produced is

$$1 + \tfrac{1}{3} + \tfrac{1}{5} + \tfrac{1}{7} + \cdots$$

which term by term is greater than the comparison series

$$\tfrac{1}{2} + \tfrac{1}{4} + \tfrac{1}{6} + \cdots$$

which is recognized immediately as divergent, since it is merely the harmonic series

$$1 + \tfrac{1}{2} + \tfrac{1}{3} + \cdots$$

multiplied by the constant $\tfrac{1}{2}$.

The range of convergence then is given for this series as $-1 \leqq x < 1$.

9-4 Power-series Theorems

Theorem 1. *If two power series, each a function of a variable u, are represented by*

$$S_1(u) = a_0 + a_1u + a_2u^2 + \cdots + a_{n-1}u^{n-1} + \cdots$$

and

$$S_2(u) = b_0 + b_1u + b_2u^2 + \cdots + b_{n-1}u^{n-1} + \cdots$$

and are convergent in the range $u < |\alpha|$ *for* $S_1(u)$ *and in the range* $u < |\beta|$ *for* $S_2(u)$*, respectively, then the series formed by the addition of the respective terms to form*

$$S_3(u) = c_0 + c_1u + c_2u^2 + \cdots + c_{n-1}u^{n-1} + \cdots$$

where

$$c_0 = a_0 + b_0 \qquad c_1 = a_1 + b_1 \qquad \cdots \qquad c_{n-1} = a_{n-1} + b_{n-1} \qquad \cdots$$

is convergent at least in that range for $|u|$ *less than* α *or* β, *whichever is the smaller.*

EXAMPLE. Consider the series

$$S_1(u) = e^{ju} = 1 + \frac{ju}{1!} + \frac{j^2u^2}{2!} + \frac{j^3u^3}{3!} + \frac{j^4u^4}{4!} + \cdots$$

$$S_2(u) = e^{-ju} = 1 - \frac{ju}{1!} + \frac{j^2u^2}{2!} - \frac{j^3u^3}{3!} + \frac{j^4u^4}{4!} + \cdots$$

$$S_1(u) + S_2(u) = 2\left[1 + \frac{j^2u^2}{2!} + \frac{j^4u^4}{4!} + \cdots\right]$$

$$S_3(u) = 2\left[1 - \frac{u^2}{2!} + \frac{u^4}{4!} - \cdots\right]$$

The series $S_1(u)$ and $S_2(u)$ converge for all finite values of u, as can be seen by the use of the standard test for convergence. The series $S_3(u)$ will also converge for all finite values of u. It will be recalled from the application of Euler's relation that $S_3(u)$ is equal to $2 \cos u$.

Theorem 2. *If two power series, each a function of a variable* u, *are represented by*

$$S_1(u) = a_0 + a_1u + a_2u^2 + \cdots + a_{n-1}u^{n-1} + \cdots$$

and

$$S_2(u) = b_0 + b_1u + b_2u^2 + \cdots + b_{n-1}u^{n-1} + \cdots$$

and are convergent in the range $|u| < \alpha$ *for* $S_1(u)$ *and the range* $|u| < \beta$ *for* $S_2(u)$, *respectively, then the resulting series formed by the multiplication of the respective terms to form the product series* $M(u)$,

$$M(u) = a_0b_0 + (a_0b_1 + a_1b_0)u + (a_0b_2 + a_1b_1 + a_2b_0)u^2 + \cdots$$

is convergent for at least $|u| < \alpha$ *or* $|u| < \beta$, *whichever is the smaller.*

EXAMPLE. Consider the power series for $\sin u$ and $\cos u$,

$$\sin u = u - \frac{u^3}{3!} + \frac{u^5}{5!} - \frac{u^7}{7!} + \cdots$$

$$\cos u = 1 - \frac{u^2}{2!} + \frac{u^4}{4!} - \frac{u^6}{6!} + \cdots$$

By termwise multiplication, there is obtained

$$M(u) = u - \frac{4u^3}{6} + \frac{16u^5}{120} - \cdots$$

or

$$M(u) = \frac{1}{2}\left[2u - \frac{(2u)^3}{3!} + \frac{(2u)^5}{5!} - \cdots\right]$$

which is immediately recognized as

$$M(u) = \tfrac{1}{2} \sin 2u$$

Theorem 3. If two power series, each a function of u, are represented by

$$S_1(u) = a_0 + a_1u + a_2u^2 + a_3u^3 + \cdots$$

and

$$S_2(u) = b_0 + b_1u + b_2u^2 + b_3u^3 + \cdots$$

and are convergent in the range $|u| < \alpha$ *for* $S_1(u)$ *and in the range* $|u| < \beta$ *for* $S_2(u)$*, respectively, then the quotient power series* $Q(u)$*, obtained by division of* $S_1(u)$ *by* $S_2(u)$*,*

$$Q(u) = \frac{S_1(u)}{S_2(u)} = \frac{a_0}{b_0} + \frac{a_1b_0 - a_0b_1}{b_0^2} u + \frac{a_2b_0^2 - a_1b_0b_1 + a_0b_1^2 - a_0b_0b_2}{b_0^3} u^2$$

is of such a nature that no statement can be validly made which concerns its range of convergence.

EXAMPLE

$$S_1(u) = \sin u = u - \frac{u^3}{3!} + \frac{u^5}{5!} - \cdots$$

$$S_2(u) = \cos u = 1 - \frac{u^2}{2!} + \frac{u^4}{4!} - \cdots$$

$$Q(u) = \frac{S_1(u)}{S_2(u)} = u + \frac{u^3}{3} + \frac{2u^5}{15} + \cdots$$

$Q(u)$ is readily recognized as the power series in $\tan x$. The series $S_1(u)$ and $S_2(u)$ are convergent for all finite values of the variable u, but the quotient series is convergent for $|u| < \pi/2$.

Theorem 4. If the power series

$$\theta = a_0 + a_1u + a_2u^2 + \cdots$$

converges for $|\theta| < \alpha$*, and the series*

$$u = b_0 + b_1x + b_2x^2 + \cdots$$

converges for $|u| < \beta$ *and* $|b_0| < \alpha$*, then for u substituted from the second series into the first series, there will be obtained a relation where* θ *is a power series in x. For sufficiently small values of x, the resulting series converges.*

Theorem 5. If the power series

$$S_1(u) = a_0 + a_1u + a_2u^2 + \cdots$$

converges for $|u| < \alpha$*, then the series* $S_2(u)$ *obtained by taking the derivative of* $S_1(u)$ *by termwise differentiation is also convergent for the same range of values.*

EXAMPLE

$$
\begin{aligned}
S_1(u) &= \cos u = 1 - \frac{u^2}{2!} + \frac{u^4}{4!} - \frac{u^6}{6!} + \cdots \\
S_2(u) &= \frac{dS_1}{du} = \frac{d(\cos u)}{du} = -\frac{2u}{2!} + \frac{4u^3}{4!} - \frac{6u^5}{6!} + \cdots \\
&= -\frac{u}{1!} + \frac{u^3}{3!} - \frac{u^5}{5!} + \cdots = -\left[\frac{u}{1!} - \frac{u^3}{3!} + \frac{u^5}{5!} - \cdots\right] \\
&= -\sin u
\end{aligned}
$$

Both the sin u and cos u power-series expansions are convergent for all values of u.

Theorem 6. *If the power series*

$$S_1(u) = a_0 + a_1u + a_2u^2 + \cdots$$

converges for $|u| < \alpha$, *then the series* $S_2(u)$ *obtained by integrating termwise the given power series is also convergent for* $|u| < \alpha$.

EXAMPLE. Suppose

$$S_1(u) = 1 - \frac{u^2}{2!} + \frac{u^4}{4!} - \frac{u^6}{6!} + \cdots$$

is a given power series. This will be recognized as the cos u series, which is known to converge for all finite values of u. If this series is integrated with respect to u in a termwise manner,

$$
\begin{aligned}
&\int_0^u \left(1 - \frac{u^2}{2!} + \frac{u^4}{4!} - \frac{u^6}{6!} + \cdots\right) du \\
&= \left[u - \frac{u^3}{2!(3)} + \frac{u^5}{4!(5)} - \frac{u^7}{6!(7)} + \cdots\right]_0^u = u - \frac{u^3}{3!} + \frac{u^5}{5!} - \frac{u^7}{7!} + \cdots
\end{aligned}
$$

which is the power series for the sin u. Both these series converge for all finite values of the variable u.

9-5 Applications: Series Expansion of Functions

Infinite series seldom appear in the solution of an engineering problem unless directly introduced by the engineer as an aid to analysis. Series forms are usually employed for engineering purposes in several ways. In one of these a series is substituted for a known function to make the process of solution simpler. In another, which is discussed in the chapter that follows, a set of experimental data is represented approximately by several terms of a power series.

Series expansions of functions are most conveniently secured in power-series form and are usually applied to transcendental functions. To expand some function $f(x)$ in the region of interest at which $x = a$, the power series, called a *Taylor series expansion*,

$$f(x) = b_0 + b_1(x - a) + b_2(x - a)^2 + b_3(x - a)^3 + \cdots \tag{9-65}$$

is assumed. The general coefficient b_n is

$$b_n = \frac{f^{(n)}(a)}{n!} \tag{9-66}$$

where $f^{(n)}(a)$ is the value of the nth derivative of $f(x)$ at a. When it is possible to take $a = 0$ in Eq. (9-65), the special case of the Taylor expansion of $f(x)$ is obtained, called a *Maclaurin series expansion*. The condition of $a = 0$ is obviously simpler to handle unless the form of the function prevents its use. Since the expansion given by (9-65) and (9-66) is a power series, the convergence of this form of functional representation is best established by the ratio test.

As an example of the use of the Taylor expansion, consider the formation of a series equivalent for the logarithmic function $\ln x$.

If $a = 1$, the successive derivatives and coefficients according to (9-66) are

$$\begin{aligned} f(x) &= \ln x & f(1) &= \ln 1 = 0 \\ f'(x) &= x^{-1} & f'(1) &= 1 \\ f''(x) &= -x^{-2} & f''(1) &= -1 \\ f'''(x) &= 2x^{-3} & f'''(1) &= 2 \\ f^{\text{iv}}(x) &= -6x^{-4} & f^{\text{iv}}(1) &= -6 \\ &\vdots & &\vdots \end{aligned}$$

According to Eq. (9-65), the function $\ln x$ in power-series form is

$$\ln x = (x-1) - \frac{(x-1)^2}{2} + \frac{(x-1)^3}{3} - \frac{(x-1)^4}{4} + \cdots \tag{9-67}$$

It should be noted that in this case the series could not have been expanded around the point $a = 0$, because all the coefficients in the series expansion would become infinite. Equation (9-67) is convergent in the interval $0 < x \leqq 2$ as may be shown by Sec. 9-3.

On the other hand, where the condition $a = 0$ (the Maclaurin series) can be utilized, the expansion can be secured even more simply.

For the function ϵ^x, for example, the derivatives and the coefficients are

$$\begin{aligned} f(x) &= \epsilon^x & f(0) &= 1 \\ f'(x) &= \epsilon^x & f'(0) &= 1 \\ f''(x) &= \epsilon^x & f''(0) &= \frac{1}{2!} \\ f'''(x) &= \epsilon^x & f'''(0) &= \frac{1}{3!} \\ &\vdots & &\vdots \end{aligned}$$

and from Eq. (9-65) the series expansion is simply

$$\epsilon^x = 1 + x + \frac{x^2}{2} + \frac{x^3}{6} + \cdots \tag{9-68}$$

which is convergent for all finite values of x.

As an example of an application of such an expansion, consider Eq. (9-69), which represents the rise in the voltage developed across the

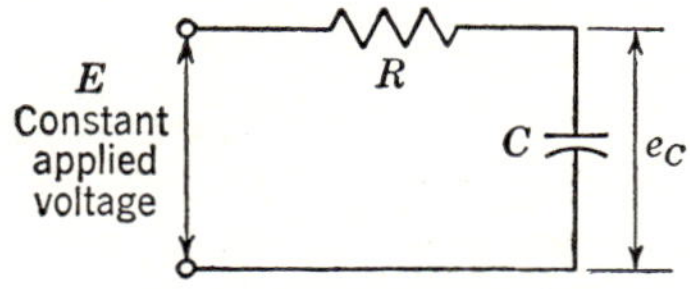

FIG. 9-2 RC network.

capacitor C in Fig. 9-2 as a function of time t when a constant voltage E is applied to the circuit:

$$e_C = E(1 - \epsilon^{-t/RC}) \tag{9-69}$$

The function $1 - \epsilon^{-t/RC}$ can be expanded into series form by means of the result already obtained in Eq. (9-68). If the independent variable x is replaced by $-t/RC$, where RC is the circuit time constant, the expansion is modified to

$$\epsilon^{-t/RC} = 1 - \frac{t}{RC} + \frac{t^2}{2R^2C^2} - \frac{t^3}{6R^3C^3} + \cdots \tag{9-70}$$

and the expression for e_C is

$$e_C = E\left[\frac{t}{RC} - \frac{t^2}{2R^2C^2} + \frac{t^3}{6R^3C^3} - \cdots\right] \tag{9-71}$$

Suppose that it is necessary to determine the value of e_C after an elapsed time of $t = 0.1RC$ sec. According to (9-71), this would be

$$\begin{aligned} e_C \text{ (at } 0.1RC) &= E\left[0.1 - \frac{0.01}{2} + \frac{0.001}{6} - \cdots\right] \\ &= E[0.1 - 0.005 + 0.000167 - \cdots] \\ &= E(0.095167) \end{aligned} \tag{9-72}$$

The accuracy of the result obtained for specific values of the independent variable depends on the number of terms to which the expansion is carried. However, the number of terms which can be evaluated is finite, and therefore some error is always introduced. The extent of the error incurred by using only the first few terms of a Taylor-series expansion can be gauged from the fact that if the terms of an alternating series decrease monotonically (single-valuedly) and approach zero, the remainder after n terms is numerically less than the $(n + 1)$th term. The greatest accuracy is obviously achieved by selecting values of a which are reasonably close to the region in which the series expansion is to be used. For example, if the value of sin 33° is to be computed by series expansion, a choice of $\pi/6$ for a will provide a much more rapid convergence and a consequent greater accuracy of approximation in fewer terms than will a choice of $a = 0$.

In many cases, the accuracy with which the variables in an engineering problem are known precludes the use of more than the first few terms

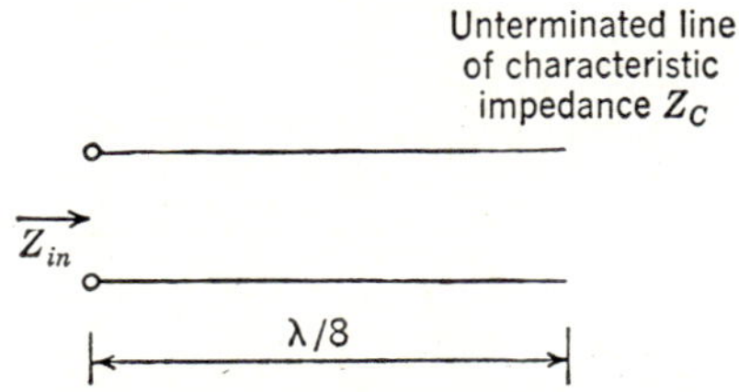

FIG. 9-3 Unterminated eighth-wave transmission line.

of a Taylor-series expansion. In these cases such an expansion proves a real time saver. For example, consider the unterminated transmission line in Fig. 9-3.

The input impedance of this line is a capacitive reactance which is equal in magnitude to the characteristic impedance Z_c of the line at that frequency f_0 at which the line length is $\lambda/8$. For any other frequency f, the input impedance is

$$z_{\text{in}} = -jz_c \cot\left(\frac{\pi}{4}\frac{f}{f_0}\right) \tag{9-73}$$

The Taylor-series expansion of the cotangent function around $\pi/4$ is

$$\begin{aligned}\cot\left(\frac{\pi}{4}\frac{f}{f_0}\right) &= \cot\frac{\pi}{4} - \frac{\pi}{4}\left(\frac{f}{f_0} - 1\right)\csc^2\frac{\pi}{4} + \cdots \\ &= 1 - \frac{2\pi}{4}\left(\frac{f}{f_0} - 1\right) + \cdots\end{aligned} \tag{9-74}$$

To a first approximation, then, the input impedance Z_{in} is

$$z_{\text{in}} = -jz_c\left(2.57 - 1.57\frac{f}{f_0}\right) \tag{9-75}$$

which is much more easily evaluated than Eq. (9-73).

The Taylor- and Maclaurin-series expansions of known functions find a wide variety of applications in electronic-circuit analysis. Some interesting cases are given by Edson and Teasdale,[1] Cohen and Bloom,[2]

[1] W. A. Edson and R. D. Teasdale, A Relation Between Susceptance Slope and Selectivity for Oscillator Design, *Proc. Natl. Electronics Conf.*, **8**: 427–434, 1952.

[2] V. W. Cohen and A. Bloom, Microphonism in a Subminiature Triode, *Proc. I.R.E.*, **36** (8): p. 1040, August, 1948.

and Vallarino and Buyer.[1] In most engineering applications which involve simple approximation, it is reasonable to assume that convergence is rapid and that higher-degree terms can be neglected. However, where a series expansion is used as the basis of theoretical solution of a complex technical problem, much care must be exercised to see that the expansion used is both correct and convergent so that the results obtained will be meaningful.

Problems

Prob. 9-1 Write the general term u_n of the following series:

(a) $$\frac{1}{1^2} + \frac{1}{3^2} + \frac{1}{5^2} + \frac{1}{7^2} + \cdots$$

(b) $$\frac{1}{1+\sqrt{1}} + \frac{2}{1+\sqrt{2}} + \frac{3}{1+\sqrt{3}} + \cdots$$

(c) $$1/4 + 1/7 + 1/10 + 1/13 + \cdots$$

(d) $$1/6 + 1/13 + 1/27 + 1/55 + \cdots$$

(e) $$\frac{7}{2 \cdot 3} - \frac{7}{3 \cdot 4} + \frac{7}{4 \cdot 5} - \frac{7}{5 \cdot 6} + \cdots$$

(f) $$\frac{1^3}{2!} - \frac{3^3}{4!} + \frac{5^3}{6!} - \frac{7^3}{8!} + \cdots$$

(g) $$\frac{3!}{2} + \frac{4!}{3} + \frac{5!}{4} + \cdots$$

Prob. 9-2 If series have the respective general, or U_n, terms as indicated below, write the first five terms of each of the corresponding series.

(a) $$U_n = \frac{1}{2n}\left(\frac{3}{2}\right)^n$$

(b) $$U_n = (-1)^{n+1}\frac{n^4}{(n-1)!}$$

(c) $$U_n = \frac{5^n}{2n}$$

(d) $$U_n = (-1)^{n+1}\frac{x^{n+1}}{n!}$$

(e) $$U_n = \frac{x^{2n-1}}{(n+1)!}$$

(f) $$U_n = \frac{3^{n+2}}{2^n(n!)}$$

(g) $$U_n = (-1)^n\frac{x^{3n-1}}{(3n-1)!}$$

[1] A. R. Vallarino and M. S. Buyer, Harmonic Distortion in Frequency-modulation Off-resonance Discriminator, *Elec. Commun.*, **26** (2): 168, June, 1949.

Prob. 9-3 Examine and investigate each of the following series to determine whether convergent or divergent:

(a) $$\frac{1}{1^2} + \frac{1}{2^2} + \frac{1}{3^2} + \cdots + \frac{1}{n^2} + \cdots$$

(b) $$\frac{1}{1\sqrt{1}} + \frac{1}{2\sqrt{2}} + \frac{1}{3\sqrt{3}} + \cdots + \frac{1}{n\sqrt{n}} + \cdots$$

(c) $$\frac{1}{2+1} + \frac{1}{2\cdot 2+1} + \frac{1}{2\cdot 3+1} + \cdots + \frac{1}{2n+1} + \cdots$$

(d) $$\tfrac{1}{3} + \tfrac{1}{6} + \tfrac{1}{9} + \cdots$$

(e) $$\frac{1}{2(\ln 2)^3} + \frac{1}{3(\ln 3)^3} + \frac{1}{4(\ln 4)^3} + \cdots + \frac{1}{(n+1)[\ln (n+1)]^3} + \cdots$$

(f) $$\frac{4}{1\cdot 2\cdot 3} + \frac{5}{2\cdot 3\cdot 4} + \frac{6}{3\cdot 4\cdot 5} + \cdots$$

(g) $$\frac{1}{2}\left(\frac{3}{2}\right) + \frac{1}{4}\left(\frac{3}{2}\right)^2 + \frac{1}{6}\left(\frac{3}{2}\right)^3 + \cdots$$

(h) $$\frac{7}{1\cdot 2} + \frac{7}{2\cdot 3} + \frac{7}{3\cdot 4} + \cdots$$

(i) $$\frac{1}{3} + \frac{1\cdot 2}{3\cdot 5} + \frac{1\cdot 2\cdot 3}{3\cdot 5\cdot 7} + \frac{1\cdot 2\cdot 3\cdot 4}{3\cdot 5\cdot 7\cdot 9} + \cdots$$

(j) $$\frac{1^3}{1!} + \frac{2^3}{2!} + \frac{3^3}{3!} + \cdots$$

Prob. 9-4 Examine for conditional and absolute convergence and divergence the following alternating series:

(a) $$\frac{1}{3} - \frac{1}{2\cdot 3^2} + \frac{1}{3\cdot 3^3} - \frac{1}{4\cdot 3^4} + \cdots$$

(b) $$\frac{1}{2} - \frac{1}{2\cdot 2^2} + \frac{1}{3\cdot 2^3} - \frac{1}{4\cdot 2^4} + \cdots$$

(c) $$1(2)^{-1} - 2(2)^{-2} + 3(2)^{-3} - 4(2)^{-4} + \cdots$$

(d) $$\frac{1}{\ln 2} - \frac{1}{\ln 3} + \frac{1}{\ln 4} - \cdots$$

Prob. 9-5 Determine the value of x for which the following series converge:

(a) $$1 - \frac{x^2}{2!} + \frac{x^4}{4!} - \frac{x^6}{6!} + \cdots$$

(b) $$\frac{(2x)}{1} - \frac{(2x)^3}{3} + \frac{(2x)^5}{5} - \frac{(2x)^7}{7} + \cdots$$

(c) $$x - \frac{x^2}{2} + \frac{x^3}{3} - \frac{x^4}{4} + \cdots$$

(d) $$1 + \frac{x^2}{2^k} + \frac{x^4}{4^k} + \cdots$$

(e) $$x + \frac{x^2}{2!} + \frac{x^4}{4!} + \frac{x^6}{6!} + \cdots$$

(f) $$\frac{1}{x} + \frac{\sqrt{2}}{x^2} + \frac{\sqrt[3]{3}}{x^3} + \frac{\sqrt[4]{4}}{x^4} + \cdots$$

(g) $$\frac{\sin x}{1^2} - \frac{\sin 2x}{2^2} + \frac{\sin 3x}{3^2} - \frac{\sin 4x}{4^2} + \cdots$$

(h) $$\frac{1}{x} + \frac{2}{x^2} + \frac{2 \cdot 3}{x^3} + \frac{2 \cdot 3 \cdot 4}{x^4} + \cdots$$

(i) $$1 - x \ln 1 + 2x^2 \ln 2 - 3x^3 \ln 3 + \cdots$$

(j) $$1 + \frac{1}{2(\ln 2)^p} + \frac{1}{3(\ln 3)^p} + \frac{1}{4(\ln 4)^p} + \cdots$$

Prob. 9-6 By the application of Taylor's or its special case, Maclaurin's, series, expand the following around the indicated value of x:

(a) $$\epsilon^x \text{ around } x = 1$$

(b) $$\sin x \text{ around } x = \frac{\pi}{3}$$

(c) $$\epsilon^{\sin x} \text{ around } x = 0$$

(d) $$\ln x \text{ around } x = 1$$

(e) $$\tan x \text{ around } x = \frac{\pi}{4}$$

Prob. 9-7 By algebraic processes, find Maclaurin's expansion to the term that contains x^5:

(a) $$\epsilon^x \sin x$$

(b) $$\ln \cos x$$

(c) $$\ln \sqrt{1 + x^2}$$

(d) $$[\ln (1 - x)]^2$$

Prob. 9-8 Prove that

(a) $$\sin^2 x + \cos^2 x = 1$$

(b) $$\epsilon^{j\theta} = \cos \theta + j \sin \theta = \text{cis } \theta$$

by using series equivalents for the sine, cosine, and exponential functions and by assuming that the series for ϵ^x is valid for complex as well as real values of x.

Prob. 9-9 Evaluate:

(a) $$\int_0^{0.1} \frac{\sin x}{x} \, dx$$

(*b*) $$\int_0^{0.2} \sin \sqrt{x}\, dx$$

(*c*) $$\int_{0.5}^{1} \frac{(\cos 2x)}{x}\, dx$$

(*d*) $$\int_0^{0.5} \sqrt{1 - x^3}\, dx$$

Prob. 9-10 Obtain the familiar binomial expansion

$$(a + x)^n = a^n + na^{n-1}x + \frac{n(n-1)}{2!} a^{n-2}x^2 + \cdots$$

from the Taylor-series expansion of $(a + x)^n$.

Prob. 9-11 Write the expression for the voltage V after the switch is momentarily closed and reopened in the sawtooth generating circuit

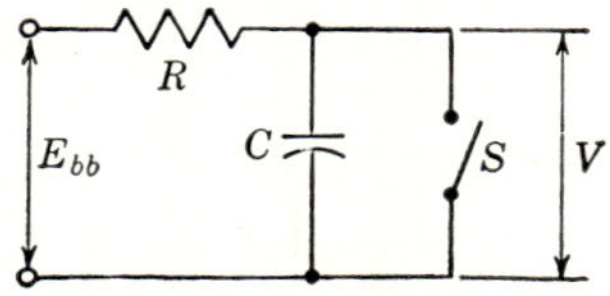

FIG. 9-4 Sawtooth-generating circuit.

shown in Fig. 9-4. By means of a power-series application:

(*a*) Show that the first term of the series is the initial rate of rise of voltage V.

(*b*) Show that the linearity ratio a, which represents the deviation of V from a linear sawtooth wave, is approximately

$$a = \tfrac{1}{2}\left(\frac{t}{T}\right)^2$$

where T is the network time constant.

(*c*) Find the time constant required for 1 per cent linearity in terms of time which elapses between switch closings.

Prob. 9-12 If the input impedance of a short-circuited transmission-line stub is

$$Z_{\text{in}} = jR_0 \tan\left(\frac{\pi}{4}\frac{f}{f_1}\right)$$

where f_1 is the frequency at which the stub is ⅛ wavelength long and R_0 is the characteristic impedance of the line, find an approximate form for Z_{in} which does not contain the tangent function.

Prob. 9-13 The sequence of numbers 17, 17, 16, 14, 12, 10, etc., can be used in the absence of trigonometric tables to provide approximate values of the sine and cosine functions. For example, for the sine of 30°, the sum of the first three numbers is obtained and multiplied by 0.01, giving the desired value of 0.50. For the sine of 60°, the sum of the first *six* numbers is used, which yields 0.86. By means of the series expansion for $\sin \theta$, compute the remainder of the sequence. Find a similar sequence by series expansion valid for the tangent function in steps of 5 to 45°.

Prob. 9-14* The operating frequency of the degenerative positive-bias multivibrator circuit shown in Fig. 9-5 is given as

$$f = \frac{1}{2R_gC_e\left[\ln \frac{b}{a} + \ln\left(\frac{1+ax}{1+bx}\right)\right]}$$

where R_g is grid-return resistance, C_e is effective circuit capacitance, and a, b, and x depend on the applied bias voltage E_c, the plate voltage E_b, and the circuit-component values. Expand the second logarithmic

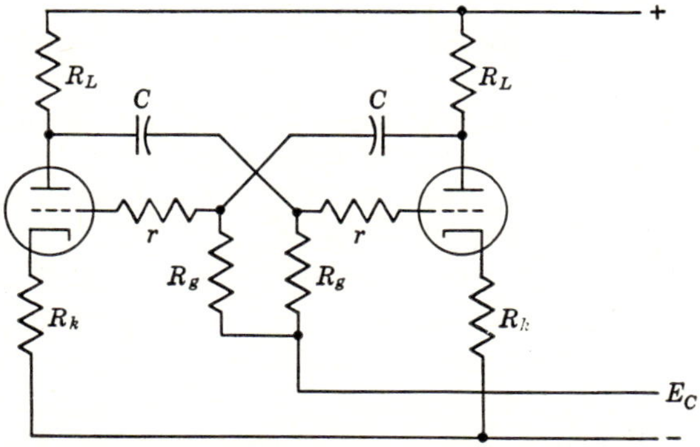

FIG. 9-5 Degenerative positive-bias multivibrator.

expression in the denominator (which contains x) in a power series, and, by developing an approximation for the expression

$$\frac{a^n + a^{n-1}b + a^{n-2}b^2 + \cdots + b^n}{n+1}$$

where $a \cong b$, show that

$$f \cong \frac{1 + \left(\frac{a+b}{2}\right)x}{2R_gC_e\left\{\ln \frac{b}{a} + x\left[\left(\frac{a+b}{2}\right)\ln \frac{b}{a} + (a-b)\right]\right\}}$$

* S. Bertram, Degenerative Positive-bias Multivibrator, *Proc. I.R.E.*, **36** (2): 280, February, 1948.

Chapter 10

NONLINEAR ELECTRONIC DEVICES

10-1 Nonlinear Characteristics

A basic concept of electric-network theory is that of linearity. The existence of a linear network permits the use of the superposition theorem. This theorem says, in effect, that in a linear system the various modes of behavior of the system that arise from various separately applied initial conditions can be added together algebraically to produce the same result as would be obtained if the initial conditions were added together algebraically and applied to the system in combination. If, for example, two sine waves of small amplitude can be applied to the resistive coupling network in an audio-amplifier phase-inverter stage, these will superpose linearly so that the output voltage of the network is just the phasor sum of the applied voltages. When the applied voltages are large, however, the resistors in the network may prove to be voltage sensitive, the net output voltage may not be simply the sum, and the assumption of linearity will not be justified. In practice, the difficulty is overcome by replacing inexpensive resistor types which are voltage sensitive with the more expensive deposited-carbon precision types which are not.

If it is desired, however, to analyze the effects of such nonlinear circuit-component behavior it is obvious that the superposition theorem cannot be used. Further, nonlinear problems cannot always be avoided by the omission of nonlinear components or by tacit assumption of linearity. For the electronics engineer may, on occasion, rely as much on circuit nonlinearities to achieve certain end effects as he does on linear circuit operation.

To analyze the operation of a circuit in which the characteristics of some component change with the amplitude of the applied signal, information on the input-output, or transfer, characteristic of the circuit is required. This information is usually available (or, if necessary, can be prepared from measurements) in graphical form, such as the data presented in electron-tube handbooks. With such graphs, it is possible to undertake many design problems such as Class-A, -B, or -C power-amplifier-performance calculations purely by geometrical constructions. If the nonlinear effect is to be treated mathematically, however, graphical information must be converted to functional form.

Functions which are obtained from experimental data rather than theoretical analysis are generally described as empirical equations, and the methods for securing such equations have been analyzed in some detail.[1]

One general approach to the problem of determining which type of empirical equation is best suited to represent a set of data is to plot the

TABLE 10-1 EMPIRICAL-EQUATION STRAIGHT-LINE TESTS

Form	Yields straight line when plotted as
1. $y = mx + b$	x vs. y
2. $y = ax^n$	$\log x$ vs. $\log y$
3. $y = b\epsilon^{ax}$	x vs. $\log y$
4. $y = \dfrac{ax + b}{x}$	x vs. xy
5. $y = \dfrac{x}{ax + b}$	x vs. $\dfrac{x}{y}$
6. $y = a + bx^2$	x^2 vs. y

data graphically as in Table 10-1. Where it is necessary to plot the log of x or y, graph paper having semilog or log-log coordinates should be used.

As an example of this procedure, the volt-ampere characteristic of the nonlinear material thyrite proves to be most nearly linear when plotted on log-log coordinates. Consequently, the best fit will be obtained with an equation of the second form, that is, $I = aE^n$, where a is the current in amperes at an applied voltage E of 1 volt and n is an exponent determined by the slope of the linear curve obtained on log-log paper (n usually has a value from 3.5 to 7.0).

It should be noted that since all the equation types included in Table 10-1 can be plotted in straight-line form on the proper graph paper, they can also be converted by algebraic manipulation to the linear form. For example, in the case of the third form in the table,

$$y = b\epsilon^{ax} \tag{10-1}$$

[1] See for example D. S. Davis, "Empirical Equations and Nomography," McGraw-Hill Book Company, Inc., 1943; R. P. Hoelscher, J. N. Arnold, and S. H. Pierce, "Graphic Aids in Engineering Computation," McGraw-Hill Book Company, Inc., New York, 1952.

taking the logarithm of both sides of the equation yields

$$\log_{10} y = \log_{10} b + ax \log_{10} \epsilon \tag{10-2}$$

which is of the form

$$y' = m'x + b' \tag{10-3}$$

where $y' = \log_{10} y$
$m' = 0.4343a$
$b' = \log_{10} b$

Thus in this case, if y' is plotted against x on linear graph paper, a straight line will result.

In many instances it will be possible to secure the values of the two parameters m and b (or m' and b' when forms 2 through 6 are converted to straight-line form) directly from the slope and intercept of the straight line which appears to fit the data best.

A simple method for determining m and b, which may be used instead of graphical inspection, is as follows. In all the forms listed in the table there are two parameters, m and b, a and n, or a and b. Therefore, if a set of data is to be represented by one of these equations, an equation should be formed for each pair of x and y values. These should be divided into two groups which correspond in number to the two parameters which must be determined. The equations in the two groups are then added, and the resulting pair of equations are solved simultaneously for the two parameters. For example, a measured set of input and output data is:

Volts in, x	*Volts out,* y
1	2.3
2	3.6
3	5.1
4	6.0
5	7.5
6	8.9

A test plot indicates that these data can be approximately represented by the first form in Table 10-1, $y = mx + b$. Forming two groups of equations corresponding to the number of parameters to be determined, and adding these together, yields

Group 1	*Group* 2
$2.3 = m + b$	$3.6 = 2m + b$
$5.1 = 3m + b$	$6.0 = 4m + b$
$7.5 = 5m + b$	$8.9 = 6m + b$
$14.9 = 9m + 3b$	$18.5 = 12m + 3b$

The simultaneous solution of these equations gives the values $m = 1.2$

and $b = 1.37$ so that the input-output equation is

$$y = 1.2x + 1.37 \tag{10-4}$$

This equation should be plotted and compared with the actual measured points. In some cases, a slight alteration in the slope or intercept value will improve the fit.

10-2 Power-series Representation of Nonlinear Characteristics

The preceding methods for determining empirical input-output characteristics are dependent to a great extent on the judgment of the investigator. It is usually possible to achieve more precise equations for experimental data by the use of power-series forms. Normally, the use of more than the first three or four terms of a power series to fit a set of data accurately is unnecessary, so that the series forms may actually be represented by third- or fourth-degree polynomials.

TABLE 10-2 DIFFERENCE TABLE WHICH CONTAINS EXPERIMENTAL DATA IN FIRST TWO COLUMNS AND FIRST, SECOND, AND THIRD DIFFERENCES IN REMAINING COLUMNS

x	y	ΔY	$\Delta^2 Y$	$\Delta^3 Y$
0	2.0			
		0.6		
1	2.6		0.4	
		1.0		0
2	3.6		0.4	
		1.4		0
3	5.0		0.4	
		1.8		0
4	6.8		0.4	
		2.2		
5	9.0			

It can be shown that (*a*) if a succession of values for y, the dependent variable, is obtained by increasing the independent variable x in equal steps, and (*b*) if the nth differences of y are constant, then the relation between x and y can be represented by a polynomial of degree n, or

$$y = a_0 + a_1x + a_2x^2 + \cdots + a_nx^n \tag{10-5}$$

What is meant by the term *differences* is best illustrated by Table 10-2. The input-output data from an experiment is tabulated in the first two columns of the table. The third column from the left represents the first differences, or ΔY, between the y values in the second column. The fourth column represents the second differences, or $\Delta^2 Y$, between the ΔY values in the third column, etc. This presentation of successive differences is called a *difference table*, and the notation ΔY, $\Delta^2 Y$, $\Delta^3 Y$, etc., is called the *advancing difference notation.*

According to the rule stated above, the relationship between x and y in this case must be of the form

$$Y = a_0 + a_1x + a_2x^2 \tag{10-6}$$

because the second differences of y are constant and all higher-order differences are necessarily zero.

Once the polynomial form has been determined by checking differences, the constants a_0, a_1, a_2, etc., and the complete equation can be evaluated by one of the following methods:

A. Forward Gregory-Newton Formula. The Gregory-Newton formula is a polynomial which is satisfied by the successive pairs of tabular values given in the first two columns of the difference table. The formula itself is

$$y = y_0 + r\,\Delta y_0 + \frac{r(r-1)}{2!}\,\Delta^2 y_0 + \frac{r(r-1)(r-2)}{3!}\,\Delta^3 y_0 + \frac{r(r-1)(r-2)(r-3)}{4!}\,\Delta^4 y_0 + \cdots \tag{10-7}$$

where y_0 is the initial value of y, and Δy_0 is the first difference between y_0 and y_1, $\Delta^2 y_0$ is the second difference between Δy_0 and Δy_1, etc. The constant r is taken as

$$r = \frac{x - x_0}{h} \tag{10-8}$$

where x_0 is the value of x at y_0, and h is the interval between successive values of x. It is obviously desirable, if possible, to select a value for y_0 at which $x_0 = 0$, for in that case $r = x/h$.

To illustrate the application of this formula, consider the data in Table 10-2. For these data, $x_0 = 0$, $y_0 = 2.0$, $\Delta y_0 = 0.6$, $\Delta^2 y_0 = 0.4$, and $h = 1.0$,

$$r = \frac{x - x_0}{h} = \frac{x - 0}{1.0} = x$$

and, from the formula, the relationship is

$$y = 2.0 + x(0.6) + \frac{x(x-1)(0.4)}{2} = 2.0 + 0.6x + \frac{0.4x^2 - 0.4x}{2} \tag{10-9}$$

and

$$y = 2.0 + 0.4x + 0.2x^2 \tag{10-10}$$

which is the desired equation.

B. Method of Least Squares. The method just described provides a function which will pass through each point of a set of data. In many instances, the difference table may not provide the firm and conclusive indications (i.e., constant differences) on which the method must be based. It may, therefore, be desirable to weaken the requirements so that the function ultimately obtained will not pass exactly through each point in the data, but will rather straddle the data and come as "close as possible" to each point. In this way, a linear equation may be found which, for the number of parameters it contains, best fits a given set of data. In many cases, a straight line which falls as close as possible to each point in a set may be more useful than a complex curve which passes exactly through each point, particularly if there has been an appreciable amount of experimental error in the determination of each point. The *method of least squares* permits the evaluation of such a curve of best fit, and the procedure for its use is as follows:[1]

Suppose it is desired to find the best straight-line curve to represent the data in Table 10-2. It is already known that this data can be fitted exactly by a second-degree equation. However, in this case, a simpler, linear expression, of the form $y = a + bx$, is desired. The input-output data from the first two columns is first tabulated as shown in Table 10-3.

TABLE 10-3

x	y	x^2	xy
0	2.0	0	0
1	2.6	1	2.6
2	3.6	4	7.2
3	5.0	9	15.0
4	6.8	16	27.2
5	9.0	25	45.0
$\Sigma x = 15$	$\Sigma y = 29.0$	$\Sigma x^2 = 55.0$	$\Sigma xy = 97.0$

Additional columns representing x^2 and xy are then added as shown, and the sums Σx, Σy, Σx^2, and Σxy are evaluated. The two equations which must be solved to evaluate the constants a_0 and a_1 are

$$\Sigma y = na_0 + a_1 \Sigma x \tag{10-11}$$
$$\Sigma(xy) = a_0 \Sigma x + a_1 \Sigma x^2 \tag{10-12}$$

where n is the number of pairs of x and y values in the first two columns. In the case of the data in Table 10-3, these equations are

$$\begin{aligned} 29.0 &= 6a_0 + 15a_1 \\ 97.0 &= 15a_0 + 55a_1 \end{aligned} \tag{10-13}$$

Solving these simultaneously yields

$$a_0 = 1.33 \qquad a_1 = 1.4$$

[1] For the theory underlying this method, see a text in advanced mathematics, such as C. R. Wylie, "Advanced Engineering Mathematics," McGraw-Hill Book Company, Inc., New York, 1951.

and the required straight line is

$$y = 1.33 + 1.4x \tag{10-14}$$

To fit the same data to a parabolic function by this method requires that additional columns be added to Table 10-3. These are shown in Table 10-3A.

TABLE 10-3A

x^3	x^2y	x^4
0	0	0
1	2.6	1
8	14.4	16
27	45.0	81
64	108.8	256
125	225.0	625
$\Sigma x^3 = 225$	$\Sigma x^2y = 395.8$	$\Sigma x^4 = 979$

The three equations which must be solved to evaluate the constants a_0, a_1, and a_2 in the parabolic form are

$$\Sigma y = na_0 + a_1\Sigma x + a_2\Sigma x^2 \tag{10-15}$$
$$\Sigma xy = a_0\Sigma x + a_1\Sigma x^2 + a_2\Sigma x^3 \tag{10-16}$$
$$\Sigma x^2y = a_0\Sigma x^2 + a_1\Sigma x^3 + a_2\Sigma x^4 \tag{10-17}$$

For the data in Table 10-3, these equations are

$$\begin{aligned} 29.0 &= 6a_0 + 15a_1 + 55a_2 \\ 97.0 &= 15a_0 + 55a_1 + 225a_2 \\ 395.8 &= 55a_0 + 225a_1 + 979a_2 \end{aligned} \tag{10-18}$$

Solving these simultaneously by means of third-order determinants gives

$$a_0 = 2.0 \qquad a_1 = 0.4 \qquad a_2 = 0.2$$

which were the values secured by means of method (A). It is evident that the method of least squares will provide exactly the same functional form when the difference table indicates constant differences of some order. However, it is most valuable when the difference table does not yield exactly constant differences in any column, for in that case method (A) cannot be applied.

If a sufficiently close fit is not secured by the parabolic form, a third-order least-squares polynomial can be obtained with the following four equations:

$$\Sigma y = na_0 + a_1\Sigma x + a_2\Sigma x^2 + a_3\Sigma x^3 \tag{10-19}$$
$$\Sigma xy = a_0\Sigma x + a_1\Sigma x^2 + a_2\Sigma x^3 + a_3\Sigma x^4 \tag{10-20}$$
$$\Sigma x^2y = a_0\Sigma x^2 + a_1\Sigma x^3 + a_2\Sigma x^4 + a_3\Sigma x^5 \tag{10-21}$$
$$\Sigma x^3y = a_0\Sigma x^3 + a_1\Sigma x^4 + a_2\Sigma x^5 + a_3\Sigma x^6 \tag{10-22}$$

In this case, columns for $\Sigma x^3 y$, Σx^5, and Σx^6 as well as for $\Sigma x^2 y$, Σx^3, and Σx^4 must be added to Table 10-3.

The method of least squares can be abbreviated considerably by deliberately restricting the number of pairs of data taken to form the first two columns of Table 10-3 to an *odd* number. If this is done, the median or central value in the x column can be taken as the origin, counting both ways from this point to determine the x values, as shown in Table 10-4. The great advantage in this method is that, as can be seen

TABLE 10-4

x values used	*x values revised to place origin at median*
0	−3
1	−2
2	−1
3	0
4	1
5	2
6	3
$\Sigma x = 21$	$\Sigma x = 0$

from the right-hand column, $\Sigma x = 0$, and all products of Σx, Σx^3, Σx^5, etc., are also zero.

This condition simplifies the equations given previously as follows:

Linear case:

$$\Sigma y = na_0 \tag{10-23}$$
$$\Sigma xy = a_1 \Sigma x^2 \tag{10-24}$$

In this case the necessity for simultaneous solution is eliminated.

Second-degree case:

$$\Sigma y = na_0 + a_2 \Sigma x^2 \tag{10-25}$$
$$\Sigma xy = a_1 \Sigma x^2 \tag{10-26}$$
$$\Sigma x^2 y = a_0 \Sigma x^2 + a_2 \Sigma x^4 \tag{10-27}$$

Third-degree case:

$$\Sigma y = na_0 + a_2 \Sigma x^2 \tag{10-28}$$
$$\Sigma xy = a_1 \Sigma x^2 + a_3 \Sigma x^4 \tag{10-29}$$
$$\Sigma x^2 y = a_0 \Sigma x^2 + a_2 \Sigma x^4 \tag{10-30}$$
$$\Sigma x^3 y = a_1 \Sigma x^4 + a_3 \Sigma x^6 \tag{10-31}$$

10-3 Harmonic Production

The signal wave forms applied to various electronic devices can be synthesized by the Fourier analysis from properly selected sine-wave components. It is therefore customary to analyze the effects of the nonlinear input-output (transfer) characteristics of these devices in terms of the modifications produced during the transmission of one or more sine- (or cosine-) wave signals. If such modifications are undesirable, they are characterized as distortion. Where they are desired, as in harmonic generators of various kinds, they become the output product of interest.

The practical method for determining the harmonic output of an amplifier is the use of a wave analyzer or distortion meter. However, much can be learned from an analytic study of the input-output characteristic of the amplifier provided that this has been converted to finite power-series form, as described in the previous section. In this case, an input-output characteristic of the form

$$y = a_0 + a_1x + a_2x^2 + a_3x^3 + a_4x^4 \tag{10-32}$$

will have been obtained, where y is the output voltage or current of the device, and x is the input voltage or current. The number of terms to which the series is limited depends on the complexity of the input-output characteristic as was seen in the previous section. In most cases, a limit of six terms will prove adequate and will not invalidate the result.

The effects of the input-output characteristic on a sine- (or cosine-) wave signal or signals are calculated by substituting, for the independent variable x, voltage (or currents) of the form $E_\omega \cos \omega t$, where E_ω is the peak value of an input voltage of frequency $\omega/2\pi$. This produces a series of the form

$$y = a_0 + a_1E_\omega \cos \omega t + a_2E_\omega{}^2 \cos^2 \omega t + a_3E_\omega{}^3 \cos^3 \omega t + \cdots \tag{10-33}$$

The terms of form $\cos^n \theta$ can be expanded into Fourier series by means of the sum and difference formulas used in trigonometry. These expansions yield the following multiple-angle formulas:

$$\cos^2 \theta = \frac{1}{2}(1 + \cos 2\theta) \tag{10-34}$$

$$\cos^3 \theta = \frac{3}{4}\cos \theta + \frac{1}{4}\cos 3\theta \tag{10-35}$$

$$\cos^4 \theta = \frac{3}{8} + \frac{1}{2}\cos 2\theta + \frac{1}{8}\cos 4\theta \tag{10-36}$$

$$\vdots$$

The use of the cosine form has an advantage in that the expansion of the $\cos^n \theta$ forms always yields positive cosine terms, which is not always the case for the corresponding $\sin^n \theta$ expansions.

The full evaluation of Eq. (10-33) by means of (10-34) to (10-36) is rather a time-consuming task. Fortunately, much work has already been expended on this general problem for both single- and multiple-cosine-wave inputs, and comprehensive results have been published by Salz,[1] Callendar and Matthews,[2] and Wass.[3]

Table 10-5, compiled by Salz, presents a tabulation of the output products, harmonics, and sidebands which result when the input-output characteristic is carried to six terms, the last being a_5x^5, and the applied signal is of the form

$$x = E_0 + E_\omega \cos \omega t + E_\rho \cos \rho t \tag{10-37}$$

i.e., it consists of two cosine waves of frequency $\omega/2\pi$ and $\rho/2\pi$ superimposed on a constant bias voltage E_0. The table is obtained by substituting Eq. (10-37) into (10-32), expanding into single-frequency components, and grouping the terms.

(*a*) *The product of the information contained in each square by the respective column heading represents the contribution of the series constant at the column head to the production of the frequency indicated by the row heading.* For example, the contribution of the series constant a_2 to the production of the frequency $\cos \omega t$ is $4E_0E_\omega(a_2/2)$, or $2a_2E_0E_\omega$.

(*b*) *The information contained in each row represents the total contribution to the production of the frequency of the respective row heading.* For example, the sum and difference frequencies (sidebands) $\cos(\rho + 3\omega)$ and $\cos(\rho - 3\omega)$ have as a coefficient

$$\left(\frac{a_4}{8}\right) 4E_\omega{}^3E_\rho + \left(\frac{a_5}{16}\right) 40E_0E_\omega{}^2E_\rho$$

(*c*) *The single-frequency solution originally sought from Eq.* (10-33) *can be secured by setting the conditions that* E_0 *and* E *are* 0. This yields the simplified results shown in Table 10-6, which has been extended to nine terms.

(*d*) *The special case of* $E_\omega = E_\rho = E$ *and* $E_0 = 0$ *has also been tabulated and is shown in Table* 10-7.

An examination of these tables yields some important general conclusions. For a single sine- or cosine-wave input, in general, the lower-harmonic amplitudes are always much greater than those of the higher harmonics. In all cases, the amount of nth harmonic contributed by the nth term in the power series is much less than the contribution that the same term makes to the lower-harmonic amplitudes.

[1] N. P. Salz, The Frequency Spectrum of Modulation Components, *Sylvania Technologist*, **5** (2): 44–46, April, 1952.

[2] M. V. Callendar and S. Matthews, Relations between Amplitudes of Harmonics and Intermodulation Frequencies in the Output from a Non-linear Amplifier or Mixer, *Electronic Eng.*, vol. 23, pp. 230–232, June, 1951.

[3] C. A. A. Wass, A Table of Intermodulation Products, *Proc. I.E.E.*, pt. III, **95** (1): 31–39, January, 1948.

TABLE 10-5 OUTPUT COMPONENTS WHEN INPUT IS OF FORM $E_0 + E_\omega \cos \omega t + E_\rho \cos \rho t$

Output product	Series constants					
	a_0	a_1	$a_2/2$	$a_3/4$	$a_4/8$	$a_5/16$
d-c	1	E_0	$2E_0^2 + (E_\omega^2 + E_\rho^2)$	$2E_0[2E_0^2 + 3(E_\omega^2 + E_\rho^2)]$	$[8E_0^4 + 24E_0^2(E_\omega^2 + E_\rho^2) + 3(E_\omega^4 + E_\rho^4) + 12E_\omega^2E_\rho^2]$	$2E_0[8E_0^4 + 40E_0^2(E_\omega^2 + E_\rho^2) + 15(E_\omega^4 + E_\rho^4) + 60E_\omega^2E_\rho^2)]$
$\cos \omega$	...	E_ω	$4E_0E_\omega$	$3E_\omega[4E_0^2 + (E_\omega^2 + 2E_\rho^2)]$	$8E_0E_\omega[4E_0^2 + 3(E_\omega^2 + 2E_\rho^2)]$	$10E_\omega[8E_0^4 + 12E_0^2(E_\omega^2 + 2E_\rho^2) + (E_\omega^4 + 3E_\rho^4) + 6E_\omega^2E_\rho^2]$
$\cos 2\omega$	...	...	E_ω^2	$6E_0E_\omega^2$	$4E_\omega^2[6E_0^2 + (E_\omega^2 + 3E_\rho^2)]$	$40E_0E_\omega^2[2E_0^2 + (E_\omega + 3E_\rho^2)]$
$\cos 3\omega$	...	...	...	E_ω^3	$8E_0E_\omega^3$	$5E_\omega^3[8E_0^2 + (E_\omega^2 + 4E_\rho^2)]$
$\cos 4\omega$	...	...	...	...	E_ω^4	$10E_0E_\omega^4$
$\cos 5\omega$	...	...	...	...	...	E_ω^5
$\cos \rho$	...	E_ρ	$4E_0E_\rho$	$3E_\rho[4E_0^2 + (2E_\omega^2 + E_\rho^2)]$	$8E_0E_\rho[4E_0^2 + 3(2E_\omega^2 + E_\rho^2)]$	$10E_\rho[8E_0^4 + 12E_0^2(2E_\omega^2 + E_\rho^2) + (3E_\omega^4 + E_\rho^4) + 6E_\omega^2E_\rho^2]$
$\cos (\rho \pm \omega)$	...	...	$2E_\omega E_\rho$	$12E_0E_\omega E_\rho$	$12E_\omega E_\rho[4E_0^2 + (E_\omega^2 + E_\rho^2)]$	$40E_0E_\omega E_\rho[4E_0^2 + 3(E_\omega^2 + E_\rho^2)]$
$\cos (\rho \pm 2\omega)$	...	...	...	$3E_\omega^2E_\rho$	$24E_0E_\omega^2E_\rho$	$10E_\omega^2E_\rho[12E_0^2 + (2E_\omega^2 + 3E_\rho^2)]$
$\cos (\rho \pm 3\omega)$	...	...	...	...	$4E_\omega^3E_\rho$	$40E_0E_\omega^3E_\rho$
$\cos (\rho \pm 4\omega)$	...	...	...	...	...	$5E_\omega^4E_\rho$
$\cos 2\rho$	...	...	E_ρ^2	$6E_0E_\rho^2$	$4E_\rho^2[6E_0^2 + (3E_\omega^2 + E_\rho^2)]$	$40E_0E_\rho^2[2E_0^2 + (3E_\omega^2 + E_\rho^2)]$
$\cos (2\rho \pm \omega)$	...	...	...	$3E_\omega E_\rho^2$	$24E_0E_\omega E_\rho^2$	$10E_\omega E_\rho^2[12E_0^2 + (3E_\omega^2 + 2E_\rho^2)]$
$\cos (2\rho \pm 2\omega)$	...	...	...	...	$6E_\omega^2E_\rho^2$	$60E_0E_\omega^2E_\rho^2$
$\cos (2\rho \pm 3\omega)$	...	...	...	...	...	$10E_\omega^3E_\rho^2$
$\cos 3\rho$	...	...	...	E_ρ^3	$8E_0E_\rho^3$	$5E_\rho^3[8E_0^2 + (4E_\omega^2 + E_\rho^2)]$
$\cos (3\rho \pm \omega)$	...	...	...	...	$4E_\omega E_\rho^3$	$40E_0E_\omega E_\rho^3$
$\cos (3\rho \pm 2\omega)$	...	...	...	...	...	$10E_\omega^3E_\rho^2$
$\cos 4\rho$	...	...	...	...	E_ρ^4	$10E_0E_\rho^4$
$\cos (4\rho \pm \omega)$	...	...	...	...	...	$5E_\omega E_\rho^4$
$\cos 5\rho$	...	...	...	...	...	E_ρ^5

TABLE 10-6 RESULTS FROM TABLE 10-5 FOR THE CASE $E_0 = E_\rho = 0$

Output product	Series constants								
	a_0	a_1	$a_2/2$	$a_3/4$	$a_4/8$	$a_5/16$	$a_6/32$	$a_7/64$	$a_8/128$
d-c.......	1	..	E_ω^2	.....	$3E_\omega^4$	.....	$10E_\omega^6$	.....	$35E_\omega^8$
cos ω.....	..	E_ω	...	$3E_\omega^3$	.....	$10E_\omega^5$	.....	$35E_\omega^7$	
cos 2ω....	..	..	E_ω^2	.....	$4E_\omega^4$	.....	$15E_\omega^6$	.....	$56E_\omega^8$
cos 3ω....	..	..	...	E_ω^3	.....	$5E_\omega^5$	.....	$21E_\omega^7$	
cos 4ω....	..	..	...	.....	E_ω^4	.....	$6E_\omega^6$	.....	$28E_\omega^8$
cos 5ω....	..	..	...	.....	.....	E_ω^5	.....	$7E_\omega^7$	
cos 6ω....	..	..	...	.....	.....	.....	E_ω^6	.....	$8E_\omega^8$
cos 7ω....	..	..	...	.....	.....	.....	.....	E_ω^7	
cos 8ω....	..	..	...	.....	.....	.....	.....	.....	E_ω^8

TABLE 10-7 RESULTS FROM TABLE 10-5 FOR THE CASE $E_\omega = E_\rho = E$, $E_0 = 0$

Output product	Series constants						
	a_0	a_1	$a_2/2$	$a_3/4$	$a_4/8$	$a_5/16$	$a_6/32$
d-c...............	1	..	$2E^2$	....	$18E^4$	....	$190E^6$
cos ω.............	..	E	....	$9E^3$	....	$100E^5$	
cos 2ω............	..	..	E^2	....	$16E^4$	....	$225E^6$
cos 3ω............	..	..	....	E^3	....	$25E^5$	
cos 4ω............	..	..	....	....	E^4	....	$36E^6$
cos 5ω............	..	..	....	....	....	E^5	
cos 6ω............	..	..	....	....	....	....	E^6
cos $(\rho \pm \omega)$.......	..	..	$2E^2$	....	$24E^4$	....	$300E^6$
cos $(\rho \pm 2\omega)$......	..	..	....	$3E^3$	....	$50E^5$	
cos $(2\rho \pm 2\omega)$.....	..	..	....	....	$6E^4$	....	$60E^6$
cos $(\rho \pm 3\omega)$......	..	..	....	....	$4E^4$	....	$90E^6$
cos $(3\rho \pm 3\omega)$.....	..	..	....	....	....	....	$20E^6$

For example, considering the single-frequency case shown in Table 10-6, the term a_6 will produce 15 per cent second harmonic and 6 per cent fourth harmonic if it is large enough to produce only 1 per cent sixth harmonic. It follows that an exceedingly sharp discontinuity must be introduced into the input-output characteristic to raise the values of the higher-order factors a_3, a_4, a_5, etc., sufficiently to increase the output of high-order harmonics.

It should be noted that it is customary to establish a single number to represent the over-all harmonic output in the case of a single-sine-wave input. This is the *distortion factor*, defined as the ratio of the root-mean-square voltage of the combined harmonics to the fundamental output voltage, or

$$\text{Distortion factor} = \frac{\sqrt{|E_2|^2 + |E_3|^2 + \cdots + |E_n|^2}}{|E_1|} \times 100 \qquad \text{per cent} \tag{10-38}$$

where E_2, E_3, etc., indicate the absolute peak values of the harmonic components, and E_1 is the absolute peak value of the fundamental.

The following conclusions apply in the case of *two* sine- or cosine-wave inputs:

(*a*) The amplitudes of the intermodulation products $\cos(\rho \pm \omega)$, $\cos(\rho \pm 2\omega)$, etc., caused by the nth term in the input-output characteristic are much larger than the amplitude of the nth harmonic. For example, considering Table 10-7, if $E_\omega = E_\rho$ and the term containing a_6 produces 0.1 per cent of sixth harmonic, it will also produce 30 per cent $\cos(\rho \pm \omega)$, 6 per cent $\cos(2\rho + 2\omega)$, 9 per cent $\cos(\rho \pm 3\omega)$, and 2 per cent $\cos(3\rho \pm 3\omega)$.

(*b*) If the device to which these signals are applied is a mixer (such as is used in the first detector circuit of a superheterodyne) designed to provide an output of $(\rho \pm \omega)$ only, other unwanted components can be eliminated by (1) using a characteristic as nearly square-law as possible (all factors above a_2 are made zero), (2) reducing the amplitudes of both input signals.

(*c*) When a second input E_ω is added to a first input E_ρ, the output of the lower harmonics of E_ρ is greatly altered if the higher-order terms are of significant magnitude. For example, it can be seen from Tables 10-6 and 10-7 that, if the term containing a_6 originally produced 0.1 per cent sixth harmonic ($\cos 6\rho t$) when only E_ρ was applied, the addition of the second signal E_ω equal in amplitude to E_ρ will create an additional 21 per cent of second harmonic ($\cos 2\rho t$) and 3 per cent of fourth harmonic ($\cos 4\rho t$).

In conclusion, it can be seen that for both the single- and multiple-sine-or cosine-wave inputs, *the d-c and even harmonic components of the output are determined by the even-order coefficients in the power series. The fundamental and odd-harmonic components of the output are determined by the odd-order coefficients in the power series.* This means that the even-order harmonics can be suppressed by balancing out the even-order coefficients in the series, as in the push-pull amplifier; the fundamental and odd-order harmonics can be suppressed by balancing out the odd-order coefficients in the series, as in the push-push or balanced-to-unbalanced "amplifier."

10-4 Intermodulation Testing

The same nonlinearity of input-output, or transfer, characteristic which causes harmonic distortion in a system will also cause intermodulation distortion. This consists of the spurious sum and difference products created during the transmission of two or more signals of different frequencies. If two tones of equal amplitude $E_\rho = E_\omega$ are transmitted through a system, the intermodulation products, as computed from Table 10-7, will be as shown in Table 10-8. When these are compared with the simple-harmonic products produced by the same tones, it can be seen that the measurement of the intermodulation products is a more sensitive test for nonlinearity, because these components are larger in amplitude in every case. Comparison will also show that the measurement of the $\rho \pm \omega$ and the $\rho \pm 2\omega$ components will show up high-order distortion products much better than the conventional test for second and third harmonic with a single input.

It is clear that if the first-order sum and difference frequencies only are measured, this will not take into account the distorting effect of the odd power terms in the input-output characteristic. This test would show, for example, zero distortion for a well-balanced push-pull amplifier, so that the second-order ($\rho \pm 2\omega$) components must be taken into account.

The exact method by which intermodulation testing is performed is as follows. A two-frequency signal is applied to the device under test, as shown in Fig. 10-1, and the output is fed to an analyzer which consists of a high-pass filter which eliminates the low-frequency component and

TABLE 10-8 HARMONIC AND INTERMODULATION PRODUCTS FOR $E_\rho = E_\omega$

Contribution due to:	Harmonics		Sum and difference frequencies	
	2ω	3ω	$\rho \pm \omega$	$\rho \pm 2\omega$
a_2	$\frac{a_2E^2}{2}$	....	a_2E^2	....
a_3	.....	$\frac{a_3E^3}{4}$	......	$\frac{3}{4}(a_3E^3)$
a_4	$2a_4E^4$	....	$3a_4E^4$	....
a_5	.....	$(\frac{25}{16})a_5E^5$	......	$(\frac{50}{16})a_5E^5$

passes only the high frequency and the sum and difference frequencies which are generated if the device is nonlinear. The passed signal is rectified (detected) and measured, the carrier (E_ρ) is filtered out, and the modulation (sidebands, or sum and difference frequencies) is measured and expressed as a percentage of the magnitude of the high-frequency carrier E_ρ. The frequencies indicated in Fig. 10-1 are not necessarily the only ones which may be selected. Values for the low-frequency signal of 40, 60, and 100 cps and for the high-frequency signal of 1,000, 7,000, and 12,000 cps have been used. Present practice makes the high-frequency signal 12 db lower in amplitude than the low-frequency signal (1:4 voltage ratio), and this is the scheme recommended by the Society

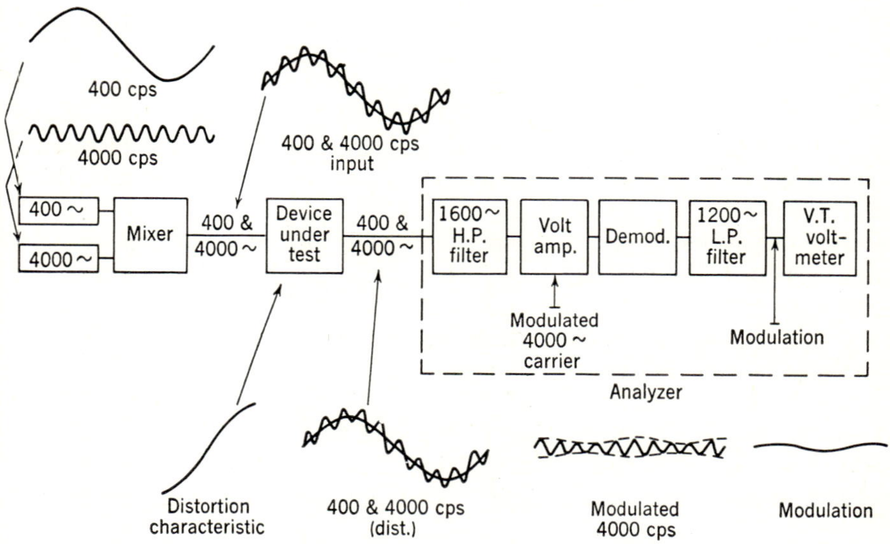

FIG. 10-1 Intermodulation test setup. (*After Roys.*)

of Motion Picture and Television Engineers (SMPTE). In Europe, the practice is to employ tones of equal intensity as recommended by the International Radio Consultative Committee (CCIR). The formal definition of intermodulation distortion is similar to that previously given for harmonic distortion:

$$\left.\begin{matrix}\text{rms intermodulation}\\ \text{distortion}\end{matrix}\right\} = \frac{\sqrt{\begin{array}{|c|}\Sigma\ (\text{sum and diff. freq.}\\ \text{voltages in output})^2\end{array}}}{\begin{array}{|c|}\text{fundamental output}\\ \text{voltage, hf signal}\end{array}} \times 100 \quad \% \qquad (10\text{-}39)$$

Warren and Hewlett[1] have analyzed the intermodulation distortion which occurs in both single-ended and push-pull amplifiers by means of series approximations for the transfer (input-output) characteristics and have determined that there are relatively fixed ratios for percentage intermodulation to percentage harmonic distortion. For the intermodu-

[1] W. J. Warren and W. R. Hewlett, An Analysis of the Intermodulation Method of Distortion Measurement, *Proc. I.R.E.*, **36** (4): 457–466, April, 1948.

lation test method, this ratio is about 3.2:1 for a single-ended amplifier and about 3.8:1 for a balanced push-pull amplifier. This paper also contains a useful table which permits direct computation of the fundamental and second-harmonic carrier terms and all prominent sideband terms directly from current values taken from an amplifier load line for corresponding, equal intervals of total peak-signal voltage. The intermodulation test procedures peculiar to film recording and phonograph recording have been discussed in detail by Frayne and Scoville,[1] and Roys.[2]

It is worthwhile to note, in conclusion, that the performance of various electron-tube types can be analyzed very quickly by using a series approximation for the plate-current characteristic, which permits rapid estimates of the performance of a particular tube as an amplifier, modulator, converter, or rectifier from a brief study of the transconductance (g_m) vs. control-grid-voltage characteristic curves.[3]

Problems

Prob. 10-1 The gamma exponent of a video system can be defined as the slope of the input-output characteristic when plotted on log-log paper. Determine the gamma exponent for the following systems:

Input	System 1 output	System 2 output	System 3 output	System 4 output	System 5 output
0.0010	0.002	0.06	0.81	2.00	1.24
0.0050	0.005	0.19	0.90	2.21	1.35
0.0025	0.010	0.61	1.00	2.42	1.45
0.0120	0.021	1.80	1.10	2.66	1.57
0.0630	0.045	5.50	1.21	2.93	1.64
0.3130	0.110	....	1.32	3.22	1.73
1.5650	0.230	....	1.44	....	1.76

Prob. 10-2 Construct a difference table for the following functions:

(a) $$y = 1.0 + 2.5x + 0.1x^2 + 10x^3$$

(b) $$y = 2.5x + 4x^3$$

(c) $$y = 0.2 + 0.3x - 0.4x^2 + 0.5x^3 + x^4$$

[1] J. G. Frayne and R. R. Scoville, Analysis and Measurements of Distortion in Variable-density Recording, *J. Soc. Motion Picture Television Engrs.*, **32**: 648, June, 1939.

[2] H. E. Roys, Analysis by the Two-frequency Intermodulation Method of Tracing Distortion Encountered in Phonograph Reproduction, *RCA Rev.*, **10** (2): 254–269, June, 1949.

[3] T. Roddam, How to Choose a Valve, *Wireless World*, **28** (6): 222–223, June, 1951, and **29** (10): 409–411, October, 1952; T. Slonczewski, Transconductance as a Criterion of Electron Tube Performance, *Bell System Tech. J.*, **28** (2): 315–328, April, 1949.

Prob. 10-3 Determine the order of polynomial which will best fit the following input-output data:

Input	System 1 output	System 2 output	System 3 output	System 4 output
1	15.0	3.1	2.1	3.20
2	18.0	3.8	2.8	3.07
3	21.5	6.9	3.6	3.10
4	25.5	12.6	4.6	3.70
5	29.9	20.9	5.8	5.50
6	34.8	31.5	7.5	9.00
7	....	....	9.5	15.00
8	....	....	12.0	24.10

Prob. 10-4 Obtain the tube characteristics for the type 6AL5 twin diode. Replot the rectified output current vs. the rms signal input voltage for load resistances of 2,000, 3,000, 4,000, 5,000, 10,000, 25,000, and 100,000 ohms. Develop equations for these input-output characteristics. In each case, develop a separate expression for the input voltage range 0 to 10 volts rms and for the over-all range of 0 to 30 volts rms.

Prob. 10-5 Evaluate the constants in the polynomial forms obtained in Prob. 10-3 by method A of Sec. 10-2.

Prob. 10-6 Fit the following input-output data by a linear equation using the method of least squares. Plot both the data and the resulting equation:

x	y
0	1.97
1	1.14
2	1.57
3	2.18
4	3.25
5	3.67
6	3.92
7	2.00
8	2.87
9	3.69

Prob. 10-7 Fit the following input-output data by the method of least squares, using an equation type determined from a difference table:

x	y
0	2
1	1.7
2	1.8
3	2.3
4	3.2

Prob. 10-8 (*a*) Construct load lines for plate load resistances of 2,500, 25,000, and 250,000 ohms for the type 6J5 triode operated with a plate supply voltage of 300 volts.

(*b*) Tabulate instantaneous plate voltage e_p vs. grid voltage e_g for these values of load resistance.

(*c*) Obtain series expressions for the input-output characteristics in (*b*).

(*d*) Determine the output components obtained when the following grid voltages are applied:

1. $E_0 = -4$ volts
2. $E_0 = -6$ volts
3. $E_0 = -8$ volts
4. $E_0 = -4$ volts; $E_\omega = 2$ volts peak at 400 cps
5. $E_0 = -8$ volts; $E_\omega = 2$ volts peak at 400 cps
6. $E_0 = -8$ volts; $E_\omega = 6$ volts peak at 400 cps
7. $E_0 = -6$ volts; $E_\omega = 2$ volts peak at 400 cps
   $E_\rho = 2$ volts peak at 4,000 cps

(*e*) Compute the distortion factor in 4, 5, 6, and 7.

Prob. 10-9 Establish from Table 10-5 that the maximum value of the eighth harmonic is not likely to exceed about 0.8 per cent for any characteristic in which the term a_8E^8 is equal to or less than the term a_1E. Find the corresponding maxima for the seventh, sixth, and fifth harmonics.

Prob. 10-10 In a balanced modulator shown in Fig. 10-2, the modulating voltage E_ω is applied at a point *a*, and the carrier voltage E_ρ at point *b*. What will be the nature of the output at *c* for the frequencies ρ and $\rho \pm \omega$?

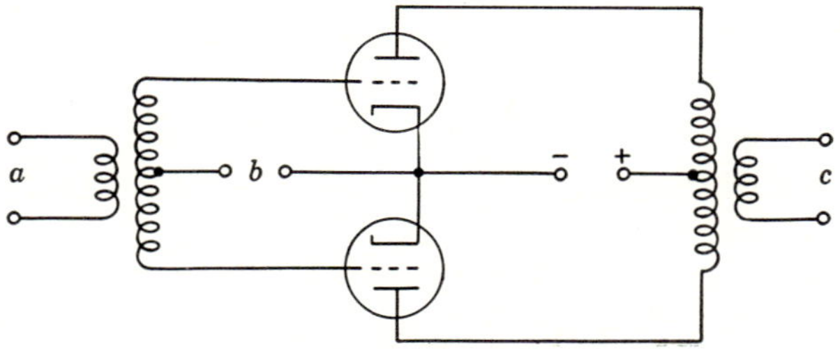

FIG. 10-2 Balanced modulator.

Prob. 10-11* In an electronic wattmeter circuit, a carrier is modulated with the current wave form under measurement and is then again modulated with the voltage wave form under measurement. The result is demodulated to yield the instantaneous voltage-and-current product. After the carrier has been modulated with the current wave form i, it has the form $i \cos \omega t$. This signal is applied in combination with the

* D. E. Garrett and F. G. Cole, A General-purpose Electronic Wattmeter, *Proc. I.R.E.*, **40** (2): 165–171, February, 1952.

voltage wave form e to two type 6BA6 variable-gain pentodes connected so that their plate currents subtract and their grid voltages are

$$e_{g_1} = i \cos \omega t + e \tag{1}$$

$$e_{g_2} = i \cos \omega t - e \tag{2}$$

respectively. If the plate current of each of these tubes is representable as a power series in e_g which contains at least four terms, show that, by substituting (1) and (2) in the series and subtracting, the circuit will yield a carrier-frequency term proportional to the instantaneous product ie.

Prob. 10-12 The Hewlett-Packard wideband amplifier, Model 460-B, has a gain of 15 db when operated with a 200-ohm load. From the data furnished for this amplifier on page 43 of the 1952 Catalog No. 21-A, the approximate input-output characteristic is:

Volts in	*Volts out*	*Volts in*	*Volts out*
0	0	5	68.0
1	7.5	6	87.0
2	18.0	7	108.0
3	33.0	8	132.0
4	48.0		

Evaluate all intermodulation products resulting when the following composite signal is applied:

$$E_\omega = 4 \text{ volts peak} \qquad E_\rho = 1 \text{ volt peak}$$

$$\frac{\omega}{2\pi} = 100 \text{ cps} \qquad \frac{\rho}{2\pi} = 12{,}000 \text{ cps}$$

Prob. 10-13 (*a*) The input-output characteristic of an amplifier is given as[1]

$$y = 250x + 2x^2$$

Determine the harmonic distortion which results when a 400-cps signal of 5 volts peak amplitude is applied to the unit.

Determine the intermodulation distortion when the applied signal is 4 volts peak at 400 cps and 1 volt peak at 4,000 cps.

(*b*) The input-output characteristic of a push-pull amplifier is given as

$$y = 250x - 2x^3$$

Determine the harmonic distortion when the applied signal is 5 volts peak at 400 cps.

Determine the intermodulation distortion when the applied signal is 4 volts peak at 400 cps and 1 volt peak at 4,000 cps.

[1] E. W. Berth-Jones, Intermodulation Testing, *Wireless World*, **28** (6): 233–236, June, 1951.

(*c*) Sketch a graph of the above input-output characteristics and note the differences in shape.

Prob. 10-14 The approximate modulation characteristic of the Hewlett-Packard Model 490-A traveling-wave tube amplifier, which operates between 2 and 4 kilomegacycles, is given as follows:[1]

Grid bias, volts	*Relative voltage output*
−25	0.0294
−20	0.0765
−15	0.1470
−10	0.2647
− 5	0.5293
0	0.9880

(*a*) Compute an input-output function for this modulation characteristic.

(*b*) For a sine-wave modulating voltage of 10 volts peak to peak at 200 mc determine the complete set of distortion products resulting when the modulator operating point is set at −20, −15, −10, and −5 volts.

Prob. 10-15 The input-output characteristic of the typical power-amplifier stage is of the "concave-downward" variety, so that at the overload point the output fails to increase with increasing input. On the other hand, the input-output characteristic of a cathode-follower is "concave upward," so that it would be feasible to use a cathode-follower in cascade with an amplifier for the purpose of linearizing the over-all transfer characteristic and reducing distortion.[2]

(*a*) Given the transfer characteristic of an amplifier stage as

$$y_a = a_1x_a + a_2x_a^2$$

and that of a cathode-follower stage as $y_c = b_1x_c + b_2x_c^2$, where y_a is the amplifier output signal, x_a is the amplifier input signal, etc. What must be the relative polarities of a_2 and b_2 for the over-all characteristic to be linear, if a_1 and b_1 are assumed positive and the amplifier and the cathode-follower are connected in cascade so that $x_c = y_a$?

(*b*) Determine the value of b_2 necessary to eliminate any downward or upward curvature in the over-all characteristic.

(*c*) Determine the new over-all transfer characteristic for the value of b_2 obtained in (*b*).

(*d*) If, for a particular case, $a_1 = b_1$ and $a_2 = 0.052$, plot the amplifier transfer characteristic, the cathode-follower transfer characteristic, and the new over-all transfer characteristic of the two stages in tandem or cascade for the most linear condition as obtained in (*c*).

[1] *Hewlett-Packard J.*, **6** (3–4): November-December, 1954.

[2] T. J. Schultz, Triode Cathode-followers for Impedance Matching to Transformers and Filters, *I.R.E. Trans. on Audio*, **3** (2): 37, March-April, 1955.

Solid graph of frequency-intensity variations in the spoken word "five." (*Bell Telephone Laboratories.*)

PART 5

Differential Equations: Theory and Applications

Chapter 11

CLASSICAL SOLUTIONS AND NON-NETWORK APPLICATIONS

11-1 Definitions and Nomenclature

A *differential equation* is an equation which contains a derivative (or differential).

The equation

$$R\frac{dq}{dt} + \frac{q}{c} = V \sin \omega t \tag{11-1}$$

taken from circuit theory, illustrates a differential equation.

An *ordinary* differential equation is a differential equation which contains only ordinary derivatives (or differentials), i.e., it is a function of only one independent variable, e.g.,

$$R\frac{dI}{dt} + \frac{1}{c}I = \frac{dE}{dt} \tag{11-2}$$

A *partial* differential equation is one that contains a partial derivative, i.e., it is a function of several variables. The simultaneous equations from transmission-line theory,

$$\begin{aligned} L\frac{\partial I}{\partial t} + RI &= \frac{-\partial V}{\partial x} \\ C\frac{\partial V}{\partial t} + GV &= \frac{-\partial I}{\partial x} \end{aligned} \tag{11-3}$$

are examples of partial differential equations.

The *order* of a differential equation is defined as equal to the highest-order derivative that appears in the equation, e.g., the order of the differential equation

$$R\frac{d^2q}{dt^2} + \frac{1}{C}\frac{dq}{dt} = \frac{dE}{dt} \tag{11-4}$$

is equal to *two*, since d^2q/dt^2 is the highest-order derivative that appears. The number of arbitrary constants of integration which must appear in the solution of a differential equation is equal to the order of the equation.

The *degree* of a differential equation is equal to the power to which the highest-order derivative is raised. (It is understood that the equation has been cleared of all fractional exponents.) The degree of Eq. (11-4) is *one*, since d^2q/dt^2, which is the highest-order derivative that is present in the equation, is raised to the *first* power.

11-2 Verification of a Solution of a Differential Equation

The most common method in use is to substitute the expression which has been obtained for a solution of the differential equation into the original equation and obtain, if the solution is a correct one, an identity.

EXAMPLE 1. Suppose that a solution

$$y = C_1 \cos \omega t + C_2 \sin \omega t \tag{11-5}$$

is obtained for the differential equation

$$\frac{d^2y}{dt^2} + \omega^2 y = 0 \tag{11-6}$$

Substituting (11-5) in (11-6), there results

$$(-C_1\omega^2 \cos \omega t - C_2\omega^2 \sin \omega t) + \omega^2(C_1 \cos \omega t + C_2 \sin \omega t) = 0 \tag{11-7}$$
$$0 = 0$$

Since (11-7) is an identical type of equation, it follows that (11-5) satisfies the given equation (11-6) and hence is a solution of it.

EXAMPLE 2. Show that

$$x = (c_1 + c_2t)\epsilon^{-t} + 3 \tag{11-8}$$

is a solution of

$$\frac{d^2x}{dt^2} + 2\frac{dx}{dt} + x = 3 \tag{11-9}$$

Since $x = (c_1 + c_2t)\epsilon^{-t} + 3$, it follows that

$$\frac{dx}{dt} = (c_1 + c_2t)(-\epsilon^{-t}) + \epsilon^{-t}(c_2) = -(c_1 + c_2t)\epsilon^{-t} + c_2\epsilon^{-t}$$
$$= (-c_1 + c_2 - c_2t)\epsilon^{-t}$$
$$\frac{d^2x}{dt^2} = -(-c_1 + c_2 - c_2t)\epsilon^{-t} + \epsilon^{-t}(-c_2) = (c_1 - c_2 + c_2t - c_2)\epsilon^{-t}$$

Upon substitution in (11-9), there is obtained

$$(c_1 - c_2 + c_2t - c_2)\epsilon^{-t} + 2(-c_1 + c_2 - c_2t)\epsilon^{-t}(c_1 + c_2t)\epsilon^{-t} + 3 = 3$$
$$(c_1 - c_2 + c_2t - c_2 - 2c_1 + 2c_2 - 2c_2t + c_1 + c_2t)\epsilon^{-t} + 3 = 3$$

Therefore $$3 = 3$$

and (11-8) is a solution of (11-9).

EXAMPLE 3. Verify that

$$y\epsilon^{-2y} - \epsilon^{x} = 0 \tag{11-10}$$

satisfies the equation

$$\frac{dy}{dx} = \frac{y}{1 - 2y} \tag{11-11}$$

Since $y\epsilon^{-2y} - \epsilon^{x} = 0$, it is seen that

$$y = \epsilon^{x+2y} \qquad \text{or} \qquad \ln y = x + 2y$$

which gives on differentiation

$$\frac{1}{y}\frac{dy}{dx} = 1 + 2\frac{dy}{dx}$$

This reduces to

$$\left(\frac{1}{y} - 2\right)\frac{dy}{dx} = 1 \qquad \text{or} \qquad \frac{dy}{dx} = \frac{y}{1 - 2y}$$

Therefore (11-10) satisfies (11-11).

11-3 First-order Differential Equations with Variables Separable

There are numerous differential equations of first order which can be reduced to a form

$$F(x)\,dx + G(y)\,dy = 0 \tag{11-12}$$

merely by the application of algebraic operations. Such equations are said to be first-order differential equations with variables separable.

EXAMPLE. Given the differential equation

$$\epsilon^{y-x}\,dy - x\,dx = 0 \tag{11-13}$$

it is required to find a solution which satisfies this equation.

Equation (11-13) may be written as

$$\frac{\epsilon^{y}}{\epsilon^{x}}\,dy = x\,dx \tag{11-14}$$

Therefore

$$\epsilon^{y}\,dy = x\epsilon^{x}\,dx \tag{11-15}$$

and

$$\epsilon^{y} = \epsilon^{x}(x - 1) + c_1 \tag{11-16}$$

which reduces to equation

$$e^{y/x} - x = c_1 e^{-x} - 1 \tag{11-17}$$

11-4 First-order Differential Equations (Homogeneous)

Consider the differential equation

$$M(x,y)\,dx + N(x,y)\,dy = 0 \tag{11-18}$$

where the functions $M(x,y)$ and $N(x,y)$ are of such a nature that, if λx is substituted for x and λy for y, then

$$M(\lambda x,\lambda y) = \lambda^{r} M(x,y) \qquad \text{and} \qquad N(\lambda x,\lambda y) = \lambda^{r} N(x,y)$$

Under these conditions, $M(x,y)$ and $N(x,y)$ are *homogeneous functions* of the same degree r. For differential equations similar to (11-18) with the condition that $M(x,y)$ and $N(x,y)$ are homogeneous to the same degree, it has been found that the substitution of $y = vx$ (or $x = wy$) will permit the equation to become one which is readily separable by algebraic means. This procedure will reduce Eq. (11-18) to the type considered in Sec. 11-2.

EXAMPLE. Solve the differential equation

$$(x + y)\,dx + (y - x)\,dy = 0 \tag{11-19}$$

It follows that if

$$M(x,y) = x + y \qquad \text{and} \qquad N(x,y) = y - x$$

then
$$M(\lambda x,\lambda y) = \lambda(x + y) = \lambda M(x,y)$$
and
$$N(\lambda x,\lambda y) = \lambda(y - x) = \lambda N(x,y)$$

Since $M(x,y)$ and $N(x,y)$ are both homogeneous functions of the same degree, namely, *one*, it follows that for the substitution

$$y = vx \tag{11-20}$$

$$\frac{dy}{dx} = -\frac{x + y}{y - x} = \frac{x + y}{x - y} \tag{11-21}$$

or
$$v + x\frac{dv}{dx} = \frac{1 + v}{1 - v} \tag{11-22}$$

which becomes

$$x\frac{dv}{dx} = \frac{1 + v}{1 - v} - v = \frac{1 + v - v + v^2}{1 - v} \tag{11-23}$$

and reduces to the equation

$$x\frac{dv}{dx} = \frac{1 + v^2}{1 - v} \tag{11-24}$$

Equation (11-24) becomes

$$\frac{dx}{x} = \frac{dv}{1 + v^2} - \frac{v\,dv}{1 + v^2} \tag{11-25}$$

which integrates into the equation

$$\ln x = \tan^{-1} v - \tfrac{1}{2}\ln(1 + v^2) + \ln c \tag{11-26}$$

where $\ln c$ is an arbitrary constant. Resubstituting $v = y/x$, Eq. (11-26) may be reduced to the form

$$\ln\left(\frac{x^2 + y^2}{k}\right) = 2\tan^{-1}\frac{y}{x} \tag{11-27}$$

which is the desired solution of Eq. (11-19), where k is equal to $\ln\sqrt{c}$.

11-5 Linear First-order Differential Equations

A linear first-order differential equation

$$f\left(x,y,\frac{dy}{dx}\right) = 0 \tag{11-28}$$

is defined as one in which the dependent variable y and its derivative dy/dx are of the first degree. It may be represented in the form

$$\frac{dy}{dx} + Py = Q \tag{11-29}$$

where P and Q are functions of the independent variable x.

This type of equation has a solution of the form

$$y\epsilon^{\int P dx} = \int Q\epsilon^{\int P dx}\, dx + C$$

or

$$y = \epsilon^{-\int P dx} \int Q\epsilon^{\int P dx}\, dx + C\epsilon^{-\int P dx} \tag{11-30}$$

EXAMPLE. Solve the differential equation

$$\frac{dy}{dx} - 3y = \epsilon^{4x} \tag{11-31}$$

By comparing this equation with (11-29), $P = -3$ and $Q = \epsilon^{4x}$. By direct substitution into (11-30)

$$y = \epsilon^{-\int -3dx} \int \epsilon^{4x}\epsilon^{\int -3dx}\, dx + c_1\epsilon^{-\int -3dx} \tag{11-32}$$

$$y = \epsilon^{3x} \int \epsilon^{x}\, dx + c_1\epsilon^{3x} \tag{11-33}$$

which readily reduces to

$$y = \epsilon^{4x} + c_1\epsilon^{3x} \tag{11-34}$$

11-6 D Operators

To expedite the solution of linear differential equations with constant coefficients, the use of D operators is essential.

The symbol D is called a *differential operator*. It represents the operation of differentiation with respect to the independent variable x and may be defined as the following identity.

$$D \equiv \frac{d}{dx} \tag{11-35}$$

This symbol, when extended to indicate higher-order derivatives, becomes

$$D^2 \equiv \frac{d^2}{dx^2} \qquad D^3 \equiv \frac{d^3}{dx^3} \qquad D^n \equiv \frac{d^n}{dx^n} \tag{11-36}$$

It must be emphasized that the D operator is *not* an entity, but merely the *symbol* which when written on the left of the function denotes that the function is to be operated upon by the D operator.

EXAMPLE 1

$$D(5x^2) = 10x \tag{11-37}$$

since

$$\frac{d(5x^2)}{dx} = 10x$$

EXAMPLE 2

$$D(3x\epsilon^{-x}) = 3\epsilon^{-x}(1 - x) \tag{11-38}$$

EXAMPLE 3

$$D^2(\cos 2x) = -4 \cos 2x \tag{11-39}$$

D operators are not algebraic entities. It is shown in tabular form (Table 11-1) that D operators nevertheless have numerous properties attributed to algebraic quantities. These properties are isomorphic, so that their behavior is readily predictable from experience with the laws that govern algebraic entities.

Two extremely important theorems concerning D operators may be stated as follows:

Theorem 1

$$\phi(D)\epsilon^{mx}f(x) = \epsilon^{mx}\phi(D + m)f(x) \tag{11-40}$$

where (D) *is defined as*

$$\phi(D) = a_0D^n + a_1D^{n-1} + \cdots + a_n$$

for positive integral values of n.

Theorem 2

$$(D - m)^n[\epsilon^{mx}f(x)] = \epsilon^{mx}D^nf(x) \tag{11-41}$$

TABLE 11-1 ISOMORPHIC PROPERTIES BETWEEN ALGEBRAIC QUANTITIES AND D OPERATORS

ADDITION

	Algebraic quantities	D operators
Commutative law..	$a + b = b + a$	$(D^m + D^n)f(x) = (D^n + D^m)f(x)$
Associative law....	$a + (b + c) = (a + b) + c$	$[D^m + (D^n + D^p)]f(x) = [(D^m + D^n) + D^p]f(x)$

MULTIPLICATION

	Algebraic quantities	D operators
Commutative law..	$ab = ba$	$(D^m \cdot D^n)f(x) = (D^n \cdot D^m)f(x)$
Associative law....	$a(bc) = (ab)c$	$(D^mD^n)f(x) = (D^n \cdot D^m)f(x)$
Distributive law...	$a(b + c) = ab + ac$	$D^m(D^n + D^p)f(x) = (D^mD^n + D^mD^p)f(x)$
Index law.........	$a^m \cdot a^n = a^{m+n}$	$(D^m \cdot D^n)f(x) = D^{m+n}f(x)$

EXAMPLE 4. By use of theorem 1, perform the indicated operations of differentiation.

$$(D^2 + D + 1)\epsilon^x \cos x \tag{11-42}$$

Here

$$\phi(D) = D^2 + D + 1 \qquad \text{and} \qquad f(x) = \cos x \tag{11-43}$$

Hence

$$\begin{aligned}\phi(D + m) = \phi(D + 1) &= D^2 + 2D + 1 + D + 1 + 1 \\ &= D^2 + 3D + 3\end{aligned} \tag{11-44}$$

$$\begin{aligned}(D^2 + D + 1)\epsilon^x \cos x &= \epsilon^x(D^2 + 3D + 3) \cos x \\ = \epsilon^x(-\cos x - 3 \sin x + 3 \cos x) &= \epsilon^x(2 \cos x - 3 \sin x)\end{aligned} \tag{11-45}$$

Example 5. By the application of theorem 2, reduce the function

$$(D - \ln 5)^7 5^x$$

Since

$$5^x = \epsilon^{\ln 5^x} = \epsilon^{x \ln 5} \tag{11-46}$$

it follows that

$$m = \ln 5 \qquad \phi(D) = (D - \ln 5)^7 \qquad f(x) = 1$$

and upon substitution in (11-40),

$$\phi(D)\epsilon^{mx}f(x) = \epsilon^{mx}\phi(D + m)f(x)$$

there is obtained

$$(D - \ln 5)^7\epsilon^{x \ln 5} = \epsilon^{x \ln 5}D^7(1) = 0 \tag{11-47}$$

11-7 Linear Differential Equation of Second Order with Constant Coefficients

A very common type of differential equation that is encountered in all branches of engineering and physics is the linear differential equation of second order with constant coefficients. For example, in electrical work, the elements resistance R, inductance L, and capacitance C are usually known fixed constants. The usual variable consists of charge q and its first and second derivatives, $dq/dt = i$ and $d^2q/dt^2 = di/dt$, respectively. In mechanics, the compliance k, the friction c, and the mass m are usually constant, and the variables are the displacement x and its first and second derivatives, $dx/dt = v$, the velocity, and $d^2x/dt^2 = dv/dt = a$, the acceleration. Therefore, the equation to be considered here will be of the form

$$a_0 \frac{d^2y}{dx^2} + a_1 \frac{dy}{dx} + a_2 y = f(x) \tag{11-48}$$

or

$$(a_0D^2 + a_1D + a_2)y = f(x) \tag{11-49}$$

where a_0, a_1, and a_2 are constants. In electricity and applied mathematics, these a's are *real* numbers. It is the assumption here that this is true, although the discussion holds for complex numbers.

It can be shown that the solution of an equation of this type will always be of the form

$$y = y_c + y_p \tag{11-50}$$

where y_c is the solution of the homogeneous equation corresponding to (11-49) [i.e., the complementary equation of (11-49)],

$$(a_0D^2 + a_1D + a_2)y_c = 0$$

and y_p is a particular integral of the identical equation

$$(a_0D^2 + a_1D + a_2)y_p = f(x) \tag{11-51}$$

Since the number of arbitrary constants of integration which are required in the solution of a second-order differential equation is equal to two, the function y_c will contribute these, and since a solution must satisfy the given equation (11-48), y_p will satisfy the $f(x)$ of the given equation; hence

$$y = y_c + y_p \tag{11-52}$$

will be a solution of the given equation.

Suppose that

$$y = \epsilon^{mx} \tag{11-53}$$

is a solution of

$$(a_0D^2 + a_1D + a_2)y = 0 \tag{11-54}$$

Then

$$(a_0m^2 + a_1m + a_2)\epsilon^{mx} = 0 \tag{11-55}$$

Since

$$\epsilon^{mx} \neq 0,$$

$$a_0m^2 + a_1m + a_2 = 0 \tag{11-56}$$

which is known as the auxiliary equation. This quadratic equation may be solved by the standard quadratic-equation formula, which will give for the roots m_1 and m_2 the respective roots:

$$m = \frac{-a_1 \pm \sqrt{a_1^2 - 4a_0a_2}}{2a_0} \tag{11-57}$$

or

$$m_1 = \frac{-a_1}{2a_0} + \frac{\sqrt{a_1^2 - 4a_0a_2}}{2a_0} \tag{11-58}$$

and

$$m_2 = \frac{-a_1}{2a_0} - \frac{\sqrt{a_1^2 - 4a_0a_2}}{2a_0} \tag{11-59}$$

The expressions $y = \epsilon^{m_1x}$ and $y = \epsilon^{m_2x}$ will satisfy the complementary equation (11-55). Hence the sum

$$y_c = c_1\epsilon^{m_1x} + c_2\epsilon^{m_2x} \tag{11-60}$$

is a complete solution for the complementary equation.

If it happens that $m_1 = m_2$, then, since any solution of the complementary equation must contain two arbitrary constants and satisfy the complementary equation, a different form for y_c in its simplest form is

$$y_c = (c_1 + c_2x)\epsilon^{m_1x} \tag{11-61}$$

If the assumption is made that a_0, a_1, and a_2 are real constants, which is the case in electric circuits, then if m_1 and m_2 are complex quantities it follows immediately that m_1 and m_2 are complex conjugates, representable as

$$m_1 = \alpha + j\beta \qquad \text{and} \qquad m_2 = \alpha - j\beta \tag{11-62}$$

where $\alpha = \dfrac{-a_1}{a_0}$

$$\beta = \frac{\sqrt{4a_0a_2 - a_1{}^2}}{2a_0}$$

When $m_1 = \alpha + j\beta$ and $m_2 = \alpha - j\beta$ are substituted into

$$y_c = c_1\epsilon^{m_1x} + c_2\epsilon^{m_2x} \tag{11-63}$$

then

$$y_c = c_1\epsilon^{(\alpha+j\beta)x} + c_2\epsilon^{(\alpha-j\beta)x}$$

$$= \epsilon^{\alpha x}(c_1\epsilon^{j\beta x} + c_2\epsilon^{-j\beta x}) \tag{11-64}$$

By Euler's relation, namely,

$$\epsilon^{j\theta} = \cos\theta + j\sin\theta \tag{11-65}$$

it follows that

$$y_c = \epsilon^{\alpha x}[c_1(\cos\beta x + j\sin\beta x) + c_2\cos(-\beta x) + j\sin(-\beta x)] \tag{11-66}$$

or

$$y_c = \epsilon^{\alpha x}[(c_1 + c_2)\cos\beta x + j(c_1 - c_2)\sin\beta x] \tag{11-67}$$

$$y_c = \epsilon^{\alpha x}(A\cos\beta x + B\sin\beta x) \tag{11-68}$$

In summary, there are three forms of solution for second-order homogeneous equations with constant coefficients, dependent upon the nature of the roots. These are

$$y_c = c_1\epsilon^{m_1x} + c_2\epsilon^{m_2x} \tag{11-60}$$

$$y_c = (c_1 + c_2x)\epsilon^{m_1x} \tag{11-61}$$

$$y_c = \epsilon^{\alpha x}(A\cos\beta x + B\sin\beta x) \tag{11-68}$$

EXAMPLE 1. Given the equation (which is nonhomogeneous)

$$(D^2 + 3D + 2)y = f(x)$$

find the complementary function y_c.

The auxiliary equation is

$$m^2 + 3m + 2 = 0$$

and $m_1 = -1$ and $m_2 = -2$. Hence

$$y_c = c_1\epsilon^{-x} + c_2\epsilon^{-2x}$$

EXAMPLE 2

$$(D^2 + 5D + 4)y = f(x)$$

The auxiliary equation becomes

$$m^2 + 5m + 4 = 0$$

and $m_1 = -1$ and $m_2 = -4$, so that y_c takes on the value

$$y_c = c_1\epsilon^{-x} + c_2\epsilon^{-4x}$$

EXAMPLE 3

$$(D^2 + 4D + 4)y = f(x)$$

The auxiliary equation is

$$m^2 + 4m + 4 = 0$$

and $m_1 = m_2 = -2$. The solution of the complementary equation becomes

$$y_c = (C_1 + C_2x)\epsilon^{-2x}$$

It is now necessary to investigate the form which y_p, the particular integral, may assume to satisfy the identical equation

$$(a_0D^2 + a_1D + a_2)y_p \equiv f(x) \tag{11-69}$$

If $f(x)$ is not contained in y_c, then assume for y_p a function which represents the sum of a function of the form of $f(x)$ plus the form of the first derivative of $f(x)$. If, however, $f(x)$ or some part of $f(x)$, is contained in y_c, then make the same assumption as to the form as in the case when $f(x)$ is not contained in y_c, except that in this case multiply the assumed form for y_p by the independent variable x raised to the least positive integral power not contained in y_c. In general, it will be found that satisfactory choices for the form of y_p are as follows:

When $f(x) = a$ (constant), $y_p = A$.
When $f(x) = ax^n$, $y_p = Ax^n + Bx^{n-1}$.
When $f(x) = a\epsilon^{\alpha x}$, $y_p = A\epsilon^{\alpha x}$.
When $f(x) = a \cos px$ and/or $a \sin px$, $y_p = A \cos px + B \sin px$.

The constants in y_p are then evaluated by the method of *undetermined coefficients*, which is best demonstrated by an example.

EXAMPLE 4. Find the complete solution of the differential equation

$$(D^2 + 3D + 2)y = 3\epsilon^x + 2 \sin x \tag{11-70}$$

The complementary equation

$$(D^2 + 3D + 2)y_c = 0 \tag{11-71}$$

will first be solved to give the complementary solution, which is

$$y_c = c_1\epsilon^{-x} + c_2\epsilon^{-2x} \tag{11-72}$$

Since it is not possible to find values for c_1 and c_2 which will make y_c or part of it identically equal to $3\epsilon^x$ or $2 \sin x$, then assume

$$y_p = A\epsilon^x + B \sin x + C \cos x \tag{11-73}$$

where A, B, and C are to be determined by the method mentioned above.

This method is known as the *method of undetermined coefficients.* This yields the *identical* equation

$$(D^2 + 3D + 2)y_p = 3\epsilon^x + 2 \sin x \tag{11-74}$$

whereas in Eq. (11-73),

$$y_p = A\epsilon^x + B \sin x + C \cos x$$

It follows that

$$\begin{aligned} 2y_p &= 2A\epsilon^x + 2B \sin x + 2C \cos x \\ 3Dy_p &= 3A\epsilon^x + 3B \cos x - 3C \sin x \\ D^2y_p &= A\epsilon^x - B \sin x - C \cos x \end{aligned} \tag{11-75}$$

which gives immediately the identity

$$3\epsilon^x + 2 \sin x \equiv 6A\epsilon^x + (B - 3C) \sin x + (3B + C) \cos x \tag{11-76}$$

This yields the simultaneous linear algebraic set of equations

$$\begin{aligned} 6A &= 3 \\ B - 3C &= 2 \\ 3B + C &= 0 \end{aligned} \tag{11-77}$$

Upon solving Eq. (11-77), the values for A, B, and C are $\frac{1}{2}$, $+\frac{1}{5}$, and $-\frac{3}{5}$, respectively. Hence

$$y_p = \tfrac{1}{2}\epsilon^x + \tfrac{1}{5} \sin x - \tfrac{3}{5} \cos x \tag{11-78}$$

The complete solution is then

$$y = y_c + y_p$$

or

$$y = c_1\epsilon^{-x} + c_2\epsilon^{-2x} + \tfrac{1}{2}\epsilon^x + \tfrac{1}{5} \sin x - \tfrac{3}{5} \cos x \tag{11-79}$$

EXAMPLE 5. A slight change in the differential equation may change the category of the particular solution, as in the equation

$$(D^2 + 3D + 2)y = 3\epsilon^{-x} + 2 \sin x \tag{11-80}$$

Again, as in the previous case, by Eq. (11-72),

$$y_c = c_1\epsilon^{-x} + c_2\epsilon^{-2x}$$

where c_1 and c_2 are arbitrary constants, but $3\epsilon^{-x}$ is already contained in y_c, since y_c becomes $3\epsilon^{-x}$ for $c_1 = 3$ and $c_2 = 0$. Therefore, for this part of the particular solution, assume $A'\epsilon^{-x}$ and multiply it by x as outlined above. Since $2 \sin x$ is not contained in y, then to $A'x\epsilon^{-x}$ there will be added the function

$$B' \cos x + C' \sin x$$

where A', B', and C' are constants that must be determined. Then

$$y_p = A'x\epsilon^{-x} + B' \cos x + C' \sin x \tag{11-81}$$

and

$$y_p' = A'(\epsilon^{-x} - x\epsilon^{-x}) - B' \sin x + C' \cos x \tag{11-82}$$

and

$$y_p'' = A'[-\epsilon^{-x} - (\epsilon^{-x} - x\epsilon^{-x})] - B' \cos x - C' \sin x \tag{11-83}$$

By setting up the identical relationship,

$$(D^2 + 3D + 2)y_p = 3\epsilon^{-x} + 2 \sin x$$

where y_p is defined as shown above in Eq. (11-80). It is then possible to determine the constants A', B', and C'.

$$\begin{aligned} D^2y_p &= -2A'\epsilon^{-x} + A'x\epsilon^{-x} - B' \cos x - C' \sin x \\ 3Dy_p &= 3A'\epsilon^{-x} - 3A'x\epsilon^{-x} + 3C' \cos x - 3B' \sin x \\ 2y_p &= + 2A'x\epsilon^{-x} + 2B' \cos x + 2C' \sin x \\ 3\epsilon^{-x} + 2 \sin x &= A'\epsilon^{-x} + (B' + 3C') \cos x \\ &\qquad + (C' - 3B') \sin x \end{aligned} \tag{11-84}$$

from which

$$A' = 3 \qquad B' + 3C' = 0 \qquad -3B' + C' = 2$$

or

$$A' = 3 \qquad B' = -\tfrac{3}{5} \qquad C' = \tfrac{1}{5}$$

A complete solution for Eq. (11-80) is given by

$$y = c_1\epsilon^{-x} + c_2\epsilon^{-2x} + 3x\epsilon^{-x} - \tfrac{3}{5} \cos x + \tfrac{1}{5} \sin x \tag{11-85}$$

An alternate method to the *undetermined-coefficient* approach of solving by ascertaining y_c and y_p is the method of *reduction of order* where a single operation yields a complete solution of a linear differential equation with constant coefficients.

Consider this method as applied to the second-order equation

$$(a_0D^2 + a_1D + a_2)y = f(x) \tag{11-86}$$

where a_0, a_1, and a_2 are constants and $a_0 \neq 0$.

Since the D operator is known to possess the same properties as algebraic quantities which obey the commutative, associative, and distributive laws, it is possible to treat the polynomial in D which appears in the left-hand member of Eq. (11-86) as follows:

Since $a_0 \neq 0$, factor $(a_0D^2 + a_1D + a_2)$, so that

$$a_0\left(D^2 + \frac{a_1}{a_0} D + \frac{a_2}{a_0}\right) = a_0(D - \alpha_1)(D - \alpha_2)$$

Then

$$a_0(D - \alpha_1)(D - \alpha_2)y = f(x) \tag{11-87}$$

or

$$(D - \alpha_1)(D - \alpha_2)y = \frac{f(x)}{a_0} \tag{11-88}$$

Replace $f(x)/a_0$ by $F(x)$; then

$$(D - \alpha_1)(D - \alpha_2)y = F(x) \tag{11-89}$$

Let

$$(D - \alpha_2)y = u \tag{11-90}$$

Then

$$(D - \alpha_1)u = F(x) \tag{11-91}$$

This is a linear differential equation of the type discussed in Sec. 11-5, hence

$$u = \epsilon^{-\int -\alpha_1 dx} \int F(x) \epsilon^{\int -\alpha_1 dx}\, dx + c_1 \epsilon^{-\int -\alpha_1 dx}$$

or

$$u = \epsilon^{\alpha_1 x} \int F(x) \epsilon^{-\alpha_1 x}\, dx + c_1 \epsilon^{\alpha_1 x} \tag{11-92}$$

But $(D - \alpha_2)y = u$, and therefore

$$(D - \alpha_2)y = \epsilon^{\alpha_1 x} \int F(x) \epsilon^{-\alpha_1 x}\, dx + c_1 \epsilon^{\alpha_1 x} \tag{11-93}$$

Again using the standard solution procedure for linear differential equations of first order, there is obtained

$$y = \epsilon^{-\int -\alpha_2 dx} \int [\epsilon^{\alpha_1 x} \int F(x) \epsilon^{-\alpha_1 x}\, dx + c_1 \epsilon^{\alpha_1 x}] \epsilon^{\int -\alpha_2 dx}\, dx + c_2 \epsilon^{-\int -\alpha_2 dx}$$

or

$$y = \epsilon^{\alpha_2 x} \int [\epsilon^{(\alpha_1 - \alpha_2)x} \int F(x) \epsilon^{-\alpha_1 x}\, dx + c_1 \epsilon^{\alpha_1 x}]\, dx + c_2 \epsilon^{\alpha_2 x} \tag{11-94}$$

which represents a complete solution of Eq. (11-86) where

$$F(x) = \frac{f(x)}{a_0}$$

EXAMPLE 6. By the *method of reduction of order*, solve the differential equation

$$(D^2 + 3D - 4)y = 6 \tag{11-95}$$

Equation (11-95) may readily be represented by

$$(D + 4)(D - 1)y = 6 \tag{11-96}$$

Let

$$(D - 1)y = u \tag{11-97}$$

Then Eq. (11-96) becomes

$$(D + 4)u = 6 \tag{11-98}$$

Then

$$u = \epsilon^{-\int 4dx} \int 6 \epsilon^{\int 4dx}\, dx + c_1 \epsilon^{-\int 4dx}$$

$$= \epsilon^{-4x} \int 6 \epsilon^{4x}\, dx + c_1 \epsilon^{-4x} = \epsilon^{-4x}(6/4\, \epsilon^{4x}) + c_1 \epsilon^{-4x} \tag{11-99}$$

or

$$u = 3/2 + c_1 \epsilon^{-4x} \tag{11-100}$$

But from Eq. (11-97) it follows that

$$(D - 1)y = 3/2 + c_1 \epsilon^{-4x}$$

Repeating the process, there is obtained

$$y = \epsilon^{-\int -dx} \int (3/2 + c_1 \epsilon^{-4x}) \epsilon^{-\int dx}\, dx + c_2 \epsilon^{-\int -dx}$$

$$= \epsilon^{x} \int (3/2 + c_1 \epsilon^{-4x}) \epsilon^{-x}\, dx + c_2 \epsilon^{x} = \epsilon^{x}\, (-3/2\, \epsilon^{-x} - 1/5\, c_1 \epsilon^{-5x}) + c_2 \epsilon^{x}$$

$$= -3/2 - 1/5\, c_1 \epsilon^{-4x} + c_2 \epsilon^{x} \tag{11-101}$$

or

$$y = C_1 \epsilon^{-4x} + C_2 \epsilon^{x} - 3/2$$

where $C_1 = -c_1/5$ and $C_2 = c_2$.

Of the possible methods of solution presented in this section, the *method of undetermined coefficients* is the one most often used in the solution of electronic engineering problems. This is because the differentiations which it requires are usually much easier than the integrations which must be done by the *method of reduction of order*. However, the method of undetermined coefficients has certain limitations. If the function of the independent variable which renders the equation nonhomogeneous consists of an algebraic polynomial or an exponential function or involves certain of the circular functions (sine or cosines), the method of undetermined coefficients is suitable; otherwise the method of reduction of order or some other general method is applicable. A general method known as Lagrange's *variation of parameter* is also extensively used for the solution of these equations, and it may be found in any advanced text on differential equations.

11-8 Simultaneous Linear Differential Equations

On occasion, a system of two linear simultaneous differential equations will be encountered of the form

$$\begin{aligned} [\phi_1(D)x] + [\psi_1(D)]y &= \theta_1(z) \\ [\phi_2(D)x] + [\psi_2(D)]y &= \theta_2(z) \end{aligned} \tag{11-102}$$

where D is the operational notation taken with respect to z, and $\phi_1(D)$, $\phi_2(D)$, $\psi_1(D)$, and $\psi_2(D)$ are polynomials in the operator D with constant coefficients.

The method employed here is directly attributed to the algebraic properties that the D operator possesses. It is upon this fact that the method of solution is predicated and the problem reduced to solution of differential equations already handled in the previous section.

An example will demonstrate:

$$Dx + (D - 1)y = \epsilon^t \tag{11-103}$$

$$2Dx + (D + 2)y = \cos t \tag{11-104}$$

Multiply Eq. (11-102) by the constant 2 and obtain

$$2Dx + 2(D - 1)y = 2\epsilon^t \tag{11-105}$$

$$2Dx + (D + 2)y = \cos t \tag{11-106}$$

$$(2D - 2 - D - 2)y = 2\epsilon^t - \cos t \tag{11-107}$$

or

$$(D - 4)y = 2\epsilon^t - \cos t \tag{11-108}$$

Let

$$y = y_c + y_p \tag{11-109}$$

where

$$y_c = c_1\epsilon^{4t}$$

and assume a particular solution

$$y_p = A\epsilon^t + B\cos t + C\sin t$$

which must satisfy identically the equation

$$(D - 4)y_p = 2\epsilon^t - \cos t$$

Hence

$$-4y_p = -4A\epsilon^t - 4B\cos t - 4C\sin t$$
$$Dy_p = A\epsilon^t + C\cos t - B\sin t$$

Therefore

$$2\epsilon^t - \cos t \equiv -3A\epsilon^t + (-4B + C)\cos t + (-B - 4C)\sin t$$
$$-B - 4C = 0$$
$$-4B + C = -1$$
$$-3A = 2$$

which gives for the constants the values $A = -\frac{2}{3}$, $B = \frac{4}{17}$, $C = -\frac{1}{17}$. It is thus found that the solution $y = y_c + y_p$ is given by

$$y = c_1\epsilon^{4t} - \tfrac{2}{3}\epsilon^t + \tfrac{4}{17}\cos t - \tfrac{1}{17}\sin t \tag{11-110}$$

The function for x in terms of t can now be found by substituting in Eq. (11-103):

$$Dx = \epsilon^t - (D - 1)y$$
$$Dx = \epsilon^t - (D - 1)(c_1\epsilon^{4t} - \tfrac{2}{3}\epsilon^t + \tfrac{4}{17}\cos t - \tfrac{1}{17}\sin t)$$
$$x = \epsilon^t - \tfrac{3}{4}c_1\epsilon^{4t} + \tfrac{5}{17}\sin t - \tfrac{3}{17}\cos t + c_2$$

A solution is then given by

$$x = \epsilon^t - \tfrac{3}{4}c_1\epsilon^{4t} + \tfrac{5}{17}\sin t - \tfrac{3}{17}\cos t + c_2$$
$$y = c_1\epsilon^{4t} - \tfrac{2}{3}\epsilon^t + \tfrac{4}{17}\cos t - \tfrac{1}{17}\sin t$$

The number of independent arbitrary constants may be obtained by setting up the determinant symbol of the polynomials of the D operators involved in the given Eq. (11-102),

$$\begin{vmatrix} \phi_1(D) & \psi_1(D) \\ \phi_2(D) & \psi_2(D) \end{vmatrix}$$

From this expression there is obtained a polynomial in D. The degree of this polynomial in D is equal to the number of arbitrary constants which a general solution of Eq. (11-102) should contain.

It follows that Eqs. (11-103) and (11-104) use for their arbitrary constant determination

$$\begin{vmatrix} D & D - 1 \\ 2D & D + 2 \end{vmatrix} = D^2 + 2D - 2D^2 + 2D = -D^2 + 4D$$

Therefore, the number of independent arbitrary constants in the solution of these equations will be two.

EXAMPLE

$$Dy + 3z = x$$
$$D^2y + (2D + 1)z = 2$$

The determinant symbol of the coefficients in D when evaluated produces a polynomial in D of degree 2, since

$$\begin{vmatrix} D & 3 \\ D^2 & 2D + 1 \end{vmatrix} = 2D^2 + D - 3D^2 = -D^2 + D$$

This means that for this particular system of simultaneous differential equations, there are two arbitrary constants required in any expression which represents a general solution.

If the first of the above equations is operated on by D and the second is then subtracted from the resulting first equation, there is obtained:

$$\begin{aligned} D^2y + D(3z) &= Dx \\ D^2y + (2D + 1)z &= 2 \\ (3D - 2D - 1)z &= -1 \\ (D - 1)z &= -1 \end{aligned}$$

If $z = z_c + z_p$, where $z_c = c_1\epsilon^x$ and $z_p = A$, then

$$\begin{aligned} (D - 1)A &= -1 \\ -A &= -1 \\ A &= 1 \end{aligned}$$

This gives

$$z = c_1\epsilon^x + 1$$

But

$$\begin{aligned} Dy &= x - 3z \\ Dy &= x - 3c_1\epsilon^x - 3 \end{aligned}$$

Hence $\qquad z = c_1\epsilon^x + 1, \qquad y = \dfrac{x^2}{2} - 3c_1\epsilon^x - 3x + c_2$

If, however, the problem is again solved from the given equations by eliminating the terms in z, it follows that

$$\begin{aligned} (2D + 1)(D)y + 3(2D + 1)z &= (2D + 1)x \\ 3D^2y + 3(2D + 1)z &= 6 \\ \hline (-D^2 + D)y &= 2 + x - 6 \\ (-D^2 + D)y &= x - 4 \\ D(D - 1)y &= 4 - x \\ y_c &= k_1 + k_2\epsilon^x \end{aligned}$$

Let

$$y_p = Bx + Cx^2$$

Then

$$\begin{aligned} y_p' &= B + 2Cx \\ y_p'' &= 2C \end{aligned}$$

and from the identical equation

$$2C - B - 2Cx \equiv 4 - x$$

it follows that

$$2C - B = 4 \qquad -2C = -1$$

or

$$C = \tfrac{1}{2} \qquad B = -3$$

Therefore a complete solution is

$$y = k_1 + k_2\epsilon^x + \frac{x^2}{2} - 3x$$

$$z = c_1\epsilon^x + 1$$

By comparison with the previous form of solution, obviously $k_1 = c_2$, $k_2 = -3c_1$. If, however, the relationship between the k's and c's is not immediately obvious, then by substituting into either one of the given equations and making use of the fact an identity exists, it is possible to establish the relationship that exists between dependent constants.

11-9 Solutions of Differential Equations by Numerical Approximation

Suppose it is desired to obtain a particular solution to the differential equation

$$\frac{dy}{dx} = F(x,y) \tag{11-111}$$

which contains the initial values $y = b$ when $x = a$.

It often happens that Eq. (11-111) cannot be formally integrated; hence no exact solution is practicable. An approximate solution is then sought. In some instances, an approximate solution is justifiably preferred to an exact solution since the very nature of the engineering application with its laboratory experimental data does not warrant the involvement often required for an exact solution. There are several methods commonly employed for finding an approximate solution which contains the simultaneous values $x = a$ and $y = b$. Here a method attributed to Picard will be discussed and illustrated.

In view of the fact that $y = b$ when $x = a$, it is required to set up a function which satisfies this condition. Such a function may be

$$y_1 = b + \int_a^x F(x,b)\,dx \tag{11-112}$$

where y_1 is known as the *first approximation.*

With y_1, a *second approximation* can be found, namely, y_2, where

$$y_2 = b + \int_a^x F(x,y_1)\,dx \tag{11-113}$$

By continuing the process, a *third approximation* y_3 can be located, and with y_3 obtained, y_4 can be had, etc.:

$$y_3 = b + \int_a^x F(x,y_2)\,dx \tag{11-114}$$

$$y_4 = b + \int_a^x F(x,y_3)\,dx \tag{11-115}$$

$$\cdot$$
$$\cdot$$
$$\cdot$$

$$y_n = b + \int_a^x F(x,y_{n-1})\,dx \tag{11-116}$$

These expressions are not solutions to Eq. (11-111), as they would have to satisfy the basic property that

$$\frac{dy_n}{dx} = F(x,y_n) \tag{11-117}$$

for $n = 1, 2, 3, 4, \ldots$. However, it may be shown that this sequence of $y_1, y_2, y_3, \ldots$ obtained gives increasingly exact approximations for increasing values of n. It was shown by Picard that this sequence has as its limiting value an exact solution of Eq. (11-111).

EXAMPLE 1. By Picard's method find an expression which gives an approximate solution to the equation

$$\frac{dy}{dx} - y^2 = x \tag{11-118}$$

and which satisfies the initial condition $y = 0$ when $x = 0$. Also find the approximate value of y that corresponds with the independent variable $x = -0.1$.

From Eq. (11-118) it follows that

$$F(x,y) = x + y^2 \qquad \text{and} \qquad F(x,0) = x$$

Hence $$y_1 = 0 + \int_0^x x\,dx \qquad \text{or} \qquad y_1 = \frac{x^2}{2}$$

Therefore

$$y_2 = 0 + \int_0^x F(x,y_1)\,dx = 0 + \int_0^x \left[x + \left(\frac{x^2}{2}\right)^2\right] dx = \frac{x^2}{2} + \frac{x^5}{20}$$

The third approximation y_3 then is

$$y_3 = 0 + \int_0^x \left[x + \left(\frac{x^2}{2} + \frac{x^5}{20}\right)^2\right] dx = 0 + \int_0^x \left(x + \frac{x^4}{4} + \frac{x^7}{20} + \frac{x^{10}}{400}\right) dx$$
$$= \frac{x^2}{2} + \frac{x^5}{20} + \frac{x^8}{160} + \frac{x^{11}}{4{,}400}$$

The value obtained for y_3 when $x = -0.1$ is then expressed by evaluating the series

$$y_3 = \frac{(-0.1)^2}{2} + \frac{(-0.1)^5}{20} + \frac{(-0.1)^8}{160} + \frac{(0.1)^{11}}{4{,}400}$$
$$= 0.005 - 0.00000054 + \cdots = 0.0049995$$

EXAMPLE 2. Show that the Picard solution to the differential equation

$$\frac{dy}{dx} = -y \tag{11-119}$$

for the given initial conditions $y = y_0$ when $x = x_0$ has for its limiting value the solution

$$y = y_0 \epsilon^{x_0 - x}$$

Verify this solution by using a direct method of solving the given equation.

$$\frac{dy}{dx} = -y$$

$$y_1 = y_0 + \int_{x_0}^{x} (-y_0)\, dx = y_0 + y_0(x_0 - x)$$

$$y_2 = y_0 + \int_{x_0}^{x} - [y_0 + y_0(x_0 - x)]\, dx = y_0 - \left[y_0 x - y_0 \frac{(x_0 - x)^2}{2} \right]_{x_0}^{x}$$

$$= y_0 - y_0 x - \frac{y_0(x_0 - x)^2}{2} + x_0 y_0 = y_0 + y_0(x_0 - x) + \frac{(x_0 - x)^2}{2}$$

$$y_3 = y_0 + \int_{x_0}^{x} - \left[y_0 + y_0(x_0 - x) + y_0 \frac{(x_0 - x)^2}{2} \right] dx$$

$$= y_0 + y_0(x_0 - x) + \frac{y_0(x_0 - x)^2}{1 \cdot 2} + \frac{y_0(x_0 - x)^3}{1 \cdot 2 \cdot 3}$$

By mathematical induction it follows that

$$y_n = y_0 \left[1 + \frac{(x_0 - x)}{1!} + \frac{(x_0 - x)^2}{2!} + \frac{(x_0 - x)^3}{3!} + \cdots + \frac{(x_0 - x)^n}{n!} \right]$$

For n increasing without limit,

$$y = \lim_{n \to \infty} y_n = y_0 \epsilon^{-(x - x_0)}$$

since from series considerations

$$\epsilon^{\theta} = 1 + \frac{\theta}{1!} + \frac{\theta^2}{2!} + \frac{\theta^3}{3!} + \cdots$$

By direct method of variables separable,

$$\frac{dy}{y} = -dx$$

$$\ln y = -x + \ln c$$

or

$$\ln \frac{y}{c} = -x$$

where c is an arbitrary constant. Hence

$$\frac{y}{c} = \epsilon^{-x} \qquad y = c\epsilon^{-x}$$

When $x = x_0$, $y = y_0$, therefore, it follows that

$$y_0 = c\epsilon^{-x_0}$$

Dividing $y = c\epsilon^{-x}$ by $y_0 = c\epsilon^{-x_0}$, there is obtained

$$\frac{y}{y_0} = \frac{\epsilon^{-x}}{\epsilon^{-x_0}} \qquad y = y_0\epsilon^{-(x-x_0)}$$

11-10 Non-network Applications

Applications of differential equations in electronics can be mainly divided into network and non-network categories. The non-network applications occur usually in the solution of physical problems concerning the distribution of mechanical forces and temperatures or the effects of applied vibratory forces. In these applications the independent variable is usually length or time, and the differential equations encountered are unrestricted as to order or degree. In network applications the independent variable is always time, and the majority of these problems rarely require the solution of anything more complex than a linear second-order equation.

A typical example of a non-network application is that contained in a study by Pohl[1] on the effects of temperature rise on the grid wires which make up the planar grid structure of a uhf triode. Considering a single grid wire, two sources of heat will cause a temperature rise. The wire will receive power along its length when the tube is operating by electron bombardment and also by radiation of heat from the cathode. With small error, it can be assumed that the power received per unit length of the wire is constant along the wire. Considering the thermal equilibrium of an element of the wire, the relationship

$$\frac{d^2t}{dl^2} = -\frac{Q}{AK} \tag{11-120}$$

can be derived, where t is temperature, l is length, Q is the heat received per second by a unit length of the wire, A is the cross-sectional area of the wire, and K is the coefficient of thermal conductivity of the wire. If the origin is taken as the center of the wire, then the initial conditions are that $t = T_0$ when $l = \pm L/2$ and $dt/dl = 0$ when $l = 0$, where T_0 is the temperature at the wire supports or grid frame and the wire length is L, as shown in Fig. 11-1a.

[1] W. J. Pohl, Aspects in the Design and Manufacture of Planar Grids for Triodes at U.H.F., *Electronic Eng.*, **23**: 95–99, March, 1951.

Obviously, Eq. (11-120) is a simple case of variable separables which requires only two integrations and the evaluation of the constants of integration for solution. When completed, this process yields for t,

$$t = T_0 + \frac{QL^2}{8AK} - \frac{QL^2}{2AK} \tag{11-121}$$

Thus the temperature distribution is a symmetrical parabolic curve, rising from T_0 at the ends to a maximum of $QL^2/8AK + T_0$ at the center.

The design of the ring grid frames used to support the grid wires is as shown in Fig. 11-1*b*. It is desired to maintain the maximum residual

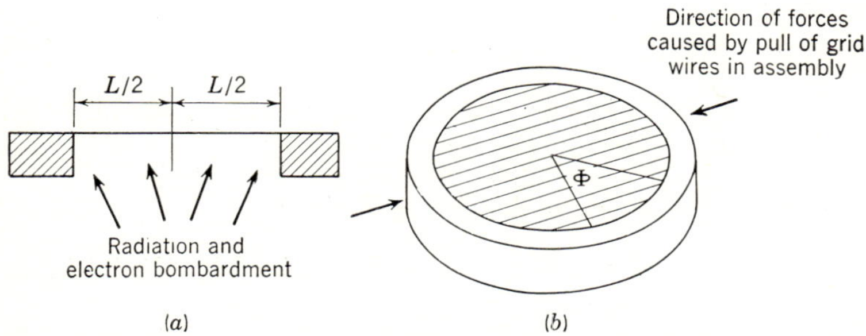

FIG. 11-1 (*a*) Grid wire and supports. (*b*) Ring grid frame of uhf triode.

tension in the grid wires. The pull of the wires tends to deform the ring into a plane oval, so that the long wires in particular will suffer a reduction in tension. The degree to which deformation of the ring will occur can be determined by estimating the deflection of the curved bar which forms the grid ring frame. For a thin ring

$$\frac{d^2\mu}{d\varphi^2} + \mu = -\frac{R^2M}{EI} \tag{11-122}$$

where μ is the radial displacement of the ring frame on deformation, φ is the polar angle specifying the section of circumference (as in Fig. 11-1*b*), R is the radius of the ring frame, M is the bending moment, E is Young's modulus of the frame, and I is the moment of inertia of a cross section of the frame.

It can be shown from mechanical considerations that M for this case is

$$M = \frac{\omega R^2}{4} \cos 2\varphi \tag{11-123}$$

where $\omega = F/p$, F is the residual force of each wire, and p is the pitch of the grid, i.e., the spacing of wires. From Eq. (11-123) it can be seen that, when $\varphi = 0$ and π, $M = \omega R^2/4$, and when $\varphi = \pi/2$ and $3\pi/2$, $M = -\omega R^2/4$. Substituting Eq. (11-123) into Eq. (11-122), we obtain

$$\frac{d^2\mu}{d\varphi^2} + \mu = -\frac{\omega R^4}{4EI} \cos 2\varphi \tag{11-124}$$

This is a complete linear differential equation of the second order with constant coefficients, which, in operational form, is

$$(D^2 + 1)\mu = f(\varphi) \tag{11-125}$$

A solution will consist of the sum of the complementary function μ_c and the particular integral μ_p. The complementary function μ_c is obtained by solving the corresponding homogeneous (or reduced) equation in which $f(\varphi) = 0$. The auxiliary equation is

$$m^2 + 1 = 0 \tag{11-126}$$

and the roots are $m = \pm j$, so that, according to the procedure outlined previously, the solution is of the form

$$\mu_c = A \cos \varphi + B \sin \varphi \tag{11-127}$$

where the constants A and B are to be determined by the initial conditions imposed upon μ and μ'.

The particular integral μ_p can be evaluated by the method of undetermined coefficients. If μ_p is assumed to be of the form

$$\mu_p = c_1 \cos 2\varphi + c_2 \sin 2\varphi \tag{11-128}$$

then the coefficients c_1 and c_2 can be evaluated by the substitution of Eq. (11-128) in Eq. (11-124). In this case

$$\mu_p' = -2c_1 \sin 2\varphi + 2c_2 \cos 2\varphi \tag{11-129}$$

and

$$\mu_p'' = -4c_1 \cos 2\varphi - 4c_2 \sin 2\varphi \tag{11-130}$$

Equating (11-130) plus (11-128) to the expression on the right in (11-124) yields

$$-3c_1 \cos 2\varphi - 3c_2 \sin 2\varphi = -\frac{\omega R^4}{4EI} \cos 2\varphi \tag{11-131}$$

According to the method of undetermined coefficients, the coefficients of similar terms on either side of the equation must be equal, so that

$$c_1 = \frac{\omega R^4}{12EI} \tag{11-132}$$

and

$$c_2 = 0 \tag{11-133}$$

The general solution of Eq. (11-124) is then

$$\mu = A \cos \varphi + B \sin \varphi + \frac{\omega R^4}{12EI} \cos 2\varphi \tag{11-134}$$

From the initial conditions of the problem, A and B can now be determined. Since $\varphi = 0$ and $\varphi = \pi/2$, $\mu' = d\mu/d\varphi = 0$, it follows that $A = B = 0$. The general solution Eq. (11-124) is then reduced to

$$\mu = \frac{\omega R^4}{12EI} \cos 2\varphi \tag{11-135}$$

which gives the deformation of the grid ring as a function of the radial angle φ.

Problems

Prob. 11-1 By the use of formulas for D operators, perform the indicated differentiation.

(*a*) $(D^2 - 4D + 4)\epsilon^{-2x} \sin x$

(*b*) $(D^2 + 4D + 4)\epsilon^{-2x} \sin x$

(*c*) $(3D^2 + 2D + 1)\epsilon^{3x} \cos 2x$

(*d*) $(D^2 + D - 1)(\epsilon^{2x} + x)$

(*e*) $(D^n + a_1D^{n-1} + a_2D^{n-2} + \cdots + a_n)\epsilon^{mx} \sin mx$

(*f*) $(D - a)^3 a^x$

(*g*) $(D - 2)^3[\epsilon^{2x}(\csc x + \sec x)]$

(*h*) $(D - a)^n a^x$

(*i*) $(D - \ln 5)^4 5^x$

(*j*) $(D + a)^n a^x$

(*k*) $(D^2 + 8D + 16)\epsilon^{-4x} \sin^2 x$

(*l*) $(D^2 - 8D + 16)\epsilon^{4x} \cos^2 x$

(*m*) $(3D^2 + 5D + 2)x\epsilon^{-x}$

(*n*) $(D^3 - 3D^2 + 3D - 1)x^2\epsilon^x$

(*o*) $(D^4 - 4D^3 + 6D^2 - 4D + 1)\epsilon^x \tan x$

Prob. 11-2 Determine whether the correspondingly numbered function that appears opposite each differential equation in the adjacent column constitutes a solution of the equation.

1. $2t^2x\,Dx - x^2 - t^2\epsilon^{(t-1/t)} = 0$	1. $x^2 - \epsilon^{t-1/t} = c\epsilon^{-1/t}$
2. $(D^2 + 9)x = 0$	2. $x = c_1 \cos 3t + c_2 \cos 3t$
3. $(D^2 - 9)x = 0$	3. $x = c_1\epsilon^{3t} + c_2\epsilon^{-3t}$
4. $(D^2 + 4)\theta = \cos t$	4. $\theta = A + B\epsilon^{-4t} + \frac{4}{17} \sin t - \frac{1}{17} \cos t$
5. $(D^2 + 6D + 5)u = 0$	5. $u = A\epsilon^{-5t} + B\epsilon^{-t}$
6. $(D^2 - \omega^2)\psi = \omega^2 + \sin \omega t$	6. $\psi = A\epsilon^{\omega t} + B\epsilon^{-\omega t} - \dfrac{\sin \omega t}{2\omega^2} - 1$
7. $(t^2 - x^2)\,Dx - 3tx = 0$	7. $(t^2 - 2x^2)^3 - ct^2 = 0$
8. $\begin{cases} t\,Dx = t - 2x \\ t\,Dy = t(x + y - 1) + 2x \end{cases}$	8. $\begin{cases} x = \dfrac{t}{3} + \dfrac{A}{t^2} \\ y = B\epsilon^t - \dfrac{t}{3} - \dfrac{A}{t^2} \end{cases}$

9. $\dfrac{dt}{d\theta} = \dfrac{11t + 4\theta - 11}{8t + 25\theta - 62}$ — 9. $4t - \theta + 2 = k\sqrt{t + 2\theta - 5}$

10. $\dfrac{2x(1 + x - \cos^2 y)\,dy}{1 - \cos^2 y} + \sec y \sin y\,dx = 0$ — 10. $x \sin^2 y - c \csc^2 y(2 + x - 2\cos^2 y) = 0$

11. $(LD + R)i = E_0 \cos \omega t$ — 11. $i = \dfrac{E_0}{R^2 + \omega^2 L^2}(R \cos \omega t + \omega L \sin \omega t) + c\epsilon^{-Rt/L}$

12. $Dx + \dfrac{\tan x}{t + x} = 0$ — 12. $(t + x)\tan x + 1 = c \sec x$

13. $Dx = 10^{tx}$ — 13. $x = c\epsilon^{5t^2}$

14. $Dx = \sec^2 x \cos t$ — 14. $4 \sin t = 2x + \sin 2x + c$

15. $t\dfrac{dx}{dt} = 2t\epsilon^{3t} + x(3t - 1)$ — 15. $tx = \epsilon^{3t}(t^2 + c)$

Prob. 11-3 The following equations are solvable by the process of variable separation. Solve and verify the answer obtained in each case.

(a) $$\frac{dy}{dx} = \sin x \sec y$$

(b) $$\frac{dy}{dx} = y^2\epsilon^{-x}$$

(c) $$\frac{dy}{dx} = \frac{(1 - x)^2}{\sqrt{y}}$$

(d) $$\frac{dy}{dx} = \frac{\sqrt{25 - 16x^2}}{9y^2}$$

(e) $$\frac{dy}{dx} = \frac{\sqrt{x + 3}}{y}$$

(f) $$\frac{dt}{dz} = \sin t \sec^3 z$$

(g) $$2y^2\,dx + y^2\,dy = 3dy$$

(h) $$dr = r \cot 2\theta\,d\theta$$

(i) $$\frac{dy}{dx} - 2y = y^2$$

(j) $$3\epsilon^{x+y}\,dy + 5\epsilon^{-2y}\,dx = 0$$

(k) $$3y\,dx = 2\sqrt{x - 1}\,dy$$

(l) $$x^3(25 - 9y^2)^{3/2}\,dx - (x^3 - 1)\,dy = 0$$

(m) $$(2\epsilon^{2x} + y)\,dx + x\,dy = 0$$

(n) $$\frac{dt}{dx} = \sec x \sec^2 t$$

(o) $$2dy + 3y\epsilon^{3x}\,dx = 2xy\,dx$$

Prob. 11-4 (a) Investigate each of the following equations for properties of homogeneous functions. (b) If homogeneous properties are possessed and also the criteria of homogeneous-function differential equations as discussed in this chapter are satisfied, then solve

(a) $$\frac{dy}{dx} = \frac{x + y}{x - y}$$

(b) $$\frac{dy}{dx} = \frac{y + \sqrt{x^2 + y^2}}{x}$$

(c) $$14tx\,dx + (7x^2 - 5t^2)\,dt = 0$$

(d) $$t(t + x)\,dx - x(t - x)\,dt = 0$$

(e) $$(x + 2y)\,dx + 2(x - 3y)\,dy = 0$$

(f) $$(y - 2x)\,dy + (2y + x)\,dx = 0$$

(g) $$(z^2\epsilon^{-2y/z} - y^2)\,dz + yz\,dy = 0$$

(h) $$w(9v - 2w)\,dv + v(w - 6v)\,dw = 0$$

(i) $$\sqrt{x^2 + y^2}\,dx - \frac{x^2}{y}\,dy = 0$$

(j) $$(4x^2 + y^2)\,dx - xy\,dy = 0$$

Prob. 11-5 By the use of the formula for the solution of linear differential equations, solve each of the following:

(a) $$\frac{dy}{dx} + \frac{y}{x} = x^3 + 2$$

(b) $$\frac{dy}{dx} + \frac{3y}{x} = x^2 - 5$$

(c) $$\frac{dy}{dx} = \frac{2}{y - x}$$

(d) $$\frac{dy}{dx} = xy + x\epsilon^{x^2}$$

(e) $$\frac{dy}{dx} = \frac{2}{1 + 4x \cot y}$$

(f) $$\frac{dr}{d\theta} + \cos\theta(2r - \sin\theta) = 0$$

(g) $$\frac{dx}{dy} = \frac{x}{xy + 4}$$

(h) $$\frac{dx}{dy} = \frac{2x\cos y + \sin y}{\sin y}$$

(i) $$\frac{dy}{dx} - \frac{2y}{x} - x = 0$$

(j) $$\frac{dy}{dx} - 3\epsilon^{-\sin x} = y\tan x$$

Prob. 11-6 Solve the following homogeneous differential equations with constant coefficients:

(a) $$(D^2 - 9D + 52)x = 0$$

(b) $$(D^2 - 5D + 6)x = 0$$

(c) $(D^2 - 8D)x = 0$

(d) $(2D^2 + 3D - 35)x = 0$

(e) $(6D^2 - 13D + 5)x = 0$

(f) $(D^2 - 16)x = 0$

(g) $(D^2 + 16)x = 0$

(h) $(D^2 - 2D + 1)x = 0$

(i) $(9D^2 + 10D + 1)x = 0$

(j) $(D^2 + 10D + 9)x = 0$

Prob. 11-7 By the method of reduction of order, solve the equations below completely. Also solve by the method of complementary solution and particular integral.

(a) $(D^2 + 4D + 4)x = 3\epsilon^{-t}$

(b) $(D^2 + 4D + 4)x = 5\epsilon^{-2t}$

(c) $(D^2 + 4D + 4)y = 2\epsilon^{2x} + 3\epsilon^{-2x} + 3x - 5$

(d) $(D^2 + 16)x = 3 \cos 2t + 5 \sin 4t$

(e) $(D^2 - 16)x = 3 + 2t + 4t^2$

(f) $(D^2 + 16)x = \cos t$

(g) $(D^2 + 16)x = 2 \cos 4t$

(h) $(D^2 - 8D)x = 3$

(i) $(D^2 + 7D + 6)x = \sec t$

(j) $(D^2 + 25)x = \cos 5t + \epsilon^{-t}$

(k) $(D^2 + 16D + 64)x = 3t\epsilon^{-8t}$

(l) $(D^2 + 5D + 4)x = \tan t$

(m) $(D^2 - 7D - 8)x = 3$

(n) $(D^2 + 14D + 49)y = 2\epsilon^{-7x}$

(o) $(3D^2 + D - 10)x = \epsilon^{-x}$

(p) $(12D^2 - 11D - 56)x = 2\epsilon^{7t/4} + 3t + 2$

(q) $(16D^2 + 14D - 2)x = \sin 2t$

(r) $(12D^2 + 7D + 1)x = \cos 3t + \tan 3t$

(s) $(9D^2 + 6D - 1)y = 3\epsilon^{2x} + 4 + \cos 2x$

(t) $(10D^2 + 8D - 2)x = 16$

Prob. 11-8 Solve the following simultaneous differential equations:

(a)
$$(D^2 + 3)x - (D + 1)y = \epsilon^{2t}$$
$$(D + 1)x - Dy = \sin t$$

(b)
$$Dx = t$$
$$x + Dy = 2t^2$$

(c)
$$Dx + 5y = 0$$
$$Dy - 5x = 0$$

(d)
$$Dx - y = 0$$
$$(D + 1)y = 3\epsilon^t$$

(e) $$(D + 1)x - Dy = 0$$
$$Dx - 4y = 2\epsilon^{4t}$$

(f) $$(D^2 - 3)x - 4y = 3$$
$$x + (D^2 - 8)y = 5$$

(g) $$x + Dy = 0$$
$$Dx - Dy = 2$$

Prob. 11-9* In a study of high-frequency gas discharges and the motion of a charged particle in an electric field, the differential equation

$$m \frac{d^2x}{dt^2} + g \frac{dx}{dt} = eE \sin \omega t$$

is obtained, where m is the particle mass, e is the particle charge, the electric field is represented as $E \sin \omega t$, g is the "coefficient of friction" between the charged particle (electron) and the gas molecules in the region of its travel, and x is the direction of the applied field. Solve this equation for the particle velocity $v = dx/dt$ if the initial velocity is v_0.

Prob. 11-10† In an analysis of debunching in a klystron the differential equation

$$\frac{d^2X}{dx^2} + h^2X = 0$$

is obtained. If it is known that $X(0) = 0$ and $dX(0)/dx = V_\omega/2V_0v_0$, solve this equation for X.

Prob. 11-11‡ The differential equation for the voltage V_2 in an electronic band-pass filter is given as

$$\frac{d^2V_2}{dt^2} + \frac{3-k}{RC}\frac{dV_2}{dt} + \frac{V_2}{R^2C^2} = \frac{k-1}{RC} E\omega \cos(\omega t + \alpha)$$

when $V_1 = E \sin(\omega t + \alpha)$, R and C are network resistance and capacitance values as shown in Fig. (11-2), $\omega = 2\pi f$, and α is the phase angle of

* L. D. Smullin and C. G. Montgomery, "Microwave Duplexers," Radiation Laboratory Series, vol. 14, sec. 5.2, p. 145, McGraw-Hill Book Company, Inc., New York, 1948.

† D. R. Hamilton, J. K. Knipp, and J. B. H. Kuper, "Klystrons and Microwave Triodes," Radiation Laboratory Series, vol. 7, p. 216, McGraw-Hill Book Company, Inc., New York, 1948.

‡ A. H. Silcocks, correspondence on "A Band Pass Filter for Low Frequencies," by G. W. Morris and P. G. M. Dowe, *Electronic Eng.*, **25** (310): 534–535, December, 1953.

V_1. Show that the solution of this is

$$V_2 = \epsilon^{-(3-k)t/2RC}[A\epsilon^{(jt/2RC)\sqrt{6k-k^2-5}} + B\epsilon^{-(jt/2RC)\sqrt{6k-k^2-5}}] + \frac{(k-1)E\cos(\omega t + \alpha - \varphi)}{\sqrt{x^2 + (3-k)^2}}$$

where $x = \dfrac{1}{\omega RC} - \omega RC$ and $\varphi = \tan^{-1}\dfrac{3-k}{x}$

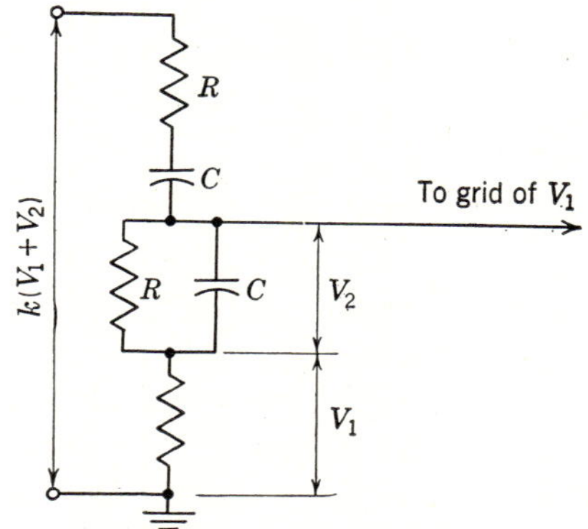

FIG. 11-2 Input circuit of electronic band-pass filter.

Prob. 11-12 The characteristic equation of an electronic weighing system is given as[1]

$$M\frac{d^2x}{dt^2} + Kx = W$$

where M is the total mass of the system, K is the spring constant (stiffness), W is the weight, and x is the scale deflection.

(a) Show that a solution of this equation is

$$x = \frac{W}{K}(1 - \cos\omega t)$$

where $\omega = 2\pi f$

(b) Show that the maximum value of x is $x_m = 2W/K$ when $t = \pi/\omega$

(c) If the weighing time T is the time required for one cycle of scale oscillation, show that $T = 2\pi\sqrt{M/K}$

(d) What are the specifications of a spring to be used in a scale with a total mass when loaded of 4 lb if the maximum weight to be measured is 100 lb and the peak deflection is to be 1 in.?

[1] R. E. Bell and J. A. Ferstle, Electronic Weighing on the Production Line, *Electronics*, **28** (6): 152–155, June, 1955.

Chapter 12

TRANSIENT SOLUTIONS AND ELEMENTARY LAPLACE TRANSFORMS

12-1 Differential Equations in Network Analysis

As suggested previously, the differential equations normally encountered in network applications are of the first or second order, are linear, and employ time as the independent variable. The fact that they are linear is no accident, since (in network analysis) linearity is almost always assumed for the sake of convenience as described in some detail in Chap. 10. It should be understood by all who make this assumption that most circuit components do not truly behave in linear fashion over more than a limited range of voltages or currents. However, the advantages that accrue by assuming linear performance usually outweigh the errors which result.

In addition to linearity, it is also necessary to assume, in setting up ordinary differential equations, that the circuit components are lumped rather than distributed. Where the assumption is not warranted, as in the case of the coaxial transmission line, which depends on distributed inductance and capacitance for its mode of operation, ordinary differential equations will not suffice and partial differential equations must be developed.

If the circuit to be analyzed consists of a single mesh or node pair, the operation of the circuit can be described in terms of a single current or voltage. Under these conditions, the system is said to possess one degree of freedom. Where multimesh or multimode networks are involved, the degree of freedom will be two or more and a set of simultaneous differential equations must be established.

Network differential equations can, therefore, be written in terms of voltages or currents. In the first and most widely used case, every term in the equation will have the dimensions of voltage. From basic physical relationships, voltage drops in transient form are represented as

$$\text{Voltage developed across } R = iR \tag{12-1}$$

$$\text{Voltage developed across } C = \frac{Q}{C} = \frac{1}{C}\int^t i\,dt \tag{12-2}$$

$$\text{Voltage developed across } L = L\frac{di}{dt} \tag{12-3}$$

According to Kirchhoff's first law, then, the sum of the voltages in a mesh containing R, C, and L and a generator of voltage $E \cos \omega t$ will be at any instant

$$iR + \frac{1}{C}\int^t i\,dt + L\frac{di}{dt} = E \cos \omega t \tag{12-4}$$

where Q, the charge on the capacitor, may be used interchangeably for the quantity $\int^t i\,dt$.

This is an integro-differential equation. Differentiating the equation once, however, yields

$$R\frac{di}{dt} + \frac{i}{C} + L\frac{d^2i}{dt^2} = -\omega E \sin \omega t \tag{12-5}$$

an expression in which each term has the dimensions of voltage/time and which can be solved directly by methods previously described. Notice that the additional differentiation has provided a differential equation of the second order. It can be concluded that networks containing all three types of circuit element (R, C, and L) will produce differential equations of the second order. Combinations of R and L or R and C alone can always be solved by means of first-order equations and can be directly integrated.

If the differential equation is to be written in terms of the currents at a single node pair, then each term in the initial equation will have the dimensions of current. The current terms, in transient forms, are

$$\text{Current in } R = \frac{e}{R} \tag{12-6}$$

$$\text{Current in } C = C\frac{de}{dt} \tag{12-7}$$

$$\text{Current in } L = \frac{1}{L}\int^t e\,dt \tag{12-8}$$

According to Kirchhoff's second law, then, the sum of the currents at a node, including an entering current from a generator of $I \cos \omega t$, will be, at any instant,

$$\frac{e}{R} + C\frac{de}{dt} + \frac{1}{L}\int^t e\,dt = I \cos \omega t \tag{12-9}$$

Once more, an additional differentiation is used to eliminate the integral, which yields an equation in which each term has the dimensions of current/time:

$$\frac{1}{R}\frac{de}{dt} + C\frac{d^2e}{dt^2} + \frac{e}{L} = -\omega I \sin \omega t \tag{12-10}$$

Note that a second-order equation is again produced.

The solution of these equations is quite straightforward and merely involves the classical procedures described previously for the handling of linear equations with constant coefficients.[1] In other words, the complementary function and the particular integral are obtained, and, after this, the initial physical conditions are used to evaluate the constants of integration.

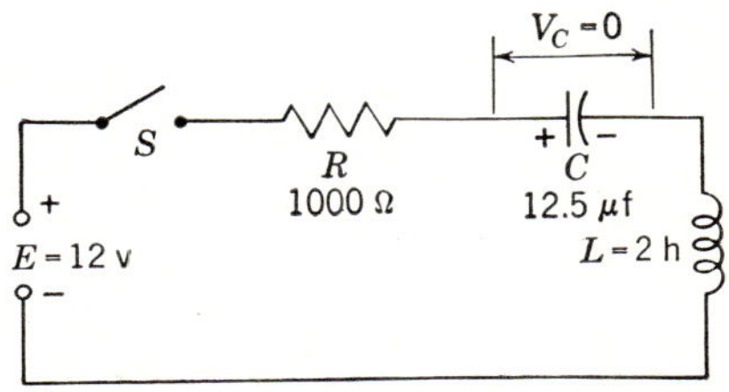

FIG. 12-1 Single-mesh circuit which contains R, C, and L.

This general technique can be illustrated by the detailed solution of Eq. (12-5) for the current in the series circuit shown in Fig. 12-1. As indicated in the diagram, the initial voltage to which C is charged is zero, and the initial current in the circuit is also zero. When the switch is closed, a constant emf of 12 volts is applied to the circuit. With $L = 2$ henrys, $C = 12.5$ μf, and $R = 1{,}000$ ohms, the differential equation is

$$2\frac{d^2i}{dt^2} + 1{,}000\frac{di}{dt} + \frac{i}{12.5 \times 10^{-6}} = 0 \tag{12-11}$$

Note that the term on the right, $-\omega E_0 \sin \omega t$, is zero in this case because the applied emf is constant, i.e., $\omega = 0$. The auxiliary equation for this case is

$$2m^2 + 1{,}000m + \frac{10^6}{12.5} = 0 \tag{12-12}$$

which has as roots $m_1 = -100$, $m_2 = -400$. The complete solution is therefore

$$i = C_1\epsilon^{-100t} + C_2\epsilon^{-400t} \tag{12-13}$$

Two initial conditions are known. The first is that, at $t = 0$, $i = 0$. Substituted in the equation above, this condition yields

$$C_1 + C_2 = 0$$

[1] Extensive treatments of network transients using classical methods have been published elsewhere. See H. H. Skilling, "Transient Electrical Currents," McGraw-Hill Book Company, Inc., New York, 1952; W. R. LePage and S. Seely, "General Network Analysis," McGraw-Hill Book Company, Inc., New York, 1952; E. Frank, "Pulsed Linear Networks," McGraw-Hill Book Company, Inc., New York, 1945.

The second is that the capacitor C is uncharged at $t = 0$, so that $V_C = 0$ and $Q = 0$. If Eq. (12-4), which also applies to this circuit, is solved for di/dt, the result is

$$\frac{di}{dt} = \frac{1}{L}\left[E \cos \omega t - iR - \frac{Q}{C}\right] \tag{12-14}$$

where Q has replaced $\int^t i\,dt$. At $t = 0$, $Q = 0$, $i = 0$, $E = 12$ volts d-c, and

$$\frac{di}{dt} = \frac{E}{L} = \frac{12}{2} = 6$$

Differentiation of the complete solution gives

$$\frac{di}{dt} = -100C_1\epsilon^{-100t} - 400C_2\epsilon^{-400t} \tag{12-15}$$

so that at $t = 0$

$$6 = -100C_1 - 400C_2$$

Solving this and the previous equation $C_1 + C_2 = 0$ simultaneously provides the values

$$C_1 = 0.02 \qquad C_2 = -0.02$$

and the final solution for the circuit current as a function of time is

$$i = 0.02(\epsilon^{-100t} - \epsilon^{-400t}) \tag{12-16}$$

12-2 Laplace Transform Method of Network Solution

From what was said in the previous section, it is clear that network analyses require only a knowledge of the manipulation of first- and second-order linear differential equations and that this manipulation is performed over and over again from one network problem to the next. This repetition obviously involves nothing in the way of mathematical novelty, and it is evident that there should be a way of systematizing the process.

To the everlasting chagrin of pure mathematicians, it was an engineer, Oliver Heaviside, who first recognized this need and developed what he called the operational method of differential-equation solution,[1] a scheme for simplifying transient analysis by reducing it to algebraic form very much as Steinmetz and others had applied the complex number and the impedance concept to reduce the labor involved in steady-state a-c network problems, as mentioned in the first chapter. Heaviside's system involved a number of mathematical short cuts which he relied upon because they worked, rather than because they could be substantiated by

[1] See E. J. Berg, "Heaviside's Operational Calculus," McGraw-Hill Book Company, Inc., New York, 1936.

formal mathematical proofs. This naturally brought him under heavy fire from the purists, but, nevertheless, working in self-elected isolation, he was able to arrive at certain correct conclusions which were considered radical at the time. He discovered, for example, that series inductive loading of telephone cables would reduce attenuation on long lines. No one listened, however, for Lord Kelvin had recommended quite a different remedy, the addition of shunt capacitance. Later on, of course, Michael Pupin settled the argument and won fame and fortune by showing that "heavification" was the answer after all. Heaviside, despite any shortcomings in his work, must be credited with the popularization of the use of differential equations in engineering.

The mathematical world has since accepted the Heaviside method but has formalized it by relating it to the parallel theoretical work done by Cauchy and Laplace. The "algebraic" method of solving differential equations is therefore referred to as the Laplace transform method. Actually, any transient problem can be solved with equal accuracy by the classical, operational method or the transform method. Recent engineering practice, however, has placed the emphasis on the transform method.

The method of transformation developed by Laplace in 1779 is somewhat analogous to the use of logarithms for the simplification of arithmetic problems. The more difficult operations, multiplication, division, and raising quantities to a power (evolution), can be replaced by the simpler processes of addition and subtraction when numbers have been transformed into logarithms. In the same way, functions of time, denoted in general by the symbol $f(t)$, are transformed into functions of a new complex variable,[1] s. When this has been accomplished, derivatives, integrals, and trigonometric functions may be expressed and manipulated algebraically. The transformation from an $f(t)$ to an $f(s)$ is accomplished by multiplying the $f(t)$ by ϵ^{-st}, where t is time measured from $t = 0$, and then integrating the product $f(t)\epsilon^{-st}$ with respect to time between the limits of 0 and ∞.

Symbolically, the Laplace transformation of $f(t)$ to $f(s)$ is written as

$$f(s) = \mathfrak{L}[f(t)] \tag{12-17a}$$

and the corresponding inverse Laplace transformation of $f(s)$ is

$$\mathfrak{L}^{-1}[f(s)] = f(t) \tag{12-17b}$$

[1] The letter p is often used to denote the complex variable s. The latter notation is preferred to prevent confusion with the Heaviside operator $p = d/dt$. The letter s generally denotes a complex number of the form $x + jy$ which can be located in the complex plane or Argand diagram (Fig. 1-1). It is subject to the one restriction that its real part is positive.

and according to the process described above

$$f(s) = \mathcal{L}[f(t)] = \int_0^\infty f(t)\epsilon^{-st}\,dt \tag{12-18}$$

From the discussion in Sec. 4-1, on the dimensions of integrals, it can be seen that the application of Eq. (12-18) to an $f(t)$ in order to convert it to an $f(s)$ involves, from a dimensional standpoint, multiplication by $[T]$. Thus, the transform of a current i, for example, will have the dimensions of $[I][T]$ or charge, $[Q]$. Note that the dimensions of the variable s, itself, are those of $[T]^{-1}$, or frequency. This suggests that under certain conditions s may be interchanged with the frequency operator $j\omega$. Use of this approach is discussed in Secs. 12-5 and 12-6.

When the operation in Eq. (12-18) has been completed, $f(t)$ will have been converted into an expression which is a function of s alone because the variable t disappears when the integration is performed. The notation used for the transform of a function $f(t)$ is $f(s)$, so that the transform for a voltage function $e(t)$ is $e(s)$. The subscript s may also be used, so that the transform of a voltage or current e or i which is a function of time can be written i_s or e_s.

The manner in which some common functions of time are transformed to functions of s by this process is as follows:

(*a*) If $f(t) = \sin \omega t$, then from Eq. (12-18)

$$f(s) = \mathcal{L}[\sin \omega t] = \int_0^\infty \epsilon^{-st} \sin \omega t\,dt \tag{12-19}$$

The exponential form of $\sin \omega t$ is

$$\sin \omega t = \frac{\epsilon^{j\omega t} - \epsilon^{-j\omega t}}{2j} \tag{12-20}$$

so that

$$\begin{aligned} f(s) &= \frac{1}{2j}\int_0^\infty (\epsilon^{j\omega t} - \epsilon^{-j\omega t})\epsilon^{-st}\,dt = \frac{1}{2j}\int_0^\infty [\epsilon^{-(s-j\omega)t} - \epsilon^{-(s+j\omega)t}]\,dt \\ &= \frac{1}{2j}\left(\frac{1}{s-j\omega} - \frac{1}{s+j\omega}\right) = \frac{\omega}{s^2+\omega^2} \end{aligned} \tag{12-21}$$

(*b*) If $f(t) = E$ (a constant direct voltage),

$$f(s) = \mathcal{L}[E] = \int_0^\infty E\epsilon^{-st}\,dt = \left[-\frac{E}{s}\epsilon^{-st}\right]_0^\infty = \frac{E}{s} \tag{12-22}$$

(*c*) If $f(t) = L\,di/dt$ (the voltage across an inductance L, where i is the current in the inductance),

$$f(s) = \mathcal{L}\left[L\frac{di}{dt}\right] = L\mathcal{L}\left[\frac{di}{dt}\right] = L\int_0^\infty \frac{di}{dt}\epsilon^{-st}\,dt = L\int_{t=0}^\infty \epsilon^{-st}\,di \tag{12-23}$$

Integrating by parts, if $u = \epsilon^{-st}$ and $dv = di$, substituting in the formula $\int u\,dv = uv - \int v\,du$ gives

$$f(s) = L\Big[i\epsilon^{-st}\Big]_{t=0}^{\infty} + Ls\int_0^{\infty} i\epsilon^{-st}\,dt \qquad (12\text{-}24)$$

The second term is, according to Eq. (12-12), simply the transform of i or $i(s)$, so that

$$f(s) = L[si(s) - i(0+)] \qquad (12\text{-}25)$$

where $i(0+)$ is the initial value[1] of the current i a very short time after $t = 0$. In general

$$\mathcal{L}[f'(t)] = s\mathcal{L}[f(t)] - f(0+) \qquad \text{or} \qquad sf(s) - f(0+) \qquad (12\text{-}26)$$

where $f(0+)$ is the initial value of $f(t)$ just after $t = 0$. And, similarly,

$$\mathcal{L}[f''(t)] = s^2 f(s) - sf(0+) - f'(0+) \qquad (12\text{-}27)$$

where $f'(0+)$ is the value of the first derivative of $f(t)$ just after $t = 0$.

The real value of the transformation method is that the above integrations need not be carried out repeatedly, because the form of $f(s)$ is always the same for a specified $f(t)$. It is, therefore, possible to employ a table of transforms of the more common functions in the same way that a table of integrals is used.

A fairly complete listing of transform pairs is given in Table 12-1. In this table, the expression shown in the left-hand column is the result obtained by transforming the corresponding $f(t)$ shown in the right-hand column.

12-3 Examples of Solutions Using the Laplace Transform

The steps to be followed in the solution of a network transient problem by means of the Laplace transformation are as follows:

1. Write the differential equation for the network.
2. Determine the initial conditions in the circuit at $t = 0$.
3. Transform the differential equation with Laplace transforms.
4. Solve for the transform of the desired unknown.
5. Obtain the inverse transform of the unknown to secure it in useful form as a function of time.

The first two steps enumerated above are approached in the manner described in Sec. 12-1 for the classical solution. Step 3 is accomplished by replacing each term in the differential equation with its own Laplace transform.

[1] The symbol $t = 0^+$ indicates that it represents the limiting value as t approaches zero from the right (i.e., through positive values of t).

Consider the following examples:

(*a*) A direct voltage of 100 volts is applied to a 1-henry inductance in series with a 100-ohm resistance. The differential equation for the network is

$$L\frac{di}{dt} + Ri = E \tag{12-28}$$

or, in this case,

$$\frac{di}{dt} + 100i = 100$$

Just after $t = 0$, $i(0+) = 0$. Using transform pair 10 for the first term on the left and transform pair 1 for the term on the right, the transformed equation is

$$[si(s) - i(0+)] + 100i(s) = \frac{100}{s} \tag{12-29}$$

Solving for $i(s)$ with $i(0+) = 0$ gives

$$i(s) = \frac{100}{s(s + 100)} \tag{12-30}$$

From transform pair 34, the corresponding inverse solution is

$$i(t) = (1 - \epsilon^{-100t}) \tag{12-31}$$

(*b*) In the same circuit, the applied voltage is of the form ϵ^{-100t}. The differential equation is

$$\frac{di}{dt} + 100i = \epsilon^{-100t} \tag{12-32}$$

Using transform pair 2 for the term on the right, the transformed equation is

$$si(s) + 100i(s) = \frac{1}{s + 100} \tag{12-33}$$

Solving for $i(s)$ gives

$$i(s) = \frac{1}{(s + 100)^2} \tag{12-34}$$

From the transform pair 9a, the corresponding inverse solution is

$$i(t) = t\epsilon^{-100t} \tag{12-35}$$

(*c*) In the same circuit, the applied voltage is a sine wave of 100 volts peak amplitude at 60 cps, so that $\omega = 377$ and $E = 100 \sin 377t$. The differential equation is

$$\frac{di}{dt} + 100i = 100 \sin 377t \tag{12-36}$$

If transform pair 3 is used for the term on the right, the transformed equation is

$$si(s) + 100i(s) = \frac{377(100)}{s^2 + 377^2} \tag{12-37}$$

Solving for $i(s)$ gives

$$i(s) = \frac{377(100)}{(s + 100)(s^2 + 377^2)} \tag{12-38}$$

From transform pair 35, the corresponding inverse solution is

$$i(t) = \frac{377(100)}{(100)^2 + (377)^2}\left(\epsilon^{-100t} + \frac{100}{377}\sin 377t - \cos 377t\right) \tag{12-39}$$

12-4 Inverse Transformations Using Partial Fractions

Usually, $f(s)$ proves to be of the form $g(s)/h(s)$, where $g(s)$ and $h(s)$ are polynomials in s which have no common factor. The fractional forms considered here are proper; i.e., the degree of the numerator must be lower than that of the denominator. If not of this form, the fraction can be reduced by division to produce the sum of an integral part and a proper fraction. When a suitable transform pair cannot be located in Table 12-1 for an $f(s)$ of this type, the inverse transformation can be accomplished by means of an expansion of $f(s)$ into partial fractions. For example, if $f(s)$ has the form

$$f(s) = \frac{2s^2 + 4s + 12}{s^3 + 8s^2 + 19s + 12} \tag{12-40}$$

the denominator can be factored so that

$$f(s) = \frac{2s^2 + 4s + 12}{(s + 1)(s + 3)(s + 4)} \tag{12-41}$$

According to the method for obtaining partial fractions, set $f(s)$ equal to three fractions which have unknown constants as numerators, and

$$f(s) = \frac{2s^2 + 4s + 12}{(s + 1)(s + 3)(s + 4)} = \frac{A_1}{(s + 1)} + \frac{A_2}{(s + 3)} + \frac{A_3}{(s + 4)} \tag{12-42}$$

A_1, A_2, and A_3 can then be evaluated by setting $s = -1, -3$, and -4, which yields

$$\begin{aligned} A_1 &= \left[\frac{2s^2 + 4s + 12}{(s + 3)(s + 4)}\right]_{S=-1} = \frac{10}{6} \\ A_2 &= \left[\frac{2s^2 + 4s + 12}{(s + 1)(s + 4)}\right]_{S=-3} = -\frac{18}{2} \\ A_3 &= \left[\frac{2s^2 + 4s + 12}{(s + 1)(s + 3)}\right]_{S=-4} = \frac{28}{3} \end{aligned} \tag{12-43}$$

TABLE 12-1 ELEMENTARY LAPLACE TRANSFORM PAIRS

$f(s) = \int_0^\infty f(t)\epsilon^{-st}\,dt$	$f(t)$ for $t > 0$ $f(t) = 0$, for $t < 0$
1. $\frac{a}{s}$	1. a
2. $\frac{1}{s+a}$	2. ϵ^{-at}
3. $\frac{\omega}{s^2+\omega^2}$	3. $\sin \omega t$
4. $\frac{s}{s^2+\omega^2}$	4. $\cos \omega t$
5. $\frac{As+B}{s^2+\omega^2}$	5. $C \cos (\omega t + \phi)$ where $C = \sqrt{A^2 + (B/\omega)^2}$ $\phi = \tan^{-1}(-B/A\omega)$
6. $\frac{\omega}{(s+a)^2+\omega^2}$	6. $\epsilon^{-at} \sin \omega t$
7. $\frac{s+a}{(s+a)^2+\omega^2}$	7. $\epsilon^{-at} \cos \omega t$
8. $\frac{n!}{s^{n+1}}$	8. t^n
8a. $\frac{1}{s^2}$	8a. t
9. $\frac{1}{(s+a)^n}$	9. $\frac{1}{(n-1)!} t^{n-1}\epsilon^{-at}$
9a. $\frac{1}{(s+a)^2}$	9a. $t\epsilon^{-at}$
10. $sf(s) - f(0+)$	10. $f'(t)$
11. $s^2f(s) - sf(0+) - f'(0+)$	11. $f''(t)$
12. $\frac{f(s)}{s} + \frac{f^{-1}(0+)}{s}$ where $f^{-1}(0+) = \int_{-\infty}^{0+} f(t)\,dt$	12. $f^{-1}(t)$ or $\int f(t)\,dt$
13. $\epsilon^{-as}f(s)$	13. $f(t-a)H(t-a)$ (See Sec. 12-8)
14. Sum of transforms of individual terms	14. Sum of terms
15. $\frac{s \sin \theta + \omega \cos \theta}{s^2+\omega^2}$	15. $\sin (\omega t + \theta)$
16. $\frac{\omega}{s^2-\omega^2}$	16. $\sinh \omega t$ or $\frac{\epsilon^{\omega t} - \epsilon^{-\omega t}}{2}$
17. $\frac{s}{s^2-\omega^2}$	17. $\cosh \omega t$ or $\frac{\epsilon^{\omega t} + \epsilon^{-\omega t}}{2}$
18. $\frac{\epsilon^{-s}}{s}$	18. $f(t) = 0,\ t < 1$ $= 1,\ t > 1$

TABLE 12-1 ELEMENTARY LAPLACE TRANSFORM PAIRS (*Continued*)

$f(s) = \int_0^\infty f(t)\epsilon^{-st}\,dt$	$f(t)$ for $t > 0$; $f(t) = 0$, for $t < 0$
19. $\dfrac{1 - \epsilon^{-s}}{s}$	19. $f(t) = 1,\ t < 1$; $= 0,\ t > 1$
20. $\dfrac{1 - (1 + s)\epsilon^{-s}}{s^2}$	20. $f(t) = t,\ t < 1$; $= 0,\ t > 1$
21. $\dfrac{\epsilon^{-s}}{s} + \dfrac{\epsilon^{-s}}{s^2}$	21. $f(t) = 0,\ t < 1$; $= t,\ t > 1$
22. $\dfrac{1 - \epsilon^{-s}}{s^2}$	22. $f(t) = t,\ t < 1$; $= 1,\ t > 1$
23. $\dfrac{(1 - \epsilon^{-s})^2}{s^2}$	23. $f(t) = t,\ t < 1$; $= 2 - t,\ 1 < t < 2$; $= 0,\ t > 2$
24. $\dfrac{1}{s} - \dfrac{(1 - \epsilon^{-s})}{s^2}$	24. $f(t) = 1 - t,\ t < 1$; $= 0,\ t > 1$
25. $\dfrac{s}{(s + a)^2}$	25. $(1 - at)\epsilon^{-at}$
26. $\dfrac{s}{(s + a)^3}$	26. $t\left(1 - \dfrac{at}{2}\right)\epsilon^{-at}$
27. $\dfrac{1}{(s + a)(s + b)}$	27. $\dfrac{\epsilon^{-bt} - \epsilon^{-at}}{a - b}$
28. $\dfrac{s}{(s + a)(s + b)}$	28. $\dfrac{a\epsilon^{-at} - b\epsilon^{-bt}}{a - b}$
29. $\dfrac{1}{(s + a)(s + b)^2}$	29. $\dfrac{\epsilon^{-at} - \epsilon^{-bt}[1 - (a - b)t]}{(a - b)^2}$
30. $\dfrac{s}{(s + a)(s + b)^2}$	30. $\dfrac{a - b(a - b)t\epsilon^{-bt} - a\epsilon^{-at}}{(a - b)^2}$
31. $\dfrac{s}{(s + a)^2 + b^2}$	31. $\epsilon^{-at}\left(\cos bt - \dfrac{a}{b}\sin bt\right)$
32. $\dfrac{1}{(s + a)(s + b)(s + c)}$	32. $\dfrac{(c - b)\epsilon^{-at} + (a - c)\epsilon^{-bt} + (b - a)\epsilon^{-ct}}{(a - b)(b - c)(c - a)}$
33. $\dfrac{s}{(s + a)(s + b)(s + c)}$	33. $\dfrac{a(b - c)\epsilon^{-at} + b(c - a)\epsilon^{-bt} + c(a - b)\epsilon^{-ct}}{(a - b)(b - c)(c - a)}$
34. $\dfrac{1}{s(s + a)}$	34. $\dfrac{1}{a}(1 - \epsilon^{-at})$
35. $\dfrac{1}{(s + a)(s^2 + \omega^2)}$	35. $\dfrac{1}{a^2 + \omega^2}\left(\epsilon^{-at} + \dfrac{a}{\omega}\sin \omega t - \cos \omega t\right)$
36. $\dfrac{1}{(s^2 + \omega^2)^2}$	36. $\dfrac{1}{2\omega^3}(\sin \omega t - \omega t \cos \omega t)$
37. $\dfrac{s}{(s^2 + \omega^2)^2}$	37. $\dfrac{t}{2\omega}\sin \omega t$

The expanded form of $f(s)$ is then

$$f(s) = \frac{10}{6(s+1)} - \frac{9}{(s+3)} + \frac{28}{3(s+4)} \tag{12-44}$$

and the solution for $f(t)$ from pair 2 is

$$f(t) = \frac{10}{6}\epsilon^{-t} - 9\epsilon^{-3t} + \frac{28}{3}\epsilon^{-4t} \tag{12-45}$$

If the denominator cannot be conveniently factored except by using a quadratic form like $s^2 + as + b$, one or more of the numerators of the partial fractions assumed must be of the form $A_1s + A_2$, as in the following example. If

$$f(s) = \frac{s+1}{s^3 + 4s^2 + 8s} \tag{12-46}$$

this can be simplified to

$$f(s) = \frac{s+1}{s(s^2 + 4s + 8)} \tag{12-47}$$

According to this rule, the partial-fraction expansion for $f(s)$ will be

$$f(s) = \frac{s+1}{s(s^2+4s+8)} = \frac{A_1}{s} + \frac{A_2s + A_3}{s^2 + 4s + 8} \tag{12-48}$$

In this case, it is more convenient to multiply both sides of the expression by the denominator on the left, which gives

$$s + 1 = A_1(s^2 + 4s + 8) + s(A_2s + A_3)$$

To solve for A_1, A_2, and A_3, equate the coefficients of like powers, so that if, in expanded form,

$$s + 1 = A_1s^2 + A_1(4s) + 8A_1 + A_2s^2 + A_3s$$

then, equating coefficients of terms with like powers, there is obtained

$$\begin{aligned} A_1 + A_2 &= 0 \\ 4A_1 + A_3 &= 1 \\ 8A_1 &= 1 \end{aligned}$$

and $A_1 = 1/8$, $A_2 = -1/8$, and $A_3 = 1/2$. The expanded form of $f(s)$ is

$$\begin{aligned} f(s) &= \frac{1}{8s} - \frac{s-4}{8(s^2+4s+8)} \\ &= \frac{1}{8s} - \left\{\frac{(s+2)}{8[(s+2)^2+4]} - \frac{(3)(2)}{8[(s+2)^2+4]}\right\} \end{aligned} \tag{12-49}$$

and the corresponding solution from pairs 1, 7, and 6 is

$$f(t) = \frac{1}{8}[1 - \epsilon^{-2t}(\cos 2t - 3 \sin 2t)] \tag{12-50}$$

It is worth while at this point to resolve the network problem treated in Sec. 12-1. It will be recalled that the differential equation developed in Eq. (12-11) was

$$2\frac{d^2i}{dt^2} + 1{,}000\frac{di}{dt} + \frac{i}{12.5 \times 10^{-6}} = 0 \tag{12-11}$$

The transformation of this equation, using pairs 11, 10, and 1, and taking the transform of i as i_s, is

$$2[s^2i_s - si(0+) - i'(0+)] + 1{,}000[si_s - i(0+)] + \frac{i_s}{12.5 \times 10^{-6}} = 0 \tag{12-51}$$

The initial conditions for the circuit are

$$i\bigg]_{t=0} = i(0+) = 0$$

$$\frac{di}{dt}\bigg]_{t=0} = i'(0+) = 6$$

Substituting these values, the transformed equation becomes

$$2s^2i_s - 12 + 1{,}000si_s + 80{,}000i_s = 0 \tag{12-52}$$

from which

$$i_s = \frac{6}{s^2 + 500s + 40{,}000} = \frac{6}{(s + 100)(s + 400)} \tag{12-53}$$

Using a partial-fraction expansion

$$\frac{6}{(s + 100)(s + 400)} = \frac{A_1}{(s + 100)} + \frac{A_2}{(s + 400)} \tag{12-54}$$

The constants A_1 and A_2 are

$$A_1 = \left[\frac{6}{s + 400}\right]_{S=-100} = 0.02$$

$$A_2 = \left[\frac{6}{s + 100}\right]_{S=-400} = -0.02$$

The expanded form of i_s is, then,

$$i_s = \frac{0.02}{(s + 100)} - \frac{0.02}{(s + 400)} \tag{12-55}$$

and the solution for $i(t)$ is

$$i = 0.02(\epsilon^{-100t} - \epsilon^{-400t}) \tag{12-56}$$

which is the result obtained before.

12-5 Network Theorems Used with Laplace Transforms

Multimesh and multinode networks will always lead to simultaneous differential equations. Quite often, however, it is possible to simplify the analysis of such networks by means of the basic network theorems. Y. P. Yu[1] has presented the following examples of such simplifications.

Using Norton's theorem, the short-circuit current between terminals 1 and 2 in Fig. 12-2a can be found from the expression

$$e = \frac{1}{C}\int_0^t i\,dt \tag{12-57}$$

From transform pair 12, the transform of i, i_s is

$$i_s = Cse_s - Q(0+) \tag{12-58}$$

and this will be the transform of the current delivered by the equivalent generator in the circuit of Fig. 12-2b, with the polarities shown.

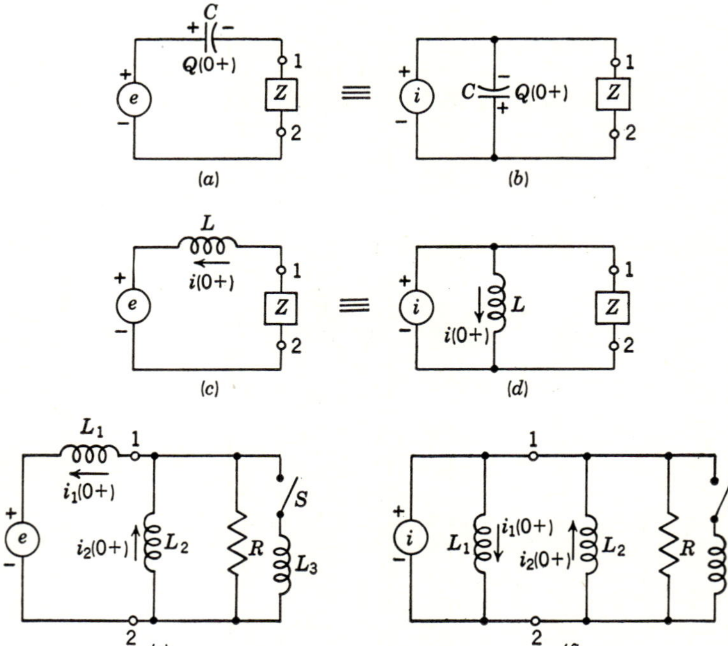

FIG. 12-2 Network simplifications by means of network theorems.

In Fig. 12-2c, the initial current in the inductance is $i(0+)$ and the equivalent constant-current generator in Fig. 12-2d will have a current

[1] Y. P. Yu, Applications of Network Theorems in Transient Analysis, *J. Franklin Inst.*, **251** (11): 381–398, November, 1949.

equal to

$$i = \frac{1}{L}\int_0^t e\,dt \tag{12-59}$$

The transform of i is

$$i_s = \frac{e_s}{sL} + \frac{i(0+)}{s} \tag{12-60}$$

As an example of this process, consider Fig. 12-2e, in which $f = 60$ cps, $\omega = 377$, $e = E \sin \omega t = 100 \sin 377t$, $L_1 = 0.1$ henry, $L_2 = 0.2$ henry, $L_3 = 0.3$ henry, and $R = 100$ ohms. The desired solution is the voltage across L_3 at $t = 0.001$ sec after the switch is closed. The transform of the short-circuit current between points 1 and 2 is given by Eq. (12-60). Now, according to pair 3, the transform e_s of $e = 100 \sin 377t$ is $(100)(377)/[s^2 + (377)^2]$ and the shunt inductance of the equivalent circuit of Fig. 12-2f is $L_1 = 0.1$ henry. Therefore

$$i_s = \frac{(100)(377)}{0.1s[s^2 + (377)^2]} + \frac{i_1(0+)}{s} \tag{12-61}$$

The initial conditions of the circuit are $i_1(0+)$ and $i_2(0+)$. Before $t = 0$, the voltage source e is shunted by an impedance z:

$$z = j\omega L_1 + \frac{j\omega L_2 R}{R + j\omega L_2} = 36.3 + j85.5 \tag{12-62}$$

Therefore i_1 is e/z, and, at $t = 0$, $i_1(0+) = -0.993$ amp. The current i_2 is

$$i_2 = \left(\frac{e}{z}\right)\left(\frac{R}{R + j\omega L_2}\right) \tag{12-63}$$

At $t = 0$, $i_2(0+) = -0.835$ amp. (Negative signs indicate current direction toward upper terminal of generator e.)

If v is taken as the voltage across L_3, then the differential equation in the equivalent network for v when the switch is closed will be (on a nodal basis) from Fig. 12-2f and Eq. (12-9)

$$i = \frac{1}{L_1}\int v\,dt + \frac{1}{L_2}\int v\,dt + \frac{v}{R} + \frac{1}{L_3}\int v\,dt \tag{12-64}$$

The transform of this equation is from Eq. (12-60)

$$i_s = \frac{v_s}{sL_1} + \frac{i_1(0+)}{s} + \frac{v_s}{sL_2} + \frac{i_2(0+)}{s} + \frac{v_s}{R} + \frac{v_s}{sL_3} + \frac{i_3(0+)}{s} \tag{12-65}$$

$$= \frac{v_s}{s}\left(\frac{1}{L_1} + \frac{1}{L_2} + \frac{1}{L_3}\right) + \frac{v_s}{R} + \frac{1}{s}[i_1(0+) + i_2(0+) + i_3(0+)] \tag{12-66}$$

The term $i_1(0+)/s$ in Eq. (12-65) and Eq. (12-66) has already been incorporated in the expression for the transform of the equivalent current generator, i_s, in Eq. (12-61). It must therefore be omitted from Eq. (12-61) when Eq. (12-61) is substituted into Eq. (12-67). The value of $i_1(0+)/s$ used in Eq. (12-66) will then be $-(-0.993)$.

In Fig. 12-2*f*, the direction of positive current is assumed downward, so that $i_2(0+) = -0.835$. At the instant $t = 0$ and the switch is closed, the current in L_3, $i_3(0+)$ will be zero. By substitution of these values, Eq. (12-66) can be solved for v_s, so that

$$v_s = \frac{100[si_s - 0.158]}{s + 1{,}833} \tag{12-67}$$

Using the value for i_s obtained previously in Eq. (12-61) but omitting $i(0+)/s$, v_s becomes

$$v_s = \frac{(10^5)(377)}{(s^2 + 377^2)(s + 1{,}833)} - \frac{15.8}{s + 1{,}833} \tag{12-68}$$

and the inverse transformation using the appropriate pairs is

$$v = 53.5[\sin\,(377t - 0.202) + (\sin\,0.202)\epsilon^{-1833t}] - 15.8\epsilon^{-1833t} \tag{12-69}$$

from which the voltage across L_3 at $t = 0.001$ is 8.47 volts. Notice that the use of the equivalent networks based on Norton's theorem has permitted the solution of a three-mesh network without the use of simultaneous equations.

It must be understood that the equivalent-network method has an important limitation. As in steady-state network analysis, the equivalent network obtained *cannot* be used to calculate operating conditions inside that portion of the network which has been converted.

In general, the equivalent current- or voltage-generator technique can be used to simplify greatly most problems. Other network theorems, such as Thévenin's theorem, the compensation theorem, and the superposition theorem, can also be employed. Since a square wave may be decomposed into several components, the last-mentioned is particularly useful in evaluating the response of a network to square pulses.[1] Consider the circuit of Fig. 12-3*a*. The charge on C_2 has reached its steady-state value before $t = 0$. The square wave is then applied to the network. A general expression for the output voltage V is desired.

Prior to $t = 0$, the voltage on C_1 is zero and across C_2 is 10 volts. After $t = 0$, the voltage V_a developed across R because of redistribution of the energy stored in C_2 can be determined from Fig. 12-3*b*. This circuit is obtained from Fig. 12-3*a* by short-circuiting all voltage generators, since they are assumed to have zero internal impedances. In transformed

[1] Y. P. Yu, *loc. cit.*

form, employing the method developed in the previous problem and Eqs. (12-9) and (12-58), V_a is

$$V_{as} = \frac{C_2}{C_1 + C_2} \cdot \frac{-10}{s + 1/R(C_1 + C_2)} \tag{12-70}$$

and conversion by the appropriate pair yields

$$V_a = -5\epsilon^{-5\times 10^3 t} \tag{12-71}$$

The voltage V_b due to the effects of the battery is, according to Fig. 12-3c,

$$V_b = -10\epsilon^{-5\times 10^3 t} \tag{12-72}$$

Finally, the pulse generator e is taken into consideration by decomposing the pulse as follows: The pulse has a duty cycle of 50 per cent and a

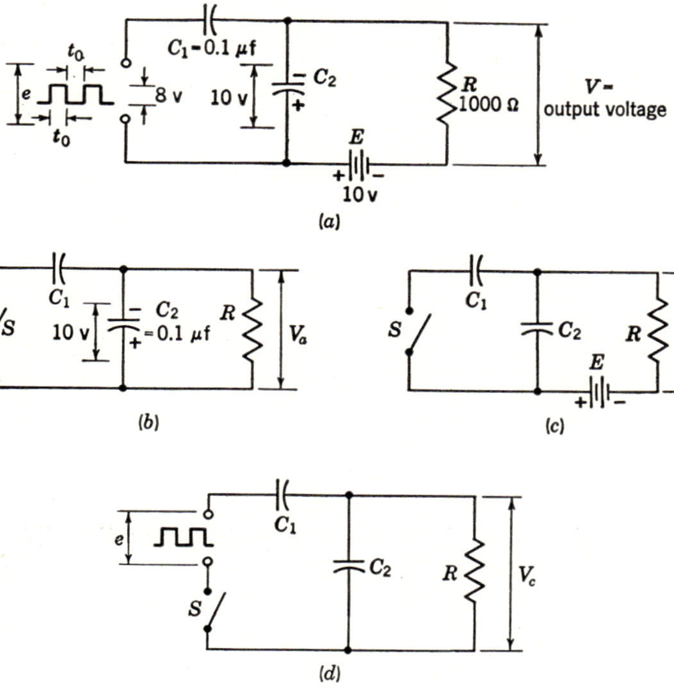

FIG. 12-3 Square wave applied to an R-C network.

length of 0. It can be considered to consist of a voltage which starts at $t = 0$ with an amplitude of +8 volts. The second component then starts at $t = t_0$ with an amplitude of −8 volts, the third component at $t = 2t_0$ with an amplitude of +8 volts, etc. Thus the superposition theorem is used by assuming that the input square wave is produced by a number of generators connected in series, each producing one of the components just described.

To compute the voltage V_c, a so-called operational-impedance method, suitable for circuits initially at rest, is used:

(*a*) In a mesh containing L, C, and R *in which there is no stored energy*, the transform of the mesh current is

$$i_s = \frac{e_s}{z_s} \tag{12-73}$$

where e_s is the transform of the generator voltage applied to the mesh and z_s is the total operational impedance, using s for $j\omega$ or

$$z_s = R + \frac{1}{sC} + sL \tag{12-74}$$

(*b*) At a node pair to which L, C, and R are connected in parallel and *in which the initial voltage across C and the initial current in L are both zero*, the transform of the node voltage is

$$e_s = i_s y_s \tag{12-75}$$

where e_s is the transform of the generator current applied to the node and Y_s is the total operational admittance, using s for $j\omega$ or

$$Y_s = \frac{1}{R} + sC + \frac{1}{sL} \tag{12-76}$$

Figure 12-3*d* can now be used along with the operational-impedance concept to evaluate V_c. Considering the first component of e mentioned above, the transform for e will be

$$e_s = \frac{8}{s} \tag{12-77}$$

Treating the circuit as if it were in steady-state operation, this voltage will appear across R reduced in magnitude by the voltage-divider effect of C_1 and $C_2 + R$ in parallel. The voltage V_c due to the first component, in operational form, is

$$V_{cs(1)} = \frac{8}{s} \cdot \frac{\dfrac{1}{1/R + sC_2}}{\dfrac{1}{sC_1} + \dfrac{1}{1/R + sC_2}} \tag{12-78}$$

and the inverse transformation gives

$$V_{c_1} = 4\epsilon^{-5\times 10^3 t} \tag{12-79}$$

V_c due to the second component will be

$$V_{c_2} = -4\epsilon^{-5\times10^3(t-t_0)} \tag{12-80}$$

The third component is

$$V_{c_3} = 4\epsilon^{-5\times10^3(t-2t_0)} \tag{12-81}$$

and that due to the nth component will be

$$V_{c_n} = (-1)^{n-1}4\epsilon^{-5\times10^3[t-(n-1)t_0]} \tag{12-82}$$

where n is an integer and is

$$n = \frac{(t - t')}{t_0} \qquad \text{where } 0 < t' < t_0 \tag{12-83}$$

The summation of all these V_c components gives

$$V_c = V_{c_1} + V_{c_2} + V_{c_3} + \cdots + V_{c_n} = 4\epsilon^{-5\times10^3t}\sum_1^n(-1)^{n-1}\epsilon^{5\times10^3(n-1)t_0} \tag{12-84}$$

$$= \begin{cases} 4\epsilon^{-5\times10^3t}\dfrac{1 - \epsilon^{5\times10^3nt_0}}{1 + \epsilon^{5\times10^3t_0}} & \text{for } n \text{ even} \\ 4\epsilon^{-5\times10^3t}\dfrac{1 + \epsilon^{-5\times10^3nt_0}}{1 + \epsilon^{-5\times10^3t_0}} & \text{for } n \text{ odd} \end{cases} \tag{12-85}$$

The final solution for the output voltage of the network after $t = 0$ is the sum of the three voltages computed in Fig. 12-3b, c, and d, or

$$V = V_a + V_b + V_c = \begin{cases} \left(4\dfrac{1 - \epsilon^{5\times10^3nt_0}}{1 + \epsilon^{5\times10^3t_0}} - 15\right)\epsilon^{-5\times10^3t} & \text{for } n \text{ even} \\ \left(4\dfrac{1 + \epsilon^{-5\times10^3nt_0}}{1 + \epsilon^{-5\times10^3t_0}} - 15\right)\epsilon^{-5\times10^3t} & \text{for } n \text{ odd} \end{cases} \tag{12-86}$$

12-6 Transfer Functions

In Sec. 12-5, certain network theorems were employed to simplify the solution of multimesh or multinode network problems. Such problems may also be simplified to a considerable degree by an extension of the operational impedance concept introduced in Sec. 12-5. The operational impedance of a mesh containing series L, C, and R elements is given in Eq. (12-74). The corresponding operational admittance of a node pair across which are connected L, C, and R elements is given in Eq. (12-76). In each case it can be seen that these expressions are the same as the usual impedance and admittance expressions except that the variable s

has been substituted for the operator $j\omega$. It can be shown that if a network contains no initial stored energy, that is, if $e_c(0+) = 0$ and $i_L(0+) = 0$ in every case, the operational impedance can be obtained directly from the steady-state impedance expression by substitution of

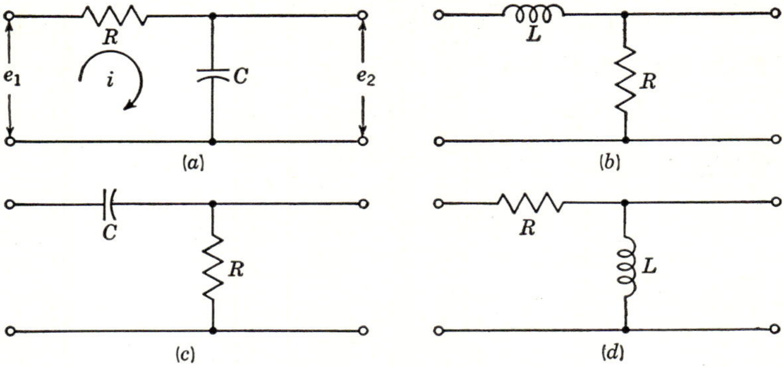

Fig. 12-4

s for $j\omega$. For example, consider the R-C network in Fig. 12-4a. The mesh equation is

$$e_1 = iR + \frac{i}{j\omega C} = i\left(R + \frac{1}{j\omega C}\right) \tag{12-87}$$

where $R + 1/j\omega C$ is the mesh impedance. The output voltage, e_2, is $i/j\omega C$, so that the substitution of $e_2(j\omega C)$ for i in Eq. (12-87) gives an expression for e_1 in terms of e_2:

$$e_1 = e_2(j\omega C)\left(R + \frac{1}{j\omega C}\right) = e_2(j\omega CR + 1) \tag{12-88}$$

The operational form of this equation may now be obtained by substituting s for $j\omega$ and taking the transform of e_1 and e_2. This yields

$$e_{1s} = e_{2s}(sT + 1) \tag{12-89}$$

where e_{1s} is the Laplace transform of e_1, e_{2s} is the transform of e_2, and T is the time constant RC. Equation (12-89) may now be used to compute the transient response of this network. If the output voltage for an applied constant voltage of $e_1 = E$ when $t > 0$ is desired, the transform of e_1 is E/s and

$$e_{2s} = \frac{E}{s} \cdot \frac{1}{(1 + sT)} = \frac{E}{T} \cdot \frac{1}{s(s + 1/T)} \tag{12-90}$$

The solution for e_2 as a function of time is the inverse transform of Eq. (12-90), so that from transform pair 34 in Table 12-1 with $a = 1/T$

$$e_s = \frac{E}{T} \cdot T(1 - \epsilon^{-t/T}) = E(1 - \epsilon^{-t/T}) \tag{12-91}$$

which is the transient response of the circuit to an applied voltage of E for $t > 0$.

The use of the operational impedance method has considerably simplified the calculation of the transient response in this case, because the transition from Eq. (12-89) to Eq. (12-90) has permitted the bypassing of Steps 1 through 3 in the procedure outlined in Sec. 12-3 for transient solutions by the transform method.

The ratio of e_{2s}/e_{1s} from Eq. (12-89) is

$$\frac{e_{2s}}{e_{1s}} = \frac{1}{sT + 1} \tag{12-92}$$

The quantity on the right is identified as the *transfer function* of the network. The transfer function of a system may be formally defined as the ratio of the transform of the output to the transform of the input, under the restriction that the initial values of both input and output quantities and all their time derivatives and integrals are zero.

The transform function for a network may be obtained by solving the network for its output/input voltage ratio on a steady-state basis and then substituting s for $j\omega$ in the result. This process may be facilitated by the use of the matrix methods described in Chap. 8. The transfer function of any network will be given by the transform of Eq. (8-41), where A and B are the general network parameters of the over-all network excluding load impedance and Z_L is the load impedance.

As an example of the manner in which a previously calculated ratio of output to input voltage may be used to determine a system transient response, consider the resistance-coupled amplifier stage shown in Fig. 6-38. The high-frequency gain of this stage is given in Eq. (6-90) as

$$\alpha_h = \frac{-g_m R_{\text{par}}}{1 + R_{\text{par}}(j\omega C_{\text{par}})} \tag{6-90}$$

The low-frequency gain, according to Eq. (6-83), is

$$\alpha_l = \frac{-g_m R_{\text{par}}}{1 - j/\omega C_c R_{\text{loop}}} \tag{6-83}$$

The mid-frequency gain, from Eq. (6-92), is

$$\alpha_m = -g_m R_{\text{par}} \tag{6-92}$$

Substituting the time constants $T_h = R_{\text{par}} C_{\text{par}}$ and $T_l = R_{\text{loop}} C_c$ and α_m from Eq. (6-92) into Eqs. (6-90) and (6-83), we obtain

$$\frac{\alpha_h}{\alpha_m} = \frac{1}{1 + j\omega T_h} \tag{12-93}$$

$$\frac{\alpha_l}{\alpha_m} = \frac{(j\omega T_l)}{1 + j\omega T_l} \tag{12-94}$$

The relative over-all gain G at high and low frequencies combined will be the product of the right-hand portions of Eqs. (12-93) and (12-94) because the relative over-all response of the stage is $\alpha_l\alpha_h/\alpha_m^2$ so that

$$G = \frac{j\omega T_l}{1 + j\omega T_l}\frac{1}{1 + j\omega T_h} \tag{12-95}$$

Assume that $T_l = 1{,}000T_h$, which would define a stage in which the high-frequency response was down 3 db at a frequency 1,000 times greater than that for which the low-frequency response was also down 3 db. For instance, if the stage is designed so that the low-frequency response at 20 cps is 3 db less than the mid-frequency response, the high-frequency response will be down 3 db from the mid-frequency response at (1,000)20 or 20,000 cps. Substituting $0.001T_l$ for T_h in Eq. (12-95) gives a relative gain G of

$$G = \frac{j\omega T_l}{(1 + j\omega T_l)(1 + 0.001 j\omega T_l)} \tag{12-96}$$

The transform of Eq. (12-96) is

$$G(s) = \frac{sT_l}{(1 + sT_l)(1 + 0.001sT_l)} \tag{12-97}$$

If a constant voltage E which has a square leading edge is applied to the circuit, the transform of E is E/s and Eq. (12-97) becomes

$$\begin{aligned} G(s) \cdot \frac{E}{s} &= \frac{E}{s} \cdot \frac{sT_l}{T_l(0.001T_l)(1/T_l + s)(1/0.001T_l + s)} \\ &= \frac{E}{0.001T_l(1/T_l + s)(1/0.001T_l + s)} \end{aligned} \tag{12-98}$$

The inverse transform of Eq. (12-98) can be obtained by means of transform pair 27 with $a = 1/T_l$ and $b = 1/0.001T_l$. The transient response of the circuit to the signal voltage E is

$$G \cdot E = \frac{E}{0.001T_l} \cdot \frac{\epsilon^{-t/0.001T_l} - \epsilon^{-t/T_l}}{(1/T_l - 1/0.001T_l)} \tag{12-99}$$

or approximately

$$G \cdot E \cong E(\epsilon^{-t/T_l} - \epsilon^{-t/0.001T_l}) \tag{12-100}$$

A plot of Eq. (12-100) will show that a stage in which T_l and T_h have been proportioned so that the -3-db points are at 20 and 20,000 cps will have a transient response for an applied square-edge wave which is a sloping pulse of amplitude 0 at $t = 0$, $0.1E$ at $t = 1$ μsec, and $0.9E$ at $t = 17$ μsec.

To facilitate the design and analysis of pulse amplifiers or amplifiers which are to be used in conjunction with servomechanisms, it is convenient to have a table of transfer functions for the various common interstage coupling and correcting networks. Chestnut[1] has prepared a detailed set of transfer functions which apply to most of the useful network configurations which can be constructed from R and C elements, and these appear in Table 12-2. The networks shown can be used to produce almost any desired frequency or transient response; the omission of R-L configurations is not limiting, because the proper R-C combination will duplicate the performance of any given R-L network. For example, the transfer function of the R-C network in Fig. 12-4*a* is identical to that of Fig. 12-4*b* and the same is true of Fig. 12-4*c* as compared with Fig. 12-4*d* when the time constants of the respective circuits are made equal.

The first column in Table 12-2 illustrates the 29 networks for which transfer functions have been calculated. Networks 1 through 11 are phase-leading networks which cause a reduction in low-frequency response, networks 12 through 18 are phase-lagging networks which cause a reduction in high-frequency response, and networks 19 through 29 produce a combination of these effects. The letters d, e, n, and v in these diagrams are the series resistance multipliers, the values of which determine the network time constants given in the last two columns of the table. Likewise, the shunt resistance multipliers are b, f, and g; the series capacitance multipliers are a and q; and the parallel capacitance multipliers are h and s.

The second column in Table 12-2 shows the idealized frequency response of each network in decibels of attenuation versus ω or frequency as plotted on semilog paper. The pertinent time constants where breaks in the attenuation-frequency diagrams occur are indicated. For the more complicated circuit configurations, the values of T_1 and T_2 are not necessarily the time constants at which breaks occur in the diagrams. The ultimate gains at zero frequency, G_0, and infinite frequency, G_∞, are also indicated in the diagrams, and the values of these are specified as functions of the resistance multipliers in the columns headed G_0 and G_∞. Note that these attenuation curves have been approximated in most cases by the use of straight-line segments which have attenuation slopes of 20 db/decade, 40 db/decade, or zero. A slope of 20 db/decade means that the attenuation increases or decreases 20 db for a decade change in frequency or, in other words, over a span in frequency of 10 to 1. A slope of 20 db/decade is the same as a slope of 6 db/octave, where an octave represents a 2-to-1 frequency change.

When two or more of these networks are so designed that they can be cascaded without unduly affecting each other's performance, the over-all transfer function of the combination will simply be the product of the

[1] H. Chestnut, Obtaining Attenuation-frequency Characteristics for Servomechanisms, *Gen. Elec. Rev.*, **50** (12): 38–44, December, 1947.

TABLE 12-2 TRANSFER FUNCTIONS OF *R-C* NETWORKS

Network	Attenuation characteristic	Transfer function	G_0	G_∞	T_1	T_2
1 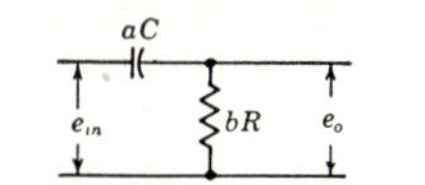	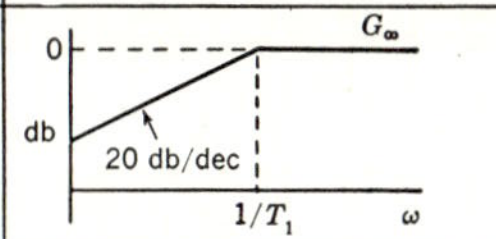	$\dfrac{T_1s}{T_1s+1}$	0	1	$abRC$	. . .
2 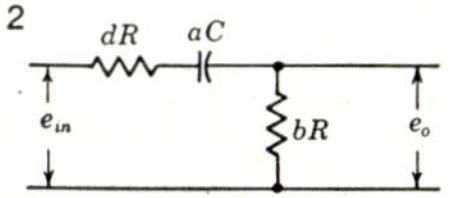	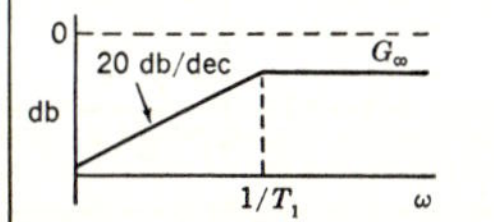	$\dfrac{T_2s}{T_1s+1}$	0	$\dfrac{1}{1+\frac{d}{b}}$	$a(b+d)RC$	$abRC$
3 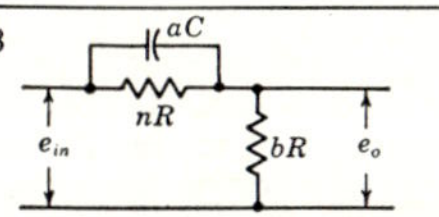	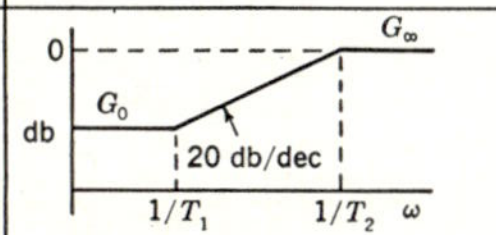	$G_0\dfrac{(T_1s+1)}{(T_2s+1)}$	$\dfrac{1}{1+\frac{n}{b}}$	1	$anRC$	$\dfrac{b}{b+n}T_1$
4 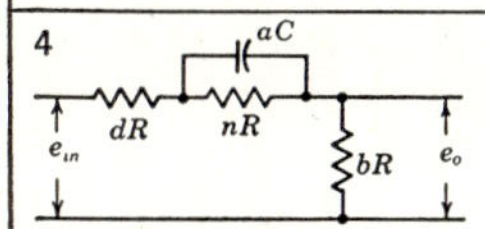	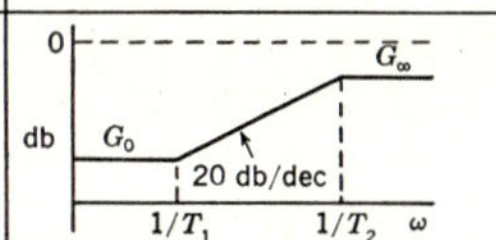	$G_0\dfrac{(T_1s+1)}{(T_2s+1)}$	$\dfrac{1}{1+\frac{d+n}{b}}$	$\dfrac{1}{1+\frac{d}{b}}$	$anRC$	$\left[\dfrac{b+d}{b+d+n}\right]T_1$
5 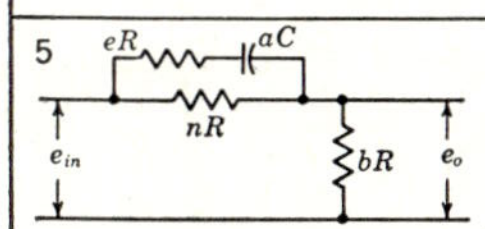	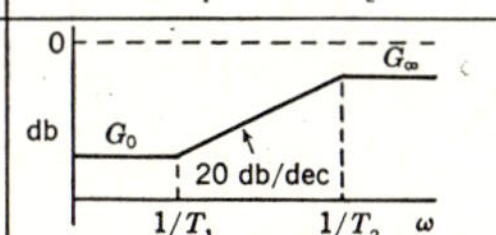	$G_0\dfrac{(T_1s+1)}{(T_2s+1)}$	$\dfrac{1}{1+\frac{n}{b}}$	$\dfrac{1}{1+\frac{en}{b(e+n)}}$	$a(e+n)RC$	$\left[\dfrac{b+\frac{en}{e+n}}{b+n}\right]T_1$
6 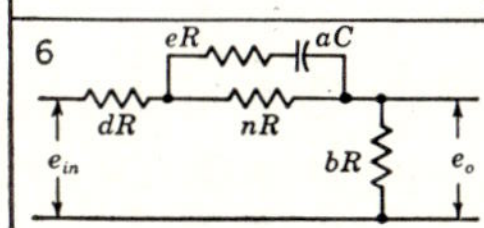	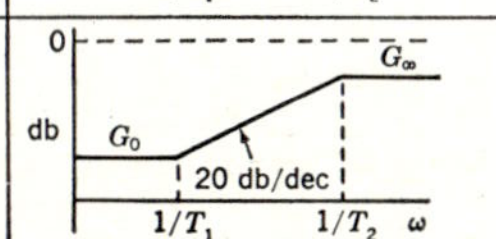	$G_0\dfrac{(T_1s+1)}{(T_2s+1)}$	$\dfrac{1}{1+\frac{d+n}{b}}$	$\dfrac{1}{1+\frac{d}{b}+\frac{en}{b(e+n)}}$	$a(e+n)RC$	$\left[\dfrac{b+d+\frac{en}{e+n}}{b+d+n}\right]T_1$

Table 12-2 Transfer Functions of *R-C* Networks (*Continued*)

Network	Attenuation characteristic	Transfer function	G_0	G_∞	T_1	T_2
7 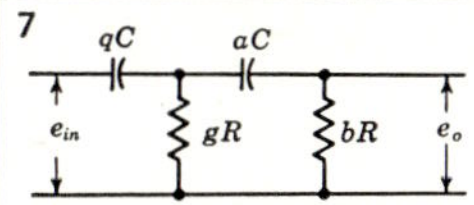	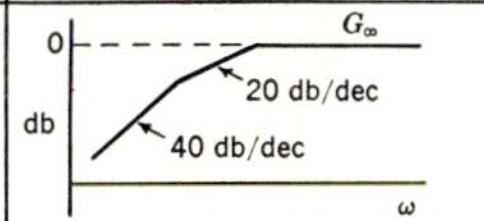	$\dfrac{T_1T_2s^2}{T_1T_2s^2+\left[T_1\left(1+\frac{g}{b}\right)+T_2\right]s+1}$	0	1	$abRC$	$gqRC$
8 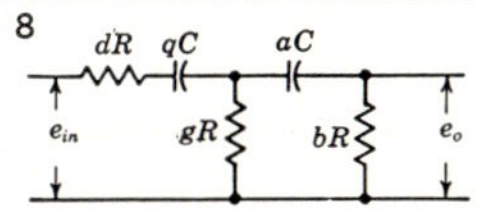	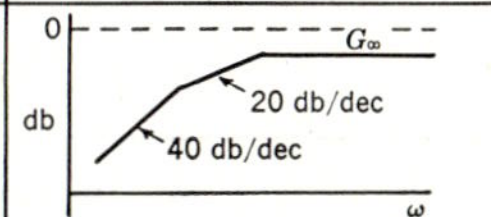	$\dfrac{T_1T_2s^2}{T_1T_2s^2\left[1+\frac{(b+g)d}{bg}\right]+\left[T_1\left(1+\frac{g}{b}\right)+T_2\left(1+\frac{d}{g}\right)\right]s+1}$	0	$\dfrac{1}{1+\frac{(b+g)d}{bg}}$	$abRC$	$gqRC$
9 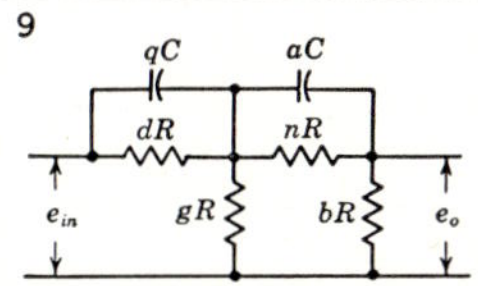	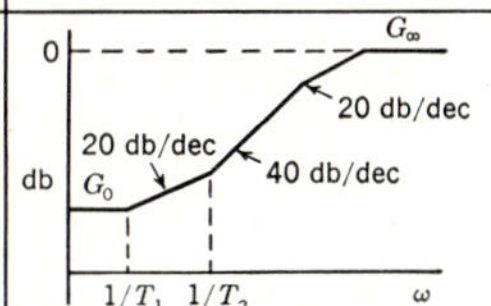	$\dfrac{(T_1s+1)(T_2s+1)}{T_1T_2s^2+\left\{T_1\left(1+\frac{n}{b}\right)+T_2\left[1+\frac{(b+g)d}{bg}\right]\right\}s+\frac{1}{G_0}}$	$\dfrac{1}{1+\frac{n}{b}+\frac{(b+g+n)d}{bg}}$	1	$dqRC$	$anRC=\dfrac{an}{dq}T_1$
10 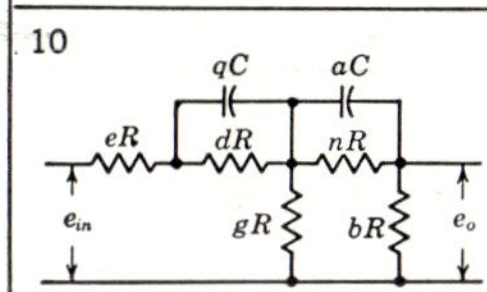	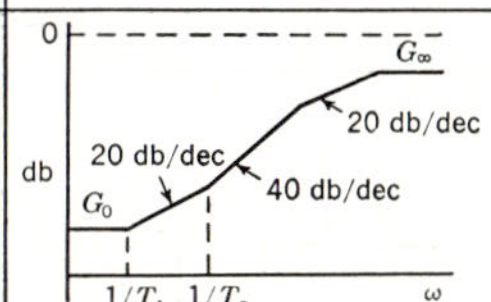	$\dfrac{(T_1s+1)(T_2s+1)}{T_1T_2\left[1+\frac{(b+g)e}{bg}\right]s^2+\left\{T_1\left[1+\frac{n}{b}+\frac{(b+g+n)e}{bg}\right]+T_2\left[1+\frac{(b+g)(d+e)}{bg}\right]\right\}s+\frac{1}{G_0}}$	$\dfrac{1}{1+\frac{n}{b}+\frac{(b+g+n)(d+e)}{bg}}$	$\dfrac{1}{1+\frac{(b+g)e}{bg}}$	$dqRC$	$anRC=\dfrac{an}{dq}T_1$
11 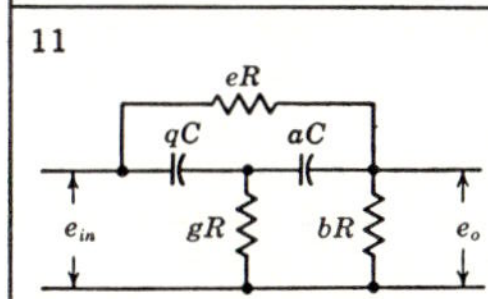	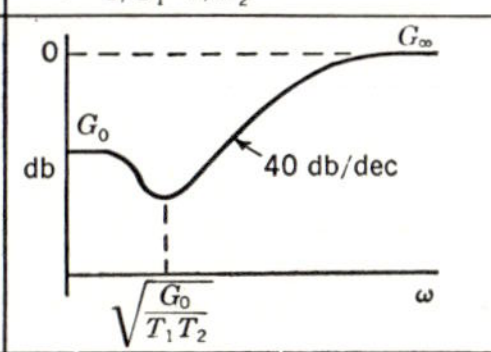	$\dfrac{T_1T_2s^2+\left[T_1\frac{g}{e}+T\left(\frac{b}{b+e}\right)\right]s+G_0}{T_1T_2s^2+\left[T_1+T_2\left(1+\frac{a}{q}\right)\right]+1}$	$\dfrac{1}{1+\frac{e}{b}}$	1	$a\dfrac{be}{b+e}RC$	gqT_1

TABLE 12-2 TRANSFER FUNCTIONS OF *R-C* NETWORKS (*Continued*)

Network	Attenuation characteristic	Transfer function	G_0	G_∞	T_1	T_2
12 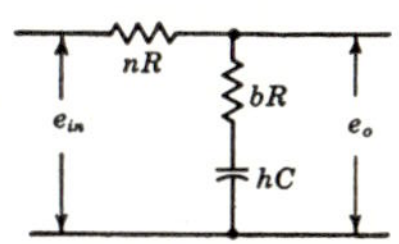	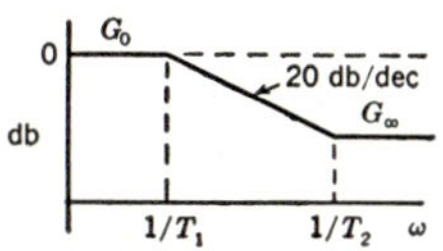	$\dfrac{T_2s+1}{T_1s+1}$	1	$\dfrac{1}{1+\frac{n}{b}}$	$\dfrac{(b+n)}{b}T_2$	$bhRC$
13 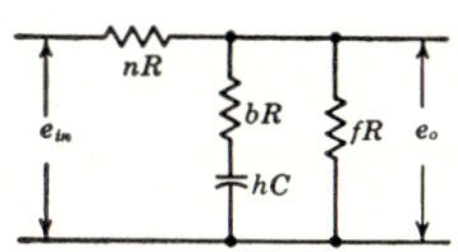	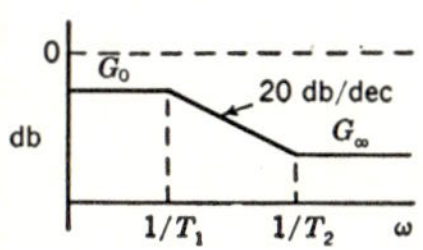	$G_0\dfrac{(T_2s+1)}{(T_1s+1)}$	$\dfrac{1}{1+\frac{n}{f}}$	$\dfrac{1}{1+\frac{n}{b}+\frac{n}{f}}$	$\left[1+\dfrac{fn}{b(f+n)}\right]T_2$	$bhRC$
14 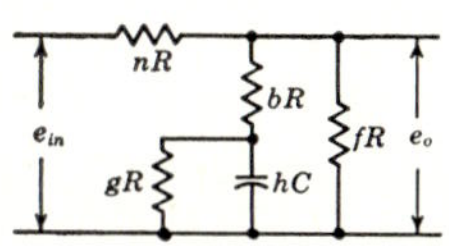	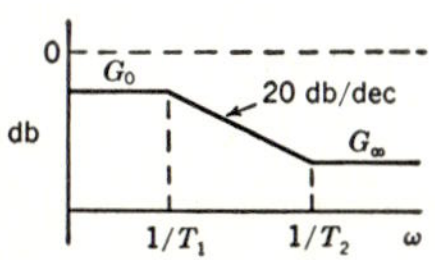	$G_0\dfrac{(T_2s+1)}{(T_1s+1)}$	$\dfrac{1}{1+\frac{n}{b+g}+\frac{n}{f}}$	$\dfrac{1}{1+\frac{n}{b}+\frac{n}{f}}$	$\left[\dfrac{1+\frac{n}{b}+\frac{n}{f}}{1+\frac{n}{b+g}+\frac{n}{f}}\right]T_2$	$\left(\dfrac{bg}{b+g}\right)hRC$

Table 12-2 Transfer Functions of R-C Networks (Continued)

Network	Attenuation characteristic	Transfer function	G_0	G_∞	T_1	T_2
15 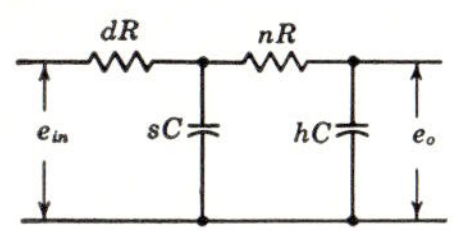	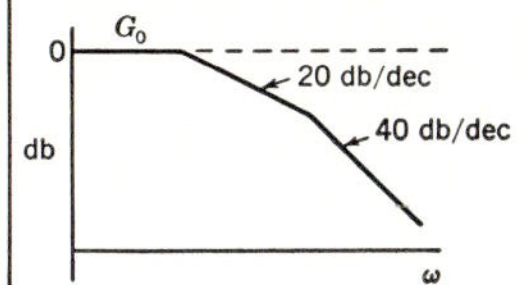	$\dfrac{1}{T_1T_2s^2+\left[T_1\left(1+\frac{d}{n}\right)+T_2\right]s+1}$	1	0	$hnRC$	$dsRC$
16 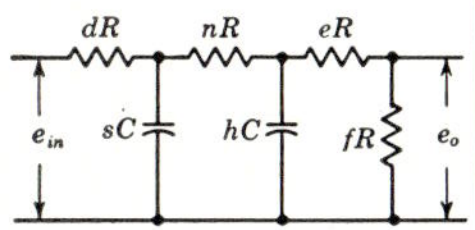	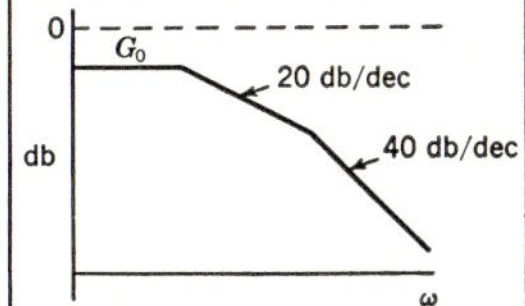	$\dfrac{1}{T_1T_2s^2+\left[T_1\left(1+\frac{d}{n}\right)+T_2\left(\frac{e+f+n}{f}\right)\right]s+\frac{1}{G_0}}$	$\dfrac{f}{d+e+f+n}$	0	$\dfrac{(e+f)}{f}hnRC$	$dsRC$
17 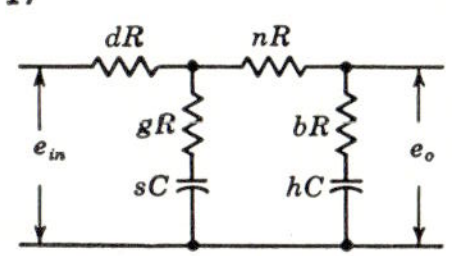	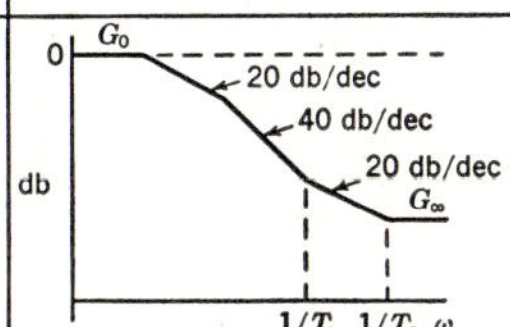 Drawn for $T_1 > T_2$	$\dfrac{(T_1s+1)(T_2s+1)}{T_1T_2\left[\left(1+\frac{d}{g}\right)\left(1+\frac{n}{b}\right)+\frac{d}{b}\right]s^2+\left[T_1\left(1+\frac{d}{b}+\frac{n}{b}\right)+T_2\left(1+\frac{d}{g}\right)\right]s+1}$	1	$\dfrac{1}{\left(1+\frac{d}{g}\right)\left(1+\frac{n}{b}\right)+\frac{d}{b}}$	$bhRC$	$gsRC$
18 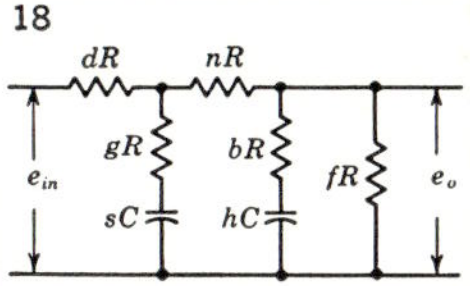	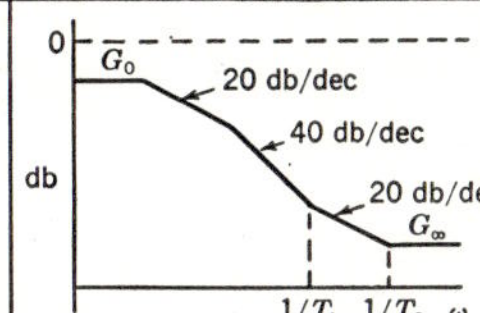 Drawn for $T_1 > T_2$	$\dfrac{\frac{f}{f+n}(T_1s+1)(T_2s+1)}{T_1T_2\left\{\left[1+\frac{fn}{b(f+n)}\right]\left[1+\frac{d}{g}\right]+\frac{d}{b}\frac{(b+f)}{(f+n)}\right\}s^2+\left\{T_1\left[1+\frac{fn}{b(f+n)}+\frac{d(b+f)}{b(f+n)}\right]+T_2\left(1+\frac{d}{g}+\frac{d}{f+n}\right)\right\}s+\frac{f}{G_0(f+n)}}$	$\dfrac{f}{d+f+n}$	$\dfrac{\frac{f}{f+n}}{\left[1+\frac{fn}{b(f+n)}\right]\left[1+\frac{d}{g}\right]+\frac{d}{b}\frac{(b+f)}{(f+n)}}$	$bhRC$	$gsRC$

Table 12-2 Transfer Functions of *R-C* Networks (*Continued*)

Network	Attenuation characteristic	Transfer function	G_0	G_∞	T_1	T_2	T_3
19 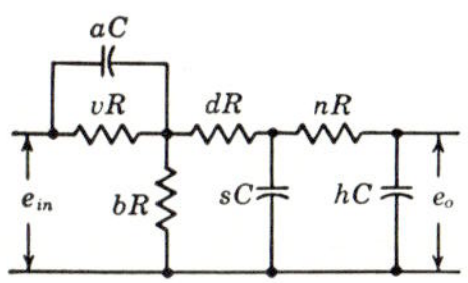	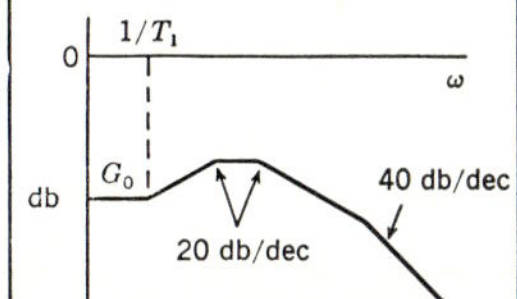 $T_1 > T_2$ and $T_1 > T_3$ for attenuation shown All factors in denominator are different.	$\dfrac{(T_1+s+1)}{T_1T_2T_3s^3 + \left\{T_1T_2\left(1+\frac{h}{s}\right)+T_1T_3+T_2T_3\left[1+\frac{v}{b}\left(1+\frac{b}{d}\right)\right]\right\}s^2 + \left\{T_1+T_2\left(1+\frac{h}{s}\right)\left[1+\frac{v}{b}\left(1+\frac{b}{d}\right)\right]+T_3\left(1+\frac{v}{b}\right)\right\}s + \frac{1}{G_0}}$	$\dfrac{1}{1+\frac{v}{b}}$	0	$avRC$	$dsRC$	$hnRC$
20 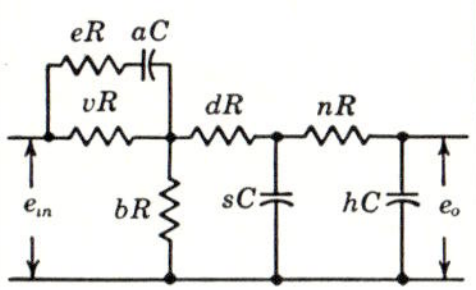	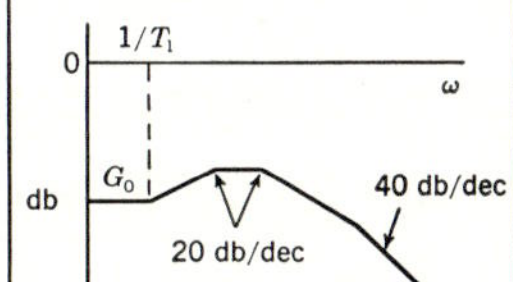 $T_1 > T_2$ and $T_1 > T_3$ for attenuation shown All factors in denominator are different.	$\dfrac{(T_1s+1)}{T_1T_2T_3\left[1+\frac{ev}{b(e+v)}\right]s^3 + \left\{T_1T_2\left(1+\frac{h}{s}\right)\left[1+\frac{ev}{b(e+v)}\left(1+\frac{b}{d}\right)\right] + T_1T_3\left[1+\frac{ev}{b(e+v)}\right]+T_2T_3\left[1+\frac{v}{b}\left(1+\frac{b}{d}\right)\right]\right\}s^2+\left\{T_1\left[1+\frac{ev}{b(e+v)}\right] + T_2\left(1+\frac{h}{s}\right)\left[1+\frac{v}{b}\left(1+\frac{b}{d}\right)\right]+T_3\left(1+\frac{v}{b}\right)\right\}s+\frac{1}{G_0}}$	$\dfrac{1}{1+\frac{v}{b}}$	0	$a(e+v)RC$	$dsRC$	$hnRC$
21 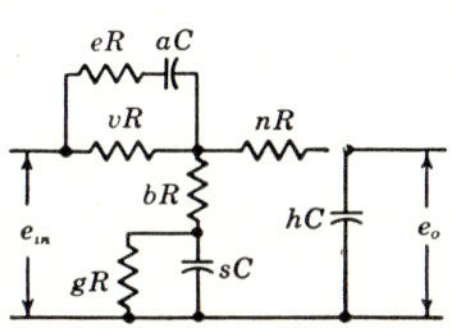	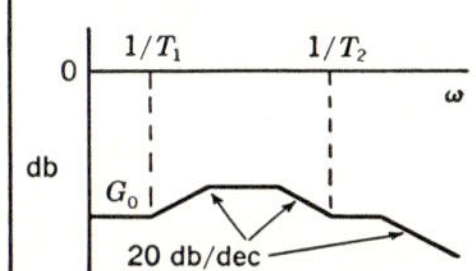 $T_1 > T_2$ and $T_1 > T_3$, G_0 small for initial lead effect. All factors in denominator are different.	$\dfrac{(T_1s+1)(T_2s+1)}{T_1T_2T_3\left[1+\left(\frac{ev}{e+v}\right)\left(\frac{b+n}{bn}\right)\right]s^3 + \left\{T_1T_2\left[1+\frac{ev}{b(e+v)}\right] + T_1T_3\left[1+\left(\frac{ev}{e+v}\right)\frac{(b+g+n)}{(b+g)n}\right]+T_2T_3\left[1+\frac{(b+n)v}{bn}\right]\right\}s^2 + \left\{T_1\left[1+\frac{ev}{(b+g)(e+v)}\right]+T_2\left(1+\frac{v}{b}\right)+T_3\left[1+\frac{(b+g+n)v}{(b+g)n}\right]\right\}s + \frac{1}{G_0}}$	$\dfrac{1}{1+\frac{v}{b+g}}$	0	$a(e+v)RC$	$\dfrac{bg}{b+g}\,sRC$	$hnRC$

TABLE 12-2 TRANSFER FUNCTIONS OF *R-C* NETWORKS (*Continued*)

Network	Attenuation characteristic	Transfer function	G_0	G_∞	T_1	T_2
22 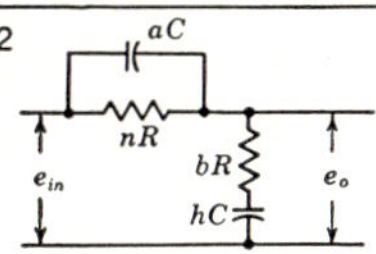	Drawn for $T_1 > T_2$ 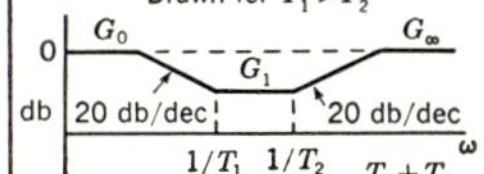 $G_1 = \frac{T_1 + T_2}{T_1(1+n/b)+T_2}$	$\frac{(T_1s+1)(T_2s+1)}{T_1T_2s^2+\left[T_1\left(1+\frac{n}{b}\right)+T_2\right]s+1}$	1	1	$bhRC$	$anRC$
23 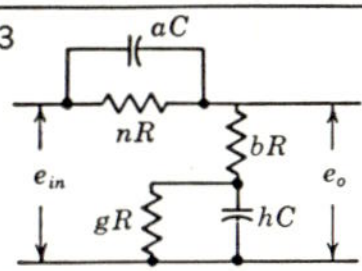	Drawn for $T_1 > T_2$ 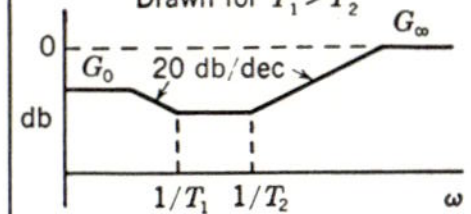	$\frac{(T_1s+1)(T_2s+1)}{T_1T_2s^2+\left[T_1\left(1+\frac{n}{b}\right)+T_2\right]s+\frac{1}{G_0}}$	$\frac{1}{1+\frac{n}{b+g}}$	1	$\frac{bg}{b+g}hRC$	$anRC$
24 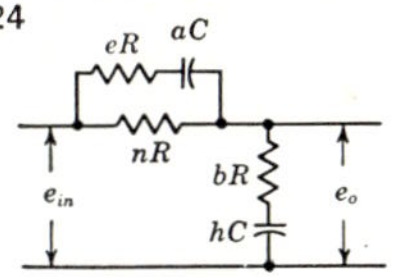	Drawn for $T_1 > T_2$ 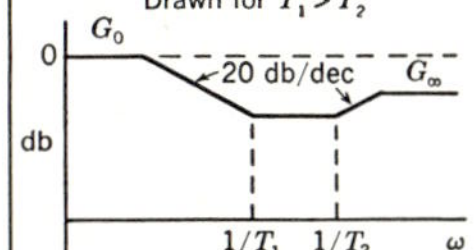	$\frac{(T_1s+1)(T_2s+1)}{T_1T_2\left[1+\frac{en}{b(e+n)}\right]s^2+\left[T_1\left(1+\frac{n}{b}\right)+T_2\right]s+1}$	1	$\frac{1}{1+\frac{en}{b(e+n)}}$	$bhRC$	$a(e+n)RC$
25 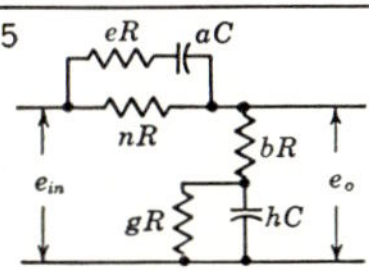	Drawn for $T_1 > T_2$ 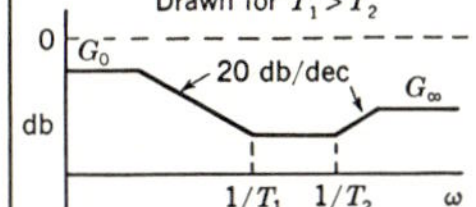	$\frac{(T_1s+1)(T_2s+1)}{T_1T_2\left[1+\frac{en}{b(e+n)}\right]s^2+\left\{T_1\left(1+\frac{n}{b}\right)+T_2\left[1+\frac{en}{(b+g)(e+n)}\right]\right\}s+\frac{1}{G_0}}$	$\frac{1}{1+\frac{n}{b+g}}$	$\frac{1}{1+\frac{en}{b(e+n)}}$	$\frac{bg}{b+g}hRC$	$a(e+n)RC$
26 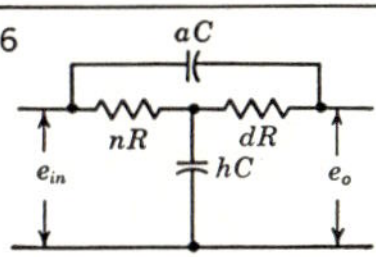	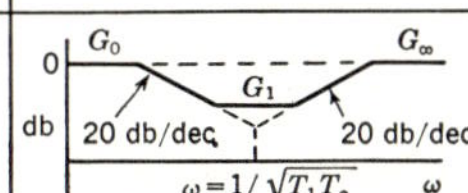 $G_1 = \left(1+\frac{d}{n}\right)\frac{T_2}{T_1}+1$	$\frac{T_1T_2s^2+T_2s+1}{T_1T_2s^2+\left[T_1\left(1+\frac{n}{d}\right)+T_2\right]s+1}$	1	1	$\frac{dn}{d+n}hRC$	$a(d+n)RC$

TABLE 12-2 TRANSFER FUNCTIONS OF *R-C* NETWORKS (*Continued*)

Network	Attenuation characteristic	Transfer function	G_0	G_∞	T_1	T_2
27 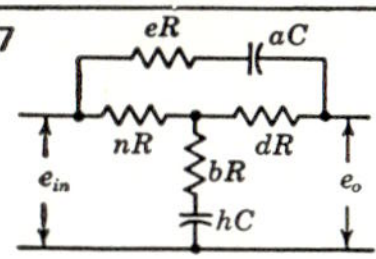	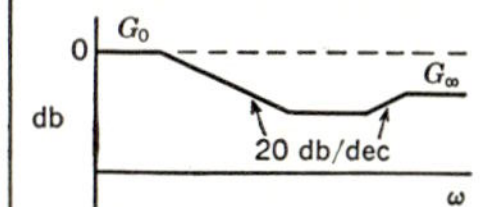	$\dfrac{T_1T_2\left[1+\dfrac{dn}{b(d+e+n)}\right]s^2+(T_1+T_2)s+1}{T_1T_2\left[1+\dfrac{n(d+e)}{b(d+e+n)}\right]s^2+\left[T_1\left(1+\dfrac{n}{b}\right)+T_2\right]s+1}$	1	$\dfrac{1+\dfrac{dn}{b(d+e+n)}}{1+\dfrac{n(d+e)}{b(d+e+n)}}$	$bhRC$	$a(d+e+n)RC$
28 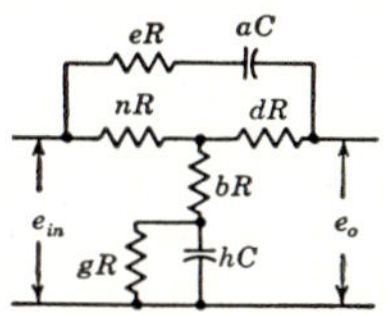	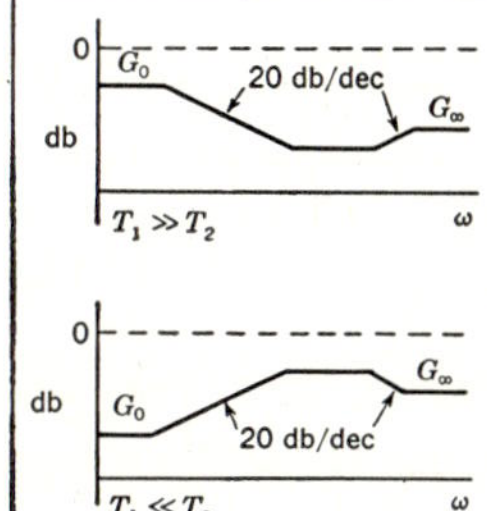	$\dfrac{T_1T_2\left[1+\dfrac{dn}{b(d+e+n)}\right]s^2+\left\{T_1+T_2\left[1+\dfrac{dn}{(b+g)(d+e+n)}\right]\right\}s+1}{T_1T_2\left[1+\dfrac{n(d+e)}{b(d+e+n)}\right]s^2+\left\{T_1\left(1+\dfrac{n}{b}\right)+T_2\left[1+\dfrac{n(d+e)}{(b+g)(d+e+n)}\right]\right\}s+\dfrac{1}{G_0}}$	$\dfrac{1}{1+\dfrac{n}{b+g}}$	$\dfrac{1+\dfrac{dn}{b(d+e+n)}}{1+\dfrac{n(d+e)}{b(d+e+n)}}$	$\dfrac{bg}{b+g}hRC$	$a(d+e+n)RC$
29 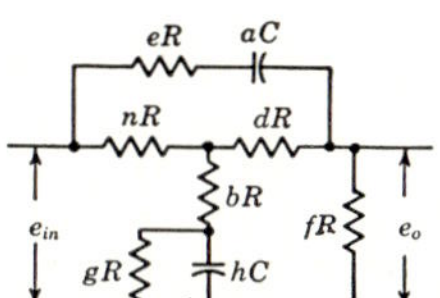	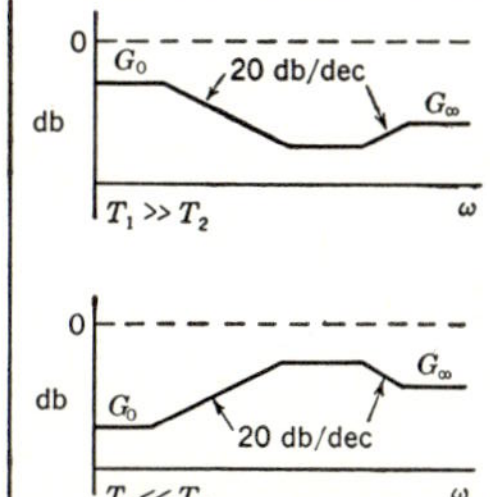 Attenuation curves are for fairly large values of *f* only.	$\dfrac{T_1T_2\left[1+\dfrac{dn}{b(d+e+n)}\right]s^2+\left\{T_1+T_2\left[1+\dfrac{dn}{(b+g)(d+e+n)}\right]\right\}s+1}{T_1T_2\left[1+\dfrac{n(d+e)}{b(d+e+n)}+\dfrac{1}{f}\left(\dfrac{e}{d+e+n}\right)\left(d+n+\dfrac{dn}{b}\right)\right]s^2+\left\{T_1\left(1+\dfrac{n}{b}\right)+\dfrac{T_1}{f}\left(d+n+\dfrac{dn}{b}\right)+T_2\left[1+\dfrac{n(d+e)}{(b+g)(d+e+n)}\right]+\dfrac{T_2}{f}\left(\dfrac{e}{d+e+n}\right)\left(d+n+\dfrac{dn}{b+g}\right)\right\}s+\dfrac{1}{G_0}}$	$\dfrac{1}{1+\dfrac{n}{b+g}+\dfrac{1}{f}\left(d+n+\dfrac{dn}{b+g}\right)}$	$\dfrac{1+\dfrac{dn}{b(d+e+n)}}{1+\dfrac{n(d+e)}{b(d+e+n)}+\dfrac{1}{f}\left(\dfrac{e}{d+e+n}\right)\times\left(d+n+\dfrac{dn}{b}\right)}$	$\dfrac{bg}{b+g}hRC$	$a(d+e+n)RC$

transfer functions of the two networks. If the networks cannot be cascaded without interaction, they must be isolated from each other by amplifier stages. As an example, the transfer function of the resistance-coupled amplifier stage given in Eq. (12-97) could have been obtained by approximating the circuit with networks 1 and 12 in Table 12-2 connected in tandem, with the multiplier b in network 12 equal to zero.

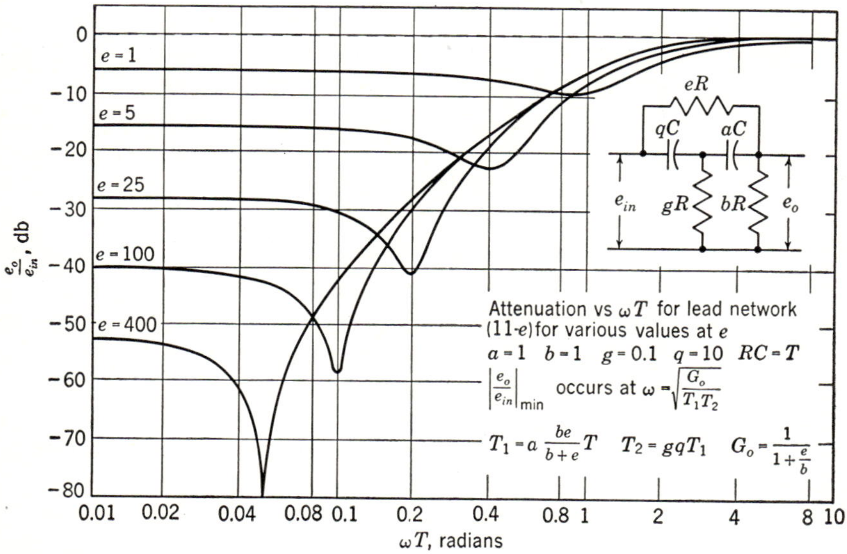

FIG. 12-5 Effect of resistor eR on attenuation-frequency characteristics of network 11.

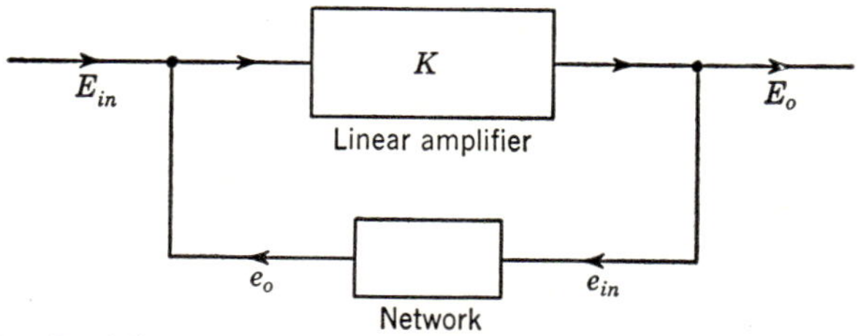

FIG. 12-6 Circuit of linear amplifier and feedback loop in which R-C networks may be used to obtain desired attenuation-frequency characteristics.

In the case of network 11, the bridging resistor eR can be adjusted to produce varying degrees of sharpness in the null shown in the response plot. The effects of such variations are shown in Fig. 12-5. When a network in Table 12-2 is used in the feedback loop around a linear amplifier as in Fig. 12-6, the over-all transfer function can be written as

$$\frac{E_o}{E_{in}} = \frac{e_{in}}{e_o}\left[\frac{Ke_o/e_{in}}{1 + Ke_o/e_{in}}\right] = \left[\frac{1}{1/K + e_o/e_{in}}\right] \tag{12-101}$$

where e_o/e_{in} is the transfer function of the network alone, K is the amplifier gain, and E_o/E_{in} is the transfer function of the amplifier with the feedback loop. When the amplifier gain is very much greater than the maximum network attenuation, so that $1/K \ll e_o/e_{in}$, the transfer function for the whole system is

$$\frac{E_o}{E_{in}} \cong \frac{e_{in}}{e_o} \tag{12-102}$$

Equation (12-102) indicates that the over-all attenuation characteristic of the system is the inverse of that shown in Table 12-2 for the network itself. Thus, the alternatives available for obtaining a desired attenuation-frequency characteristic are more numerous than is indicated by the network characteristics themselves.

12-7 Evaluation of Inverse Transforms by Residues

As mentioned in Sec. 12-4, the ultimate form of $f(s)$ is generally a proper fraction $g(s)/h(s)$. The roots of the polynomial $h(s)$ may be real, imaginary, or complex. When they are real and negative, such as $s = -a$, the transient solution will be a damped nonoscillatory wave. If the roots are purely imaginary, such as $s = +j\omega$, the solution will be a steady-state oscillation; and if they are complex, such as $s = -a \pm j\omega$, the solution will be a damped oscillation.

In the expression $f(s) = g(s)/h(s)$, those critical values of s which make $f(s) = 0$ are described as *zeros* of $f(s)$. Those values of s which make $f(s)$ infinite are identified as *poles* of $f(s)$, and these are the same as the roots of $h(s)$. Consider the function

$$f(s) = \frac{(s+1)(s+2)^2}{(s+3)^3(s+2+j2)(s+2-j2)} \tag{12-103}$$

Equation (12-103) has a *first-order zero* at $s = -1$ and a *second-order zero* at $s = -2$ and at $s = \infty$. It has a *third-order pole* at $s = -3$ and a pair of conjugate complex poles at $s = -2 \pm j2$. As another example, the $f(s)$ given in Eq. (12-92) has no zero except at $s = \infty$ and has a single pole at $s = -1/T$.

As discussed in Sec. 12-4, the inverse transform or transient solution of any $f(s)$ is obtained by first locating the poles of the function or, in other words, by finding the roots of its denominator. From these the constants of a partial fraction expansion can be determined; these constants are known as the *residues* of the poles and there is a residue associated with each pole. Note that any poles which correspond to complex values of s must occur in conjugate complex pairs, since the inverse transform of any function of s must be real.

It will be recalled that the definition of the Laplace transform of a function was given as

$$f(s) = \int_0^\infty f(t)\epsilon^{-st}\,dt \tag{12-104}$$

Now, if $f(s)$ is known and $f(t)$ is to be determined, an integral can be developed from Eq. (12-104) which is known as the inversion integral.[1] The inversion integral gives a solution for $f(t)$ when $f(s)$ is known, and can be evaluated by the method of residues. From a practical standpoint, the inversion integral need not actually be written because it can be shown that if $f(s) = g(s)/h(s)$, as defined previously, the solution of the integral by means of a Laurent series expansion yields the result that

$$f(t) = \Sigma\,\text{Res}\,\frac{g(s)\epsilon^{st}}{h(s)} \tag{12-105}$$

where Σ Res is read "the sum of the residues of." Thus, if

$$f(s) = \frac{A_1}{s} + \frac{A_2}{s^2} + \frac{A_3}{(s+\omega)} + \frac{A_4}{(s+a)^2} \tag{12-106}$$

Eq. (12-105) would be written

$$f(t) = \Sigma\,\text{Res}\left[\frac{A_1\epsilon^{st}}{s} + \frac{A_2\epsilon^{st}}{s^2} + \frac{A_3\epsilon^{st}}{(s+\omega)} + \frac{A_4\epsilon^{st}}{(s+a)^2}\right] \tag{12-107}$$

When the residue of each term in Eq. (12-107) has been evaluated, $f(t)$ will have been determined. Two formulas are available for the residues of functions. If $f(s) = g(s)/h(s)$ and the quantity $-a$ is a *first-order* pole of $f(s)$, so that $h(-a) = 0$, then

$$\text{Res}\,[f(s)]_{-a} = \frac{g(-a)}{h'(-a)} \tag{12-108}$$

where $h'(-a)$ is the value of the first derivative of $h(s)$ at $s = -a$.

[1] Further information on the inversion integral, theory of functions of a complex variable, and transform methods is given in more advanced texts. See R. V. Churchill, "Modern Operational Mathematics in Engineering," McGraw-Hill Book Company, Inc., New York, 1944; M. F. Gardner and J. L. Barnes, "Transients in Linear Systems," John Wiley & Sons, Inc., New York, 1942; K. O. Friedrichs, Integration in the Complex Plane, *Elec. Eng.*, **61** (3): 139–143, March, 1942; J. Millman, Laplacian Transform Analysis of Circuits with Linear Lumped Parameters, *Elec. Eng.*, **61** (4): 197–205, April, 1942; C. R. Wylie, Jr., "Advanced Engineering Mathematics," McGraw-Hill Book Company, Inc., New York, 1951; J. C. Jaeger, "An Introduction to the Laplace Transformation," John Wiley & Sons, Inc., New York, 1949; A. Bronwell, "Advanced Mathematics in Physics and Engineering," McGraw-Hill Book Company, Inc., New York, 1953; H. S. Carslaw and J. C. Jaeger, "Operational Methods in Applied Mathematics," Oxford University Press, New York, 1941; A. Erdelyi, "Tables of Integral Transforms," Vols. 1 and 2, McGraw-Hill Book Company, New York, 1955; and E. A. Guillemin, "The Mathematics of Circuit Analysis," John Wiley & Sons, Inc., New York, 1949.

If $f(s) = \epsilon^{st}/(s + a)^n$, where $-a$ is a pole of order n, then

$$\text{Res}\left[\frac{\epsilon^{st}}{(s+a)^n}\right]_{-a} = \frac{1}{(n-1)!}\frac{d^{n-1}\epsilon^{st}}{ds^{n-1}}\bigg|_{s=-a} \tag{12-109}$$

Consider the $f(s)$ obtained in Eq. (12-30), which was

$$f(s) = \frac{100}{s(s+100)} \tag{12-110}$$

This function has two first-order poles, one at $s = 0$ and one at $s = -100$. The solution according to Eq. (12-105) is

$$f(t) = \Sigma\, \text{Res}\left[\frac{100\epsilon^{st}}{s(s+100)}\right] \tag{12-111}$$

To find the residue at $s = 0$ by means of Eq. (12-108), let

$$g(s) = \frac{100\epsilon^{st}}{s+100}$$

and $h(s) = s$. Then $g(0) = 1$ and $h'(0) = 1$, so that the residue at $s = 0$ is $g(0)/h'(0) = 1$. To find the residue at $s = -100$, let $g(s) = 100\epsilon^{sT}/s$ and $h(s) = s + 100$. Then $g(-100) = -\epsilon^{-100t}$ and $h'(-100) = 1$, so that the residue at $s = -100$ is $-\epsilon^{-100t}$. The solution of Eq. (12-111) is the sum of the two residues, or

$$f(t) = 1 - \epsilon^{-100t} \tag{12-112}$$

which is the solution obtained in Eq. (12-31).

The residues of Eq. (12-107) may be evaluated by means of Eq. (12-109):

$$\begin{aligned}
\text{Res}\left[\frac{A_1\epsilon^{st}}{s}\right] &= \frac{A_1\epsilon^0}{0!} = A_1 \\
\text{Res}\left[\frac{A_2\epsilon^{st}}{s^2}\right] &= \frac{A_2}{1!}\frac{d\epsilon^{st}}{ds}\bigg|_{s=0} = A_2 t \\
\text{Res}\left[\frac{A_3\epsilon^{st}}{(s+\omega)}\right] &= \frac{A_3\epsilon^{-\omega t}}{0!} = A_3\epsilon^{-\omega t} \\
\text{Res}\left[\frac{A_4\epsilon^{st}}{(s+a)^2}\right] &= \frac{A_4}{1!}\frac{d\epsilon^{st}}{ds}\bigg|_{s=-a} = A_4 t\epsilon^{-at}
\end{aligned} \tag{12-113}$$

The solution of Eq. (12-106) is the sum of the residues in Eq. (12-113), or

$$f(t) = A_1 + A_2 t + A_3\epsilon^{-\omega t} + A_4 t\epsilon^{-at} \tag{12-114}$$

The formal rule for carrying out this procedure in the general case is known as the Heaviside expansion theorem.

12-8 Pulse Testing of Linear Systems

Many electronic systems are designed to transmit pulses of various shapes. Others, such as audio amplifiers, while not primarily designed for pulse transmission, can be profitably tested with pulses. If the pulse wave form is properly chosen, a transient measurement provides exactly the same information as a steady-state frequency-response measurement but provides it in a different form. Table 12-3, compiled by Oliver,[1]

TABLE 12-3 STEP-FUNCTION RESPONSE OF TYPICAL NETWORKS[1]

Unit step response | System frequency characteristic | Typical network

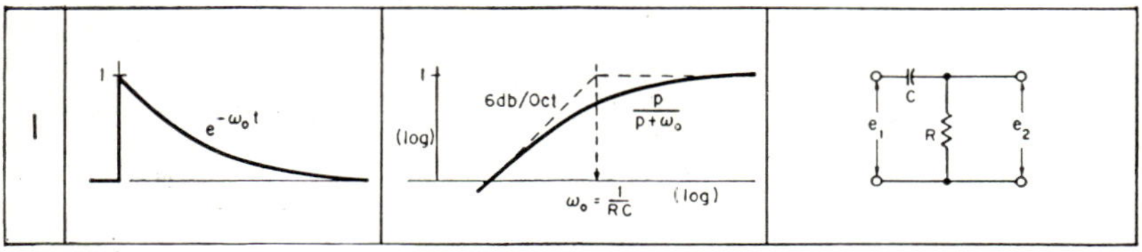

CASE 1. This is the typical simple low-frequency cutoff such as might be produced by a series capacitor–shunt resistor combination. The step response shows an abrupt rise to unity followed by an exponential decay. Usually encountered in amplifier interstages and so-called "differentiating networks." In interstages, f_0 is typically a few cycles; in differentiating networks, f_0 may be as high as several megacycles, in which case the step response is very nearly an impulse. (In this table, note that p is used instead of the s of our notation.)

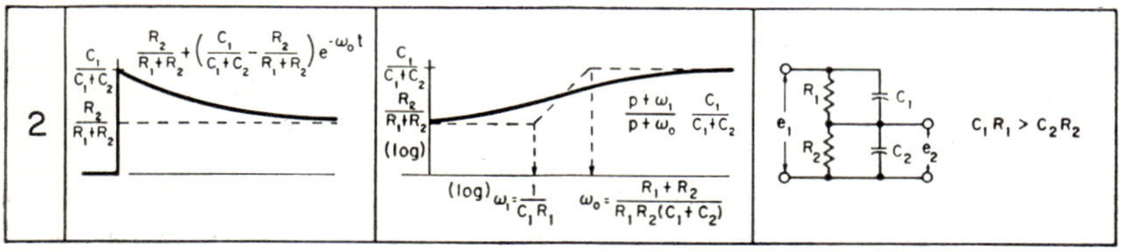

CASE 2. Rising simple step in the *frequency* characteristic. Step response rises initially to amplitude determined by high-frequency transmission, falls exponentially to level determined by low-frequency (or d-c) transmission. This is commonly encountered in improperly compensated resistance-capacitance dividers, such as scope probes, d-c amplifier interstages.

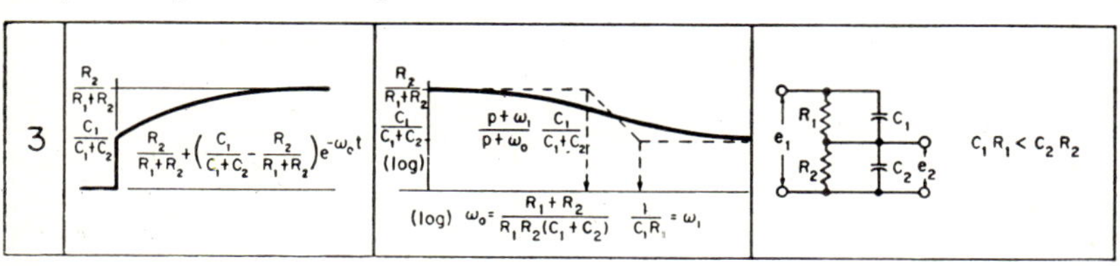

CASE 3. The counterpart of Case 2. Here it is the high-frequency transmission which is deficient.

[1] B. M. Oliver, Square Wave and Pulse Testing of Linear Systems, *Hewlett-Packard J.*, **7** (3): 1–6, November, 1955.

TABLE 12-3 STEP-FUNCTION RESPONSE OF TYPICAL NETWORKS (*Continued*)

Unit step response | System frequency characteristic | Typical network

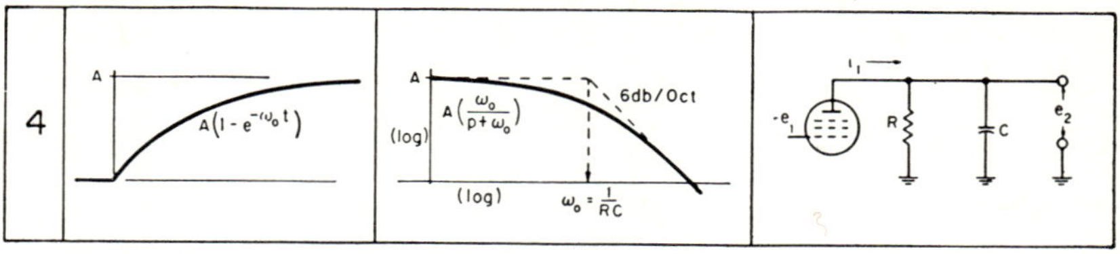

CASE 4. Typical simple high-frequency cutoff such as is produced by a parallel R-C combination. The step response rises exponentially to the final value determined by the low-frequency (or d-c) transmission. Commonly encountered in simple (not "peaked") interstages and wherever shunt capacitance (as from connecting cables) loads down a resistive source.

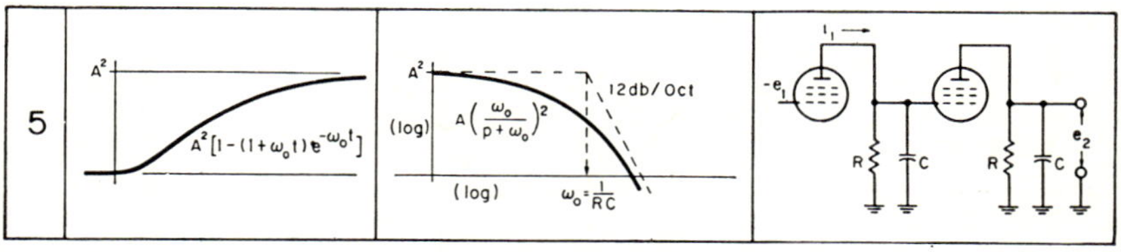

CASE 5. Two simple R-C high-frequency cutoffs in tandem. Typical rise characteristic of two-stage resistance-coupled amplifier without "peaking." Principal differences compared with Case 4: (1) longer rise time for same ω_0, (2) zero slope at $t = 0$. For each additional high-frequency cutoff, one more derivative of step response vanishes at $t = 0$. Thus, if high-frequency transmission falls (ultimately) at $6n$ db/octave, all derivatives of step response up to the nth are zero at $t = 0$.

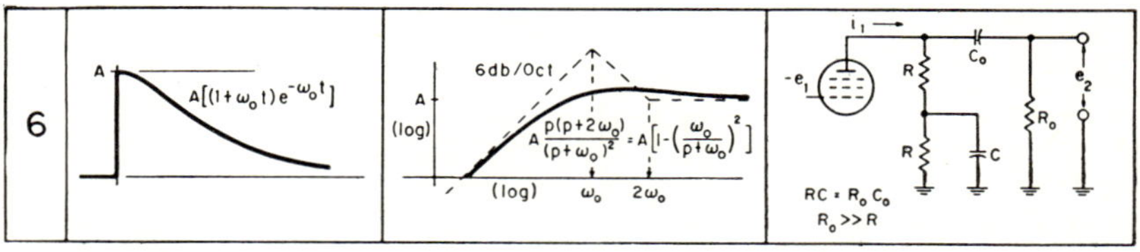

CASE 6. Phase-compensated low-end cutoff. Step-function response falls to zero eventually, but initial slope is zero. As a result, square-wave response shows little or no tilt. May be produced in a single network or by two networks (Cases 1 and 3) in tandem. Often found in video amplifiers.

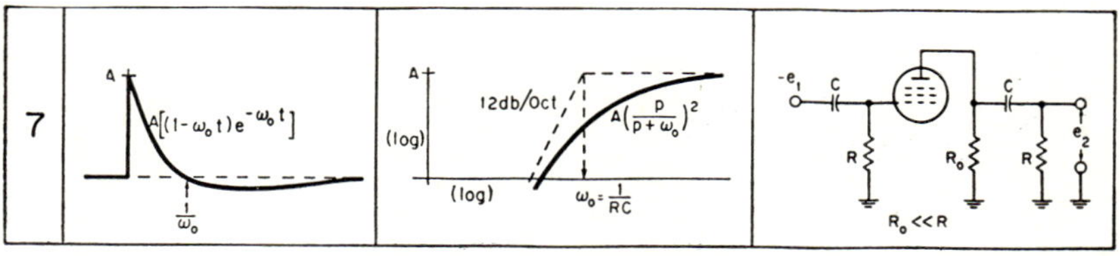

CASE 7. Two simple low-frequency cutoffs (Case 1) in tandem. Typical low-frequency transient response of single-stage resistance-coupled amplifier with input blocking capacitor, two-stage amplifier with input blocking capacitor, or two-stage amplifier with no input blocking capacitor. Principal differences compared with

TABLE 12-3 STEP-FUNCTION RESPONSE OF TYPICAL NETWORKS (*Continued*)

Case 1: (1) faster initial rate of fall for same ω_0, (2) response goes negative crossing axis at $t = 1/\omega_0$. With each additional low-end cutoff, one additional axis crossing is produced. Thus, if the low-end response falls off (ultimately) at $6n$ db/octave, there will be $n - 1$ axis crossings. They do not occur at regular intervals—each successive half-cycle takes longer.

Unit step response | System frequency characteristic | Typical network

8

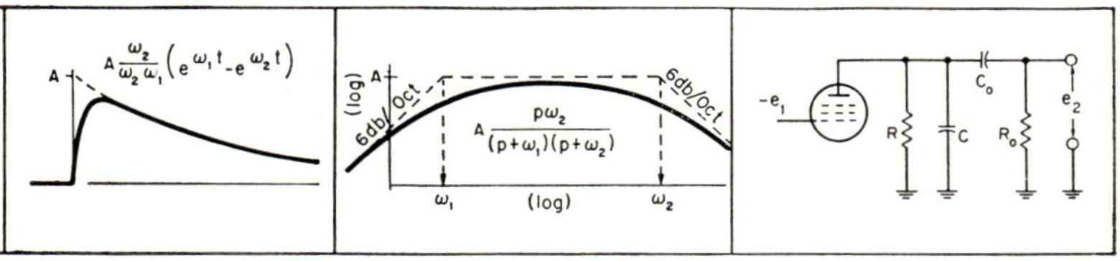

CASE 8. Simple high- and low-frequency cutoff. The step response rises exponentially at a rate determined by high-frequency cutoff, then falls exponentially at a rate determined by low-frequency cutoff. Typical complete resistance-coupled interstage response. If $\omega_2/\omega_1 \gg 1$, then on a slow time scale response looks like Case 1; on a fast time scale response looks like Case 4. If $\omega_2 = \omega_1 = \omega_0$, we have the case of a critically damped R-L-C circuit. The response then becomes $\omega_0 t\epsilon^{-\omega_0 t}$.

9

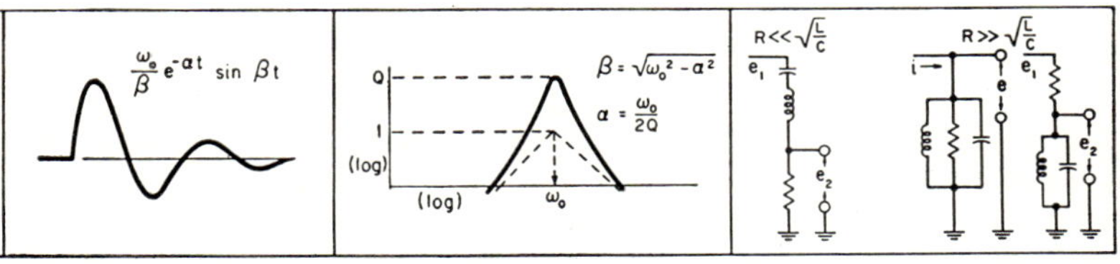

CASE 9. Typical damped oscillation. The dotted lines in the frequency characteristic are the asymptotes which the actual characteristic approaches for $\omega/\omega_0 \ll 1$ and $\omega/\omega_0 \gg 1$. The peak of the resonance curve is Q times as high as the intersection of these asymptotes. For reasonable Q's, such that $\beta \approx \omega_0$, the Q of the circuit may be readily found from the fact that the envelope of oscillation decays to $1/\epsilon$ in Q/π cycles. Thus $Q = \pi n$, where n is the number of cycles to the $1/\epsilon$ point.

10

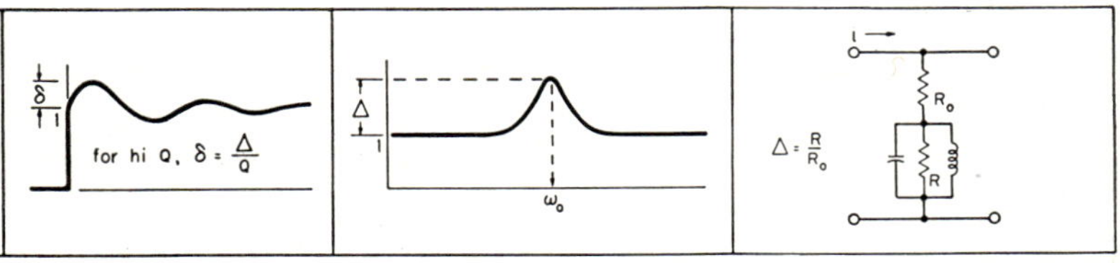

CASE 10. Small resonance in an otherwise flat characteristic. Response consists of unit step due to flat transmission plus damped oscillation due to resonance. Initial amplitude of oscillation is related to amplitude of hump in frequency characteristic as indicated in the illustration. For the same amplitude of hump, increasing the Q decreases amplitude of oscillation but oscillation persists longer. If hump is near top of band, time scale will be such that initial rise of response will not appear so abrupt, but will blend with oscillation to give response like that of overpeaked interstage. Mid-band resonances such as shown in this case often occur as a result of stray feedback paths, such as heater leads, or from attempting to bypass electrolytics with small mica capacitors. (Electrolytics become inductive at high frequencies.)

TABLE 12-3 STEP-FUNCTION RESPONSE OF TYPICAL NETWORKS (*Continued*)

Unit step response | System frequency characteristic | Typical network

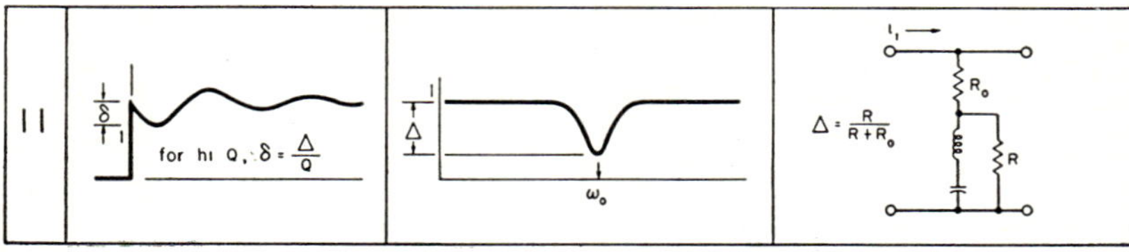

CASE 11. Similar to Case 10, but here there is a resonant dip. Note that the effect of a complete null ($\Delta = 1$) is no worse than that of a 6-db hump. The pilot separation filters used in the coaxial television system produce this type of dip—a complete null. Because their Q is so high (several thousand), the disturbance they produce, while it persists for a long time, is of such low amplitude that it is invisible in the picture.

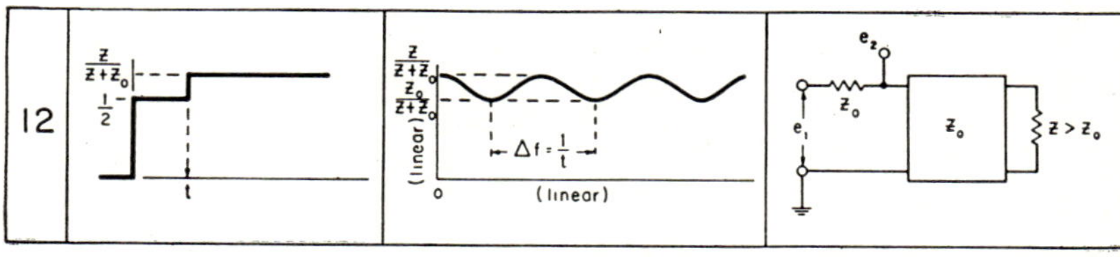

CASE 12. Positive echo. Associated frequency characteristic has nearly sinusoidal ripple in amplitude and phase. Frequency interval between successive maxima or minima is reciprocal of echo delay. The longer the delay, the closer the ripples. Commonly encountered in systems having faulty or misterminated delay lines. Also in measurements where multipath transmissions can exist such as acoustic measurements.

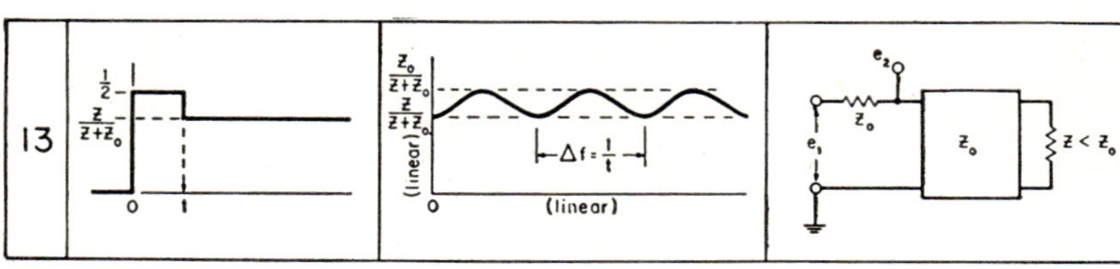

CASE 13. Negative echo. Same frequency ripples as in Case 11 but reversed 180°.

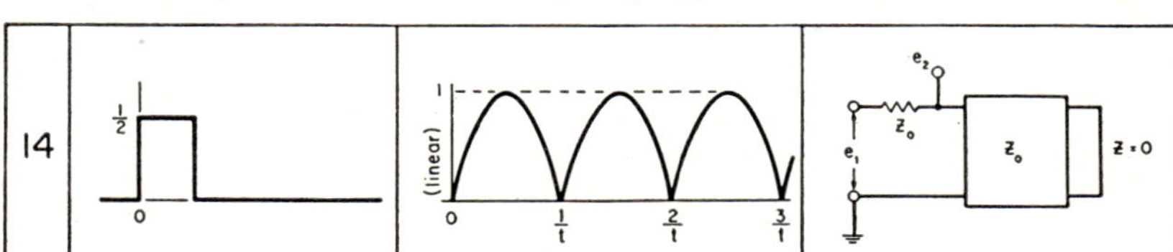

CASE 14. Rectangular pulse response. Can be considered to be a 100 per cent negative echo. Minima of frequency ripples have now become nulls. Shape of amplitude characteristic is that of rectified sine wave. Phase characteristic is sawtooth decreasing from $\pi/2$ linearly to $-\pi/2$ and jumping back to $\pi/2$ at each null. Such a characteristic can be obtained by using a delay line as shown with the near end terminated and the far end shorted.

TABLE 12-3 STEP-FUNCTION RESPONSE OF TYPICAL NETWORKS (*Continued*)

	Unit step response	System frequency characteristic	Typical network
15	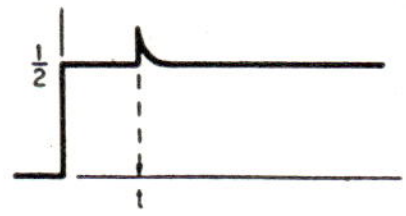	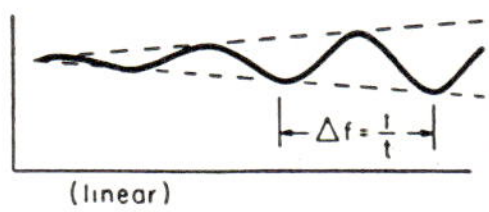	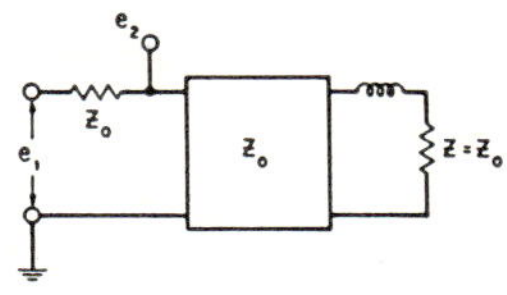

CASE 15. "Differentiated echo." This is the sort of disturbance produced when a delay line is terminated in such a way that the reflection coefficient increases with frequency. Typical causes are (1) series inductance or shunt capacitance in the termination of a smooth line, (2) termination of a constant-k filter in simple resistances. With both ends matched at low frequencies, the transmitted echo involves two reflections, both of which increase with frequency, and so tends to be "doubly differentiated" and smaller.

16	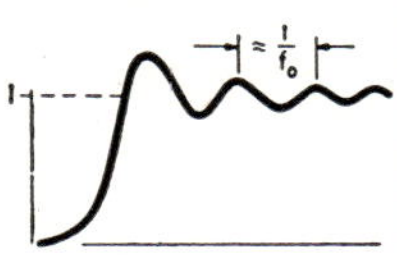	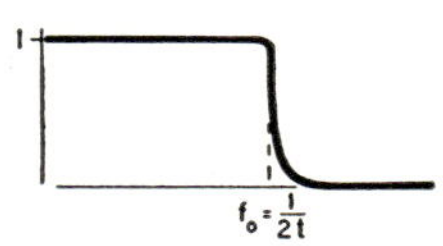	MATCHED CONSTANT-K FILTER

CASE 16. Rise characteristic (qualitative only) of a low-pass filter without phase correction. Following the initial smooth rise there is a ripple whose apparent frequency approaches the cutoff frequency after several cycles. With an increasing number of sections this ripple increased in amplitude and duration. The "ringing sound" so often attributed to sharp-cutoff filters is not due to exaggeration of frequencies near cutoff nor to the sharp cutoff per se, but rather to the delay distortion which exists near cutoff, causing those upper frequencies which are passed to arrive too late and thus be separately audible. The effect is noticeable only in extreme cases, and with proper delay equalization the effect disappears.

17	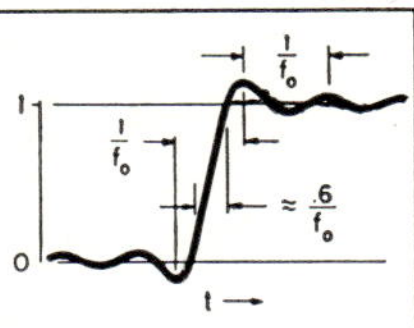	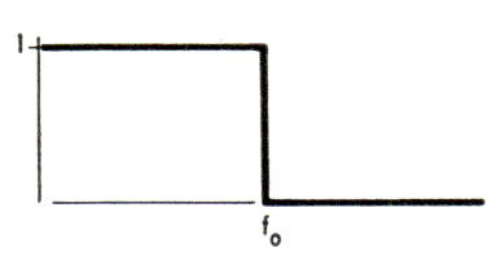	IDEAL LOW-PASS FILTER (PHASE EQUALIZED)

CASE 17. The "ideal" low-pass filter passes all frequencies below f_0 with the same amplitude and delay while attenuating completely those above f_0. Its step response is the sine integral. This function differs from zero (except at discrete points) for all $t > -\infty$. Hence the ideal filter cannot be realized without infinite delay. A practical approximation will have a finite number of wiggles before the main rise. Here again, the ripples in the step response do *not* indicate high-frequency enhancement, but are the "Gibbs effect" encountered in Fourier series, and are properly called *band elimination ripples.*

TABLE 12-3 STEP-FUNCTION RESPONSE OF TYPICAL NETWORKS (*Continued*)

	Unit step response	System frequency characterisitic	Typical network
18	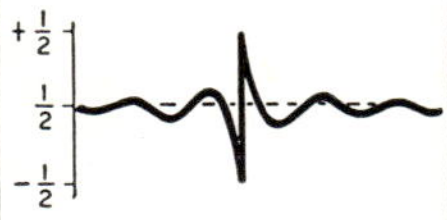	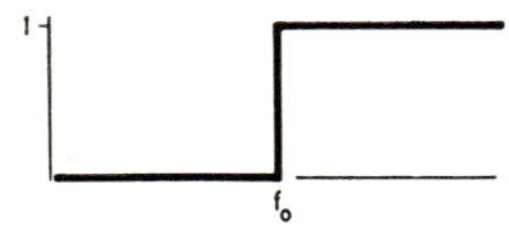	IDEAL HI-PASS FILTER (PHASE EQUALIZED)

CASE 18. The ideal high-pass filter. By superposition the response of this filter is obtained by subtracting the response of the ideal low-pass filter from an equally delayed unit step.

demonstrates the relationships which exist between transient response and frequency response. These two are inextricably interrelated because any arbitrary input pulse or transient can be represented as the sum of a number of sinusoids as determined by the Fourier analysis. If the system is linear, the output will consist of these same sinusoids modified only in amplitude and phase according to the frequency and phase response of the system. Each will be transmitted as if it alone were present (superposition). The simple sum of the output sinusoids will be the output pulse.

Table 12-3 gives the step response of 18 network configurations frequently encountered in electronic systems. A step impulse is one which assumes a value of E immediately after $t = 0$ and has a transform $f(s) = E/s$. The response of each network as shown in the first column will also be the response of the network to the positive step of a square wave, provided that the square-wave frequency is low enough so that the individual transients do not overlap. In other words, the square-wave frequency must be less than $1/2t$, where t is the time required for the step response to reach a relatively constant value. Square waves of long duration would be required, for example, to test systems in which sharp breaks in frequency response occur, such as sharp low-end cutoffs or sharp resonances within the system pass band. The complete step-function response of a system which has a low-frequency cutoff will be accurately displayed on a cathode-ray oscilloscope only if the square-wave repetition frequency is much less than the frequency of cutoff. When the required square-wave frequency is too low to be convenient, it is customary to use a frequency such that the transients from the successive positive and negative steps do not vanish but instead overlap considerably. Figure 12-7 illustrates some typical overlapped low-frequency square-wave responses.

Where the networks are similar, the frequency response data in Table 12-3 can be compared directly with that in Table 12-2. For example, the information given on network 1 is the same in each table.

The response of networks to pulses of various shapes can be computed by the transform methods discussed previously as well as by the use of the superposition theorem as described in Sec. 12-5. The transforms for special pulses or wave forms can be evaluated by means of transform

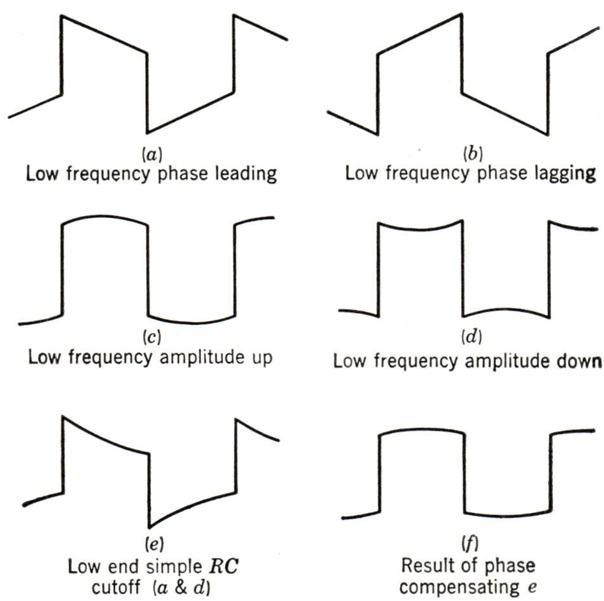

FIG. 12-7 Effects of typical low-end distortions on square waves.

pair 13 in Table 12-1. If $f(s)$ is the transform of $f(t)$, then this pair may be written in equation form as

$$\mathfrak{L}[f(t - a)H(t - a)] = \epsilon^{-sa}f(s) \tag{12-115}$$

where $a > 0$ and $H(t - a)$ is a function which equals 0 for $t < a$ and equals 1 for $t > a$. The function $H(t)$ is described as Heaviside's unit function, and Eq. (12-115) is known as Heaviside's shifting theorem. A graph of $f(t)$ will start at $t = 0$, whereas a graph of $f(t - a)H(t - a)$ will have the same shape as $f(t)$ but will have a starting point which is displaced from $t = 0$ by an amount a. As a consequence, the function $f(t)$ has been shifted to the right a units along the time axis, and this effect has led to the name "shifting theorem."

The relationship in Eq. (12-115) can be proved by taking the transform of $f(t - a)H(t - a)$. According to Eq. (12-18) this is

$$\mathfrak{L}[f(t - a)H(t - a)] = \int_0^\infty \epsilon^{-st}f(t - a)H(t - a)\,dt \tag{12-116}$$

Since $H(t - a) = 0$ for $t < a$,

$$\mathcal{L}[f(t-a)H(t-a)] = \int_a^\infty \epsilon^{-st} f(t-a)\,dt = \int_0^\infty \epsilon^{-as}\epsilon^{-st} f(t)\,dt = \epsilon^{-as} f(s) \tag{12-117}$$

The manner in which Eq. (12-115) or transform pair 13 is used to build up the transforms of step or broken functions is as follows:

(*a*) If $f(t) = E$ for $0 < t < a$
$= 0$ $\quad t > a$

then

$$f(t) = E - EH(t-a) = E[1 - H(t-a)] \tag{12-118}$$

and from Eq. (12-115)

$$f(s) = \frac{E(1 - \epsilon^{-as})}{s} \tag{12-119}$$

(*b*) If $f(t)$ is the positive half of a sine wave,

$$\begin{aligned} f(t) &= E \sin \omega t \qquad \text{for } 0 < t < \pi/\omega \\ &= 0 \qquad\qquad\quad t > \pi/\omega \end{aligned}$$

so that

$$f(t) = E \sin \omega t + E \sin \omega(t - \pi/\omega) H(t - \pi/\omega) \tag{12-120}$$

and

$$f(s) = \frac{E\omega}{s^2 + \omega^2}(1 + \epsilon^{-s\pi/\omega}) \tag{12-121}$$

(*c*) If $f(t)$ is a square wave of period of $2T$,

$$\left.\begin{aligned} f(t) &= E \qquad \text{for} \qquad nT < t < (n+1)T \\ &= -E \qquad (n+1)T < t < (n+2)T \end{aligned}\right\} n = 0, 1, 2, \ldots$$

so that

$$f(t) = E[H(t) - 2H(t-T) + 2H(t-2T) - \cdots] \tag{12-122}$$

and

$$f(s) = \frac{E}{s}(1 - 2\epsilon^{-st} + 2\epsilon^{-2st} - \cdots) = \frac{E}{s}\left(\frac{1 - \epsilon^{-st}}{1 + \epsilon^{-st}}\right) \tag{12-123}$$

(*d*) If $f(t)$ is a sawtooth wave of slope k and period T so that

$$f(t) = kT - \sum_{n=0}^{n} kT\,H(t - nT) \tag{12-124}$$

for $nT < t < (n+1)T$, where $n = 0, 1, 2, \ldots$, then

$$f(s) = \frac{k}{s^2} - \frac{kT}{s}(\epsilon^{-st} + \epsilon^{-2st} + \cdots) = \frac{k}{s^2} - \frac{kT\epsilon^{-st}}{s(1 - \epsilon^{-st})} \tag{12-125}$$

Transform pairs 18–24 in Table 12-1 may be developed by the same methods. In particular, pair 19 is equivalent to Eq. (12-119) with $E = 1$ and $a = 1$.

12-9 Evaluation of Inverse Transforms by a Numerical Method

The methods for obtaining an explicit solution for $f(t)$ when $f(s)$ is known fall into two broad classes—the analytic methods discussed in the previous sections, which yield a solution for $f(t)$ in functional form, and numerical methods, which yield a solution for $f(t)$ as a sequence of numerical values. In the simpler problems, the analytic methods have the advantage of yielding a function solution which contains the original network parameters in explicit form. As the order of the differential equation increases, this advantage is lost and numerical methods may provide a solution with less labor. Boxer and Thaler have developed a numerical method of evaluating inverse transforms which can be applied to linear, time-varying, or nonlinear systems, including those containing time lags.[1] It may be summarized as follows:

Suppose $f(s) = g(s)/h(s)$, where $g(s)$ is a polynomial of degree n, and $h(s)$ is a polynomial of degree m. The steps required to obtain the time response are as follows:

1. Express the function $f(s)$ as a rational fraction in powers of s^{-1} by dividing the numerator and denominator by s^m.

2. Substitute for s^{-K} a rational fraction in powers of z^{-1} obtained from Table 12-4 and rearrange $f(s)$ as a rational fraction in powers of z^{-1} (by definition, $z = \epsilon^{sT}$).

3. Divide the resulting expression by T, where T is the time interval between points along the time axis at which the solution is desired.

4. Expand the fraction by synthetic division into a series of the form

$$D_0 + D_1 z^{-1} + D_2 z^{-2} + \cdots + D_n z^{-n}$$

where D_n, the coefficient of z^{-n}, is the approximate value of the time response at $t = nT$.

This procedure may be illustrated by applying it to some typical problems:

(*a*) Consider the $f(s)$ of a network with a feedback loop to which a step input has been applied:

$$f(s) = \frac{1}{s^3 + s^2 + s} \tag{12-126}$$

The degree of $h(s)$ is $m = 3$, so that dividing the numerator and denominator of Eq. (12-126) by s^3 according to the first step yields

$$f(s) = \frac{s^{-3}}{1 + s^{-1} + s^{-2}} \tag{12-127}$$

[1] R. Boxer and S. Thaler, A Simplified Method of Solving Linear and Nonlinear Systems, *Proc. I.R.E.*, **44** (1): 89–101, January, 1956.

Substituting the corresponding forms for $s^{-K} = s^{-1}$, s^{-2}, and s^{-3} as obtained from Table 12-4 (Step 2) and dividing the result by T (Step 3) give

$$f(z) = \frac{6T^2(z^{-1} + z^{-2})}{(12 + 6T + T^2) - (36 + 6T - 9T^2)z^{-1} + (36 - 6T - 9T^2)z^{-2} - (12 - 6T + T^2)z^{-3}} \tag{12-128}$$

The solution is obtained by choosing a suitable value for T and dividing the denominator of Eq. (12-128) into the numerator (Step 4). Assume $T = 1.0$, causing $f(z)$ to be expressed as

$$f(z) = \frac{6z^{-1} + 6z^{-2}}{19 - 33z^{-1} + 21z^{-2} - 7z^{-3}} \tag{12-129}$$

Carrying out the division process,

$$f(z) = 0.316z^{-1} + 0.864z^{-2} + 1.15z^{-3} + 1.16z^{-4} + 1.06z^{-5} + 0.987z^{-6} + \cdots \tag{12-130}$$

According to Step 4, the coefficient D_n of the general term z^{-n} is the approximate value of the desired $f(t)$ at $t = nT$. In this case, $D_0 = 0$, $D_1 = 0.316$, $D_2 = 0.864$, $D_3 = 1.15$, etc., so that the values of $f(t)$ are

$$\begin{aligned} f(t) &= 0 && \text{at } t = 0 \\ f(t) &= 0.316 && \text{at } t = 1 \\ f(t) &= 0.864 && \text{at } t = 2, \text{ etc.} \end{aligned}$$

The accuracy of the solution will improve if smaller values of T are taken. If $T = 0.5$ is substituted into Eq. (12-128),

$$f(z) = \frac{1.5z^{-1} + 1.5z^{-2}}{15.25 - 36.75z^{-1} + 30.75z^{-2} - 9.25z^{-3}} \tag{12-131}$$

The division process yields

$$f(z) = 0.0984z^{-1} + 0.335z^{-2} + 0.610z^{-3} + 0.853z^{-4} + \cdots \tag{12-132}$$

The corresponding values of $f(t)$ are

$$\begin{aligned} f(t) &= 0 && \text{at } t = 0 \\ f(t) &= 0.0984 && \text{at } t = 0.5 \\ f(t) &= 0.335 && \text{at } t = 1.0 \\ f(t) &= 0.610 && \text{at } t = 1.5 \\ f(t) &= 0.853 && \text{at } t = 2.0, \text{ etc.} \end{aligned}$$

Note the more precise values of $f(t)$ obtained at $t = 1.0$ and $t = 2.0$ in this case as compared with the case of $T = 1.0$; for example, $f(t) = 0.335$ versus 0.316 at $t = 1.0$; $f(t) = 0.853$ versus 0.864 at $t = 2.0$, etc.

(*b*) The numerical method is equally applicable to the solution of integro-differential equations. To accomplish this, the transform of the equation is taken, including any initial conditions which exist. When the transform has been obtained, the steps enumerated above are applied in the same manner. Before proceeding, however, the transform should be manipulated according to the following rules:

1. If the degree of the denominator of the transform is higher than the degree of the numerator by two or more, no change is made in the transform.

2. If the degree of the numerator is one less than, equal to, or greater than the degree of the denominator, divide until the remainder meets the criterion of Rule 1 above.

Consider the differential equation

$$(D + 1)y = 1 \tag{12-133}$$

which has as an initial condition $y(0) = \mu$. From pair 10 in Table 12-1, the transform of Eq. (12-133) is, if the transform of y is $y(s)$,

$$s\,y(s) - \mu + y(s) = \frac{1}{s} \tag{12-134}$$

Solving for the transform of y, or $y(s)$,

$$y(s) = \frac{1 + s\,\mu}{s\,(s + 1)} = \frac{s\,\mu + 1}{s^2 + s} \tag{12-135}$$

In this case, the degree of the numerator is one less than the degree of the denominator, and, following Rule 2 above, the denominator is divided into the numerator once, which yields

$$y(s) = \frac{\mu}{s} + \frac{1 - \mu}{s^2 + s} \tag{12-136}$$

The first term of Eq. (12-136) is directly transformable, so that only the second term must be operated upon. Suppose that μ is specified as $\mu = 0.1$. Expressing the second term of Eq. (12-136) in powers of s^{-1} according to Step 1 of the numerical method,

$$y(s) = \frac{0.9s^{-2}}{1 + s^{-1}} \tag{12-137}$$

Substituting the z-forms from Table 12-4 and dividing by T, according to steps 2 and 3,

$$y(z) = \frac{0.9T(1 + 10z^{-1} + z^{-2})}{(12 + 6T) - 24z^{-1} + (12 - 6T)z^{-2}} \tag{12-138}$$

Selecting the time interval $T = 0.1$ and carrying out the long division,

$$y(z) = 0.00714 + 0.085z^{-1} + 0.163z^{-2} + 0.233z^{-3} + 0.296z^{-4} + 0.354z^{-5} \quad (12\text{-}139)$$

To Eq. (12-139) must be added the inverse transform of the first term of Eq. (12-136), which is simply μ, or in this case 0.1. From the coefficients of Eq. (12-139), the solution for y is (adding μ to each term)

$$\begin{array}{ll} t = 0 & y = 0.10714 \\ t = 0.1 & y = 0.185 \\ t = 0.2 & y = 0.263 \\ t = 0.3 & y = 0.333, \quad \text{etc.} \end{array}$$

(*c*) In a time-varying system, some network element varies with time. As a consequence, the coefficients of the dependent variable in the differential equation are no longer constants but are functions of time. An example of such an equation is

$$(D + t)y = t \quad (12\text{-}140)$$

where the value of y at $t = 0$ is $y(0) = 0$.

This type of differential equation can be solved using the z-forms of Table 12-4 by the following procedure:

1. Obtain the Laplace transform of the linear differential equation with time-varying coefficients in the usual manner, including initial conditions.
2. Proceed as in the constant-coefficient cases discussed in problems (*a*) and (*b*). In performing the long-division process, however, change the values of $C_n, C_{n-1}, \ldots, C_0$ (which are the coefficients of the differential equation and which vary with time) at each step in the division process to the values at the corresponding times of the functions.

To illustrate this procedure, consider the differential equation given by Eq. (12-140). Designate the coefficient of y which is a function of time as C, so that Eq. (12-140) may be rewritten as

$$(D + C)y = t \quad (12\text{-}141)$$

where $C = t$. Taking the Laplace transform, considering C constant and $y(0) = 0$,

$$y(s) = \frac{1}{s^3 + Cs^2} \quad (12\text{-}142)$$

Expressing Eq. (12-142) in powers of s^{-1},

$$y(s) = \frac{s^{-3}}{1 + Cs^{-1}} \quad (12\text{-}143)$$

Substituting the z-forms from Table 12-4 and dividing by T,

$$y(z) = \frac{T^2(z^{-1} + z^{-2})}{(2 + CT) - (6 + CT)z^{-1} + (6 - CT)z^{-2} - (2 - CT)z^{-3}} \tag{12-144}$$

Selecting the time interval $T = 0.4$,

$$y(z) = \frac{0.16z^{-1} + 0.16z^{-2}}{(2 + 0.4C) - (6 + 0.4C)z^{-1} + (6 - 0.4C)z^{-2} - (2 - 0.4C)z^{-3}} \tag{12-145}$$

The long-division process is now carried out, remembering that according to Rule 2 just stated the values of the coefficients which vary with time must be changed at each step in the division process to the values at the corresponding times of the functions. In this case, $C = t$, so that during successive steps of the division, $t = C = 0.4$, $t = C = 0.8$, $t = C = 1.2$, etc., yielding

$$\frac{0.16z^{-1} + 0.16z^{-2}}{(2 + 0.4C) - (6 + 0.4C)z^{-1} + (6 - 0.4C)z^{-2} - (2 - 0.4C)z^{-3}} = 0.0741z^{-1} + 0.266z^{-2} + 0.503z^{-3} + 0.714z^{-4} + \cdots \tag{12-146}$$

From Eq. (12-146), the desired solution for y is

$$\begin{array}{ll} t = 0 & y = 0 \\ t = 0.4 & y = 0.0741 \\ t = 0.8 & y = 0.266 \\ t = 1.2 & y = 0.503, \text{ etc.} \end{array}$$

(*d*) Nonlinear differential equations may be solved in a manner similar to that for the time-varying linear equation. The factors such as C_n, $C_{n-1}, \ldots, C_0$, however, must now represent functions of the dependent rather than independent variable. The rules of procedure are as follows:

1. Obtain the Laplace transform of the nonlinear differential equation in the usual manner including initial conditions.

2. Proceed as in the constant-coefficient case. In performing the long-division process, however, change the values of $C_n, C_{n-1}, \ldots, C_0$ at each step in the division process to the most recent values available for the coefficients in terms of the various functions of the dependent variable. In other words, in the first step, values of $C_n, C_{n-1}, \ldots, C_0$ are obtained from the initial conditions for y, dy/dt, ... The result of the first division is a new value of y to be used in calculating $C_n, C_{n-1}, \ldots, C_0$ for the second step in the division. New values to be used for the derivatives of y with respect to t are calculated from the new values of the dependent variable.

TABLE 12-4 z-FORMS FOR USE IN THE NUMERICAL SOLUTION OF INVERSE TRANSFORMS

	s^{-K}	z-form $= F_K(z^{-1})$
1	s^{-1}	$\frac{T}{2}\frac{1 + z^{-1}}{1 - z^{-1}}$
2	s^{-2}	$\frac{T^2}{12}\frac{1 + 10z^{-1} + z^{-2}}{(1 - z^{-1})^2}$
3	s^{-3}	$\frac{T^3}{2}\frac{z^{-1} + z^{-2}}{(1 - z^{-1})^3}$
4	s^{-4}	$\frac{T^4}{6}\frac{z^{-1} + 4z^{-2} + z^{-3}}{(1 - z^{-1})^4} - \frac{T^4}{720}$
5	s^{-5}	$\frac{T^5}{24}\frac{z^{-1} + 11z^{-2} + 11z^{-3} + z^{-4}}{(1 - z^{-1})^5}$

Consider the nonlinear differential equation

$$(D + y)y = 1 \tag{12-147}$$

where $y(0) = 0$. Carrying out the procedure as before, let $C = y$, so that Eq. (12-147) can be written

$$(D + C)y = 1 \tag{12-148}$$

Assuming C to be constant as before, the Laplace transform of Eq. (12-148) expressed in powers of s^{-1} is

$$y(s) = \frac{s^{-2}}{1 + Cs^{-1}} \tag{12-149}$$

Substituting the z-forms from Table 12-4 and dividing by T yield

$$y(z) = \frac{T(1 + 10z^{-1} + z^{-2})}{(12 + 6CT) - 24z^{-1} + (12 - 6CT)z^{-2}} \tag{12-150}$$

Selecting $T = 0.1$,

$$y(z) = \frac{0.1(1 + 10z^{-1} + z^{-2})}{(12 + 0.6C) - 24z^{-1} + (12 - 0.6C)z^{-2}} \tag{12-151}$$

Carrying out the long-division process yields

$$y(z) = 0.00833 + 0.100z^{-1} + 0.199z^{-2} + 0.295z^{-3} + 0.390z^{-4} + \cdots \tag{12-152}$$

and the desired solution for y is

$$\begin{aligned} t &= 0 & y &= 0.00833 \\ t &= 0.1 & y &= 0.100 \\ t &= 0.2 & y &= 0.199, \text{ etc.} \end{aligned}$$

According to Rule 2, the first step of the division process C was based on the initial value of y, $y(0) = 0$, so that $C = y = 0$. For the second step, the value of y obtained in the first step, $y = 0.00833$ was used, etc.

Systems which contain time lags may also be manipulated through the use of z-forms. The one requirement which must be met is that the choice of the interval T must be such that T is an integral submultiple of the time-delay.

Problems

Prob. 12-1 For the following functions, write the corresponding $\mathcal{L}[f(t)] = f(s)$ transforms. The transform table should be consulted.

$$(a)\ f(t) = \frac{1}{2}\sin 3t + \frac{1}{3}$$

$$(b)\ f(t) = \frac{3}{5}\cos 2t + \epsilon^{-t}\sin 2t$$

$$(c)\ f(t) = 6\cos t + 2\epsilon^{-t}$$

Prob. 12-2 Find the inverse transforms of the following expressions:

$$(a)\ f(s) = \frac{1+s}{s^3+2s^2}$$

$$(b)\ f(s) = \frac{1+s+s^2+3s^3}{s^4+s^3}$$

$$(c)\ f(s) = \frac{2+s+2s^2+s^3}{s^4+2s^3+2s^2}$$

$$(d)\ f(s) = \frac{1}{s(s+1)(s+2)}$$

$$(e)\ f(s) = \frac{4}{s^2+9}$$

$$(f)\ f(s) = \frac{5}{s(s+5)}$$

$$(g)\ f(s) = \frac{1}{4(s+3)}$$

$$(h)\ f(s) = \frac{1}{s(s+1)(s-1)(s+3)}$$

$$(i)\ f(s) = \frac{6}{s^2(s^2+9)(s^2+4)}$$

$$(j)\ f(s) = \frac{6+5s}{s^2+4s-5}$$

Prob. 12-3 Solve by transformation method

$$(D^2 + 5D + 4)x = 3 \qquad \text{for } t > 0$$

where $x = 2$, $dx/dt = 0$ for $t = 0$.

Prob. 12-4 Solve by transformation method the simultaneous equations

$$\left.\begin{aligned}(D^2 - 3D + 2)x + (D - 1)y &= 0\\ (D - 1)x - (D^2 - 5D + 4)y &= 0\end{aligned}\right\} t > 0$$

where $x = 0$, $y = 0$, $dx/dt = 0$, $dy/dt = 1$ for $t = 0$.

Prob. 12-5 Transform and solve the following differential equations:

(a) $(4D^2 + 2D + 1)x = 4 \cos \omega t$

(b) $(D^2 + 7D + 8)x = 10$

(c) $(D + 1)y = 2x + 2$

(d) $(3D^2 + 9)y = -\dfrac{7t}{T}\epsilon^{-t/T} \cos 5t - 4$

Prob. 12-6 In Fig. 12-8*a* is shown a choke-input filter circuit of a television-transmitter high-voltage power supply. The rectifier output

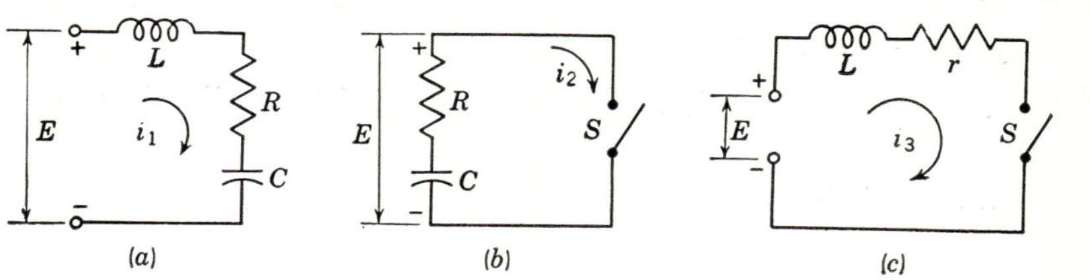

FIG. 12-8

E of 7,600 volts is applied to the filter at $t = 0$, with zero initial charge on C.

(a) Write the differential equation of the network.

(b) Solve for i_1 by classical and transform methods.

(c) If $L = 1.5$ henrys, $R = 32$ ohms, and $C = 15$ μf, determine i_1 at $t = 0.001$, 0.002, 0.004, 0.006, 0.007, 0.008, 0.01, 0.012, 0.014, 0.016, 0.018, 0.02, 0.022, 0.024, 0.026. 0.028, 0.03, 0.034, 0.036, 0.038, 0.044, 0.05, and 0.052 sec.

(d) Plot the results with time in milliseconds as the abscissa and instantaneous current in amperes as the ordinate.

(e) Determine the maximum positive and negative values of i_1 and the times at which these occur.

(f) Determine and plot the output voltage (the voltage across R and C) of the filter as a function of time.

Prob. 12-7 After the filter output voltage has been stabilized at 7,600 volts and the initial charging period has been completed, assume that a gas arc occurs in the output tube supplied by the filter. This can be represented by the closing of the switch in Fig. 12-8*b*.

(a) What is the maximum discharge current produced by C through the short circuit?

(*b*) Write the differential equation of the network.

(*c*) Solve for i_2 by transform methods.

(*d*) Evaluate i_2 for $t = 0.0001, 0.0002, 0.0004, 0.0006, 0.0008, 0.0010, 0.0012, 0.0014$, and 0.0020 sec.

(*e*) Plot the results with time in milliseconds as the abscissa and instantaneous current in amperes as the ordinate.

Prob. 12-8 If the fault mentioned in the previous problem is not cleared, the rectifier system will also begin to deliver current to the short through the input choke as in Fig. 12-8*c*. The manufacturer has determined that the power transformer associated with the rectifier will deliver a d-c short-circuit current of 211 amp. If the choke inductance remains 1.5 henrys and its series resistance r is 1.8 ohms,

(*a*) Write the differential equation of the network.

(*b*) Solve for i_3 by transform methods.

(*c*) Evaluate i_3 for $t = 0.02, 0.04, 0.06, 0.08$, and 0.10 sec.

(*d*) Plot the results with time in milliseconds as the abscissa and instantaneous current in amperes as the ordinate.

(*e*) If a 60-cycle high-speed breaker requires 4 cycles (1 cycle relay time plus 3 cycles breaker time) to operate, what current will the rectifier deliver before the breaker clears the short?

(*f*) If a more economical 6-cycle breaker is used, what current will be developed?

Note: Include power supply internal resistance in (*a*)

Prob. 12-9 In Fig. 12-9, switch 1 is closed first. When the current in the inductance has reached a steady value, switch 2 is closed.

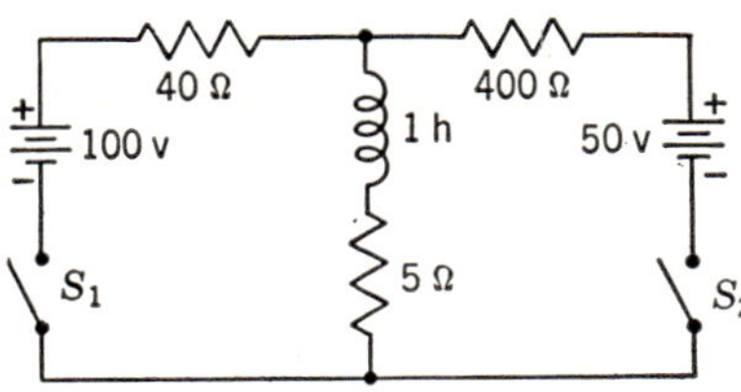

FIG. 12-9

(*a*) Write the differential equation for the network.

(*b*) Using transform methods, find the current in the inductance as a function of time after switch 2 is closed.

(*c*) What is the initial value of the voltage across the inductance when switch 2 has been closed?

Prob. 12-10 If

$$f(s) = \frac{1}{(s+1)^2(s^2+4)}$$

expand by partial fractions and solve for $f(t)$.

Prob. 12-11 If

$$f(s) = \frac{s^3 + 2s^2 + 4s + 4}{s^4 + 4s^2}$$

expand by partial fractions and the method of equating coefficients of like powers.

Prob. 12-12* Consider the multimesh network shown in Fig. 12-10*a*. Convert this complex circuit into a simple one-node network using equivalent current generators. Assume that the current in the circuit prior to the closing of switch S has reached its steady-state value. If the switch is then closed, find an expression for the voltage across R_4 at $t > 0$.

Prob. 12-13† In the circuit of Fig. 12-10*b*, the current generator i is applied to the network when the switch is opened at $t = 0$. If both

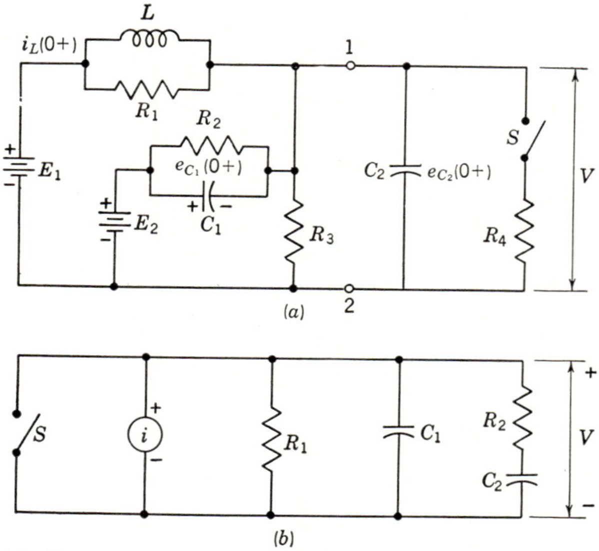

FIG. 12-10 (*a*) $E_1 = 100$ volts, $E_2 = 150$ volts, $L = 1$ henry, $C_1 = C_2 = 10\ \mu$f, $R_1 = R_2 = R_3 = 1{,}000$ ohms, $R_4 = 50$ ohms. (*b*) $i = 10 \cos 377t$ amp, $R_1 = R_2 = 100$ ohms, $C_1 = C_2 = 100\ \mu$f.

capacitors have zero charge at $t = 0$, find the voltage developed across the capacitors C_1 at 0.001 sec after the switch is opened.

Prob. 12-14

(*a*) In the circuit of Fig. 12-8*a*, the initial voltage to which C is charged is 20 volts, the upper end being positive in polarity. If $C = 1{,}110\ \mu$f, $L = 0.25$ henry, and $R = 50$ ohms, find the circuit current after $t = 0$ when the applied voltage is 50 volts d-c.

* Y. P. Yu, *op. cit.*, pp. 386–387.
† *Ibid.*, pp. 392–393.

(*b*) Find the circuit current under the same conditions except that there is no initial charge on the capacitor.

(*c*) Find the circuit current in Fig. 12-8*a* if $C = 80\ \mu f$, $L = 0.25$ henry, $R = 50$ ohms, and there is no initial charge on the capacitor.

(*d*) Find the circuit current in Fig. 12-8*c* after the switch is closed, if the applied voltage is a sine wave of 40 volts peak at 60 cps with a phase angle of 60° at $t = 0$, $L = 1$ henry, and $R = 100$ ohms.

Prob. 12-15 In the analysis of the operation of a semiconductor diode multivibrator,[1] an equivalent circuit is derived which consists of a constant voltage E applied through a switch and a series resistor R_2 to a parallel combination of a resistance R_1 and a capacitor C. At the instant the switch is closed ($t = 0$), the initial voltage V_a across C is V'_v.

(*a*) Draw the circuit described above.

(*b*) Write the voltage transform in terms of the Laplacian operator s, using the transfer function of the network.

(*c*) Solve for V_a as a function of time after $t = 0$.

Prob. 12-16 A d-c voltage of 1 volt is applied through a switch to the primary of a transformer which has an inductance of 2 henrys and a resistance of 1 ohm. The secondary has an inductance of 3 henrys and is loaded with a resistance of 2 ohms. If the mutual inductance between primary and secondary is +1 henry and there is no initial current in either primary or secondary, draw the circuit and find the primary and secondary current as a function of time after $t = 0$.

Prob. 12-17 In a beat-frequency oscillator,[2] a parallel resonant circuit is used consisting of an inductance L, a capacitance C, and a resistance R, connected in parallel. This circuit is then excited with a current $i \sin \omega t$ at its resonant frequency, ω. When steady-state conditions have been achieved, an additional capacitance, ΔC, is switched in parallel with the circuit. Solve for the resulting transient and the new steady-state condition.

Prob. 12-18 A capacitor, C_1, is charged to an initial voltage, V. It is then connected through a switch to a parallel circuit which has two branches. The first consists of resistance R_1, and the second consists of resistance R_2 in series with capacitor C_2. Draw the circuit and find the voltage v across C_2 as a function of time and the voltage V if $C_1 = 0.02\ \mu f$, $C_2 = 0.001\ \mu f$, $R_1 = 5{,}000$ ohms, and $R_2 = 1{,}000$ ohms. Plot v as a function of time from 0 to 80 μsec.

Prob. 12-19 A constant voltage V is applied through a switch and an inductance L_1 of 40 mh to a parallel combination of $L_2 = 5$ mh and

[1] J. J. Suran and E. Keonjian, A Semiconductor Diode Multivibrator, *Proc. I.R.E.*, **43** (7): 814–820, July, 1955.

[2] A. H. Taylor, Controlled Beat Frequency Oscillator, *Radio-Electronic Eng.*, **24** (3): 15–17, March, 1955.

$C = 0.01$ μf. Draw the circuit and calculate the voltage v across the parallel combination as a function of time and V. Plot v from $t = 0$ to $t = 45$ μsec. Calculate and plot the response if V is turned off after 1 μsec.

Prob. 12-20 A study by Fletcher and Munson gives the frequency response of the human ear in terms of "equal-loudness curves." This study shows that the frequency response of the ear at a fixed loudness level rises to a peak in the mid-frequency range and then falls off at the rate of 12 db/octave. If the response of the ear for frequencies above 500 cps at a typical loudness level is given by[1]

$$A = \frac{1}{1 - \omega^2T^2 - 0.5j\omega T}$$

where $T = 0.391 \times 10^{-4}$.

(*a*) Determine the absolute value of A and from this the maximum value of A and the frequency for which A is a maximum. Plot A as a function of frequency on semilog paper.

(*b*) Write the transfer function for the ear, taking $-\omega^2T^2 = (j\omega T)^2$, and calculate the transient response of the ear to an impulse with a square leading edge which might be produced by a heavy hammer striking a chime or a triangle. Plot as a function of time.

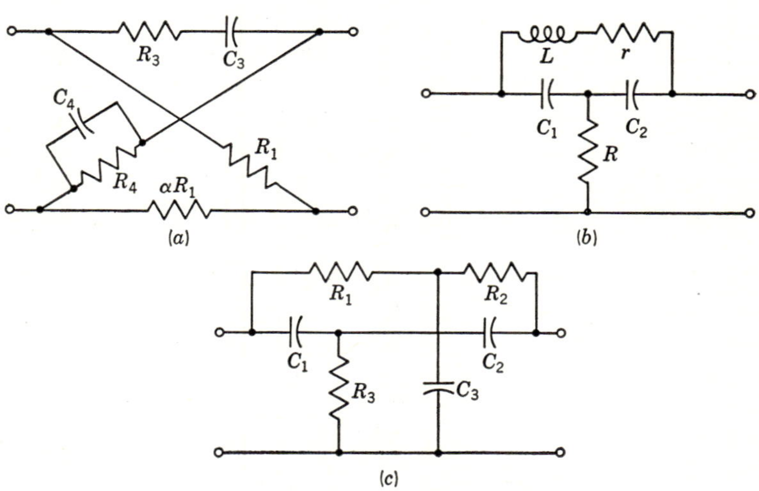

FIG. 12-11 Frequency-selective networks.

Prob. 12-21 In Figure 12-11 are shown three frequency-selective networks: (*a*) the Wien-bridge network which is a special form of lattice,

[1] C. P. Boegli, Transient and Frequency Response in Audio Equipment, *Audio Eng.*, **38** (2): 19–20, February, 1954.

(*b*) the bridged-T network in which the bridging arm is an inductance whose Q may be used as a parameter to vary selectivity, and (*c*) the twin-T network. Using matrix methods, compute the transfer functions for each of these networks and determine the conditions for an output null in each case. In case (*a*) assume $R_3 = R_4$ and $C_3 = C_4$, in (*b*) assume $C_1 = C_2$, and in (*c*) assume $R_1 = R_2$ and $C_1 = C_2$.

Prob. 12-22. Find by the method of residues the inverse transform of the function

$$f(s) = \frac{30(0.1s^2 + 25s + 10^5)}{s(1.5s^2 + 10{,}350s + 3 \times 10^6)}$$

Electronic oil prospecting by geophone. (*Philips Petroleum Co.*)

PART 6

The Direction of Mathematics in Electronics

Chapter 13

ELEMENTS OF BOOLEAN ALGEBRA

13-1 Introduction

As an example of the manner in which abstract mathematical topics can suddenly become of importance to the engineer, consider the case history of Boolean algebra. In 1847, an English mathematician named George Boole published a pamphlet entitled, "The Mathematical Analysis of Logic—Being an Essay toward a Calculus of Deductive Reasoning." This subject, now known as Boolean algebra, remained dormant with regard to application in electronics until about 1938, when Shannon utilized it in communication theory.[1] Shannon was able to demonstrate that Boolean algebra provided an extremely convenient method of notation for networks made up of contacts. Further, he was able to show with mathematical rigor that this algebra could be used to manipulate and redesign any two-terminal series, series-parallel, or parallel network of contacts.

13-2 Boolean Algebra in Terms of Logical Propositions

A simple interpretation of Boolean algebra is that given in terms of logical propositions. A letter X in the algebra corresponds to a logical proposition. The sum of two letters $(X + Y)$ represents the proposition "X or Y" and the product XY represents the proposition "X and Y." The prime mark is used to designate the negative, so that X' is the negation of X, or "not X." The constants 1 and 0 represent truth and falsity, respectively. For example, the relation $X + Y = 1$ indicates that X or Y is true, while $XY = 1$ means that X and Y are true.

[1] C. E. Shannon, A Symbolic Analysis of Relay and Switching Circuits, *Trans. A.I.E.E.*, **57**: 713, 1938. See also C. E. Shannon, The Synthesis of Two-terminal Switching Circuits, *Bell System Tech. J.*, **28** (1): 59–98, January, 1949.

13-3 Boolean Algebra in Terms of Switching Circuits

The interpretation of Boolean algebra in terms of switching circuits is similar to that given in the previous section, except that a proposition, from a switching standpoint, can be described as a certain switching operation which is to be performed within a network of contacts. The networks considered here will be those which have a single input and a single output terminal, although Boolean algebra may also be applied to the analysis of multiterminal switching networks.

The symbol X in the switching algebra is interpreted as a *make* contact on a relay or switch. A make contact is one which will close when the relay or switch of which it is a part is actuated. The negation of X, written as X', is a *break* contact on the same relay, i.e., one which will open when the relay is actuated. The constant 0 represents a *closed* or continuous circuit, and the constant 1 represents an *open* or broken connection.

In the Boolean algebra as applied to switching networks, the combining operation of addition corresponds to the *series* connection of two or more

TABLE 13-1 HINDRANCE FUNCTIONS OF SIMPLE SWITCHING CIRCUITS IN BOOLEAN ALGEBRAIC NOTATION

Network	*Hindrance function*
	0 (permanently closed circuit)
	1 (permanently open circuit)
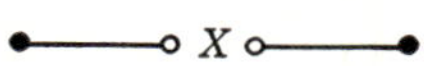	X (make contact on relay X)
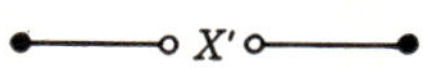	X' (break contact on relay X)
	$X + Y$ (series connection)
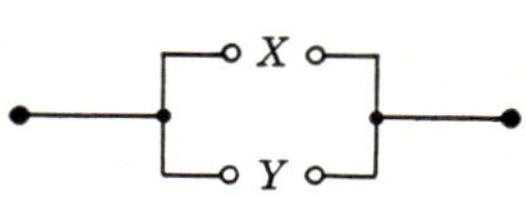	XY (parallel connection)
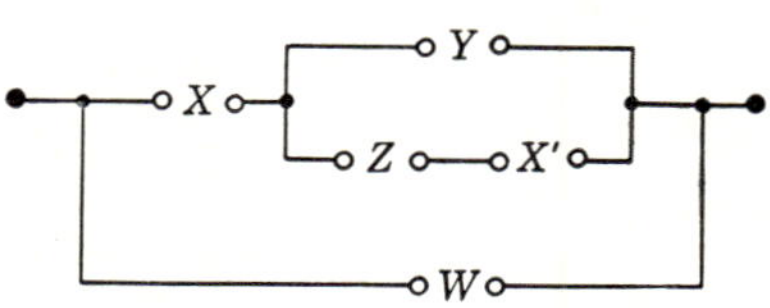	$W[X + Y(Z + X')]$

contacts. The operation of multiplication corresponds to the *parallel* connection of two or more contacts. These conventions are illustrated in Table 13-1, and make it possible to write an algebraic expression corresponding to a two-terminal network. Such an expression will include a symbol for each contact which appears in the network and is described as the *hindrance* or *hindrance function* of the network. In Table 13-1, the expression given on the right represents the hindrance of the network shown on the left.

Consider the last network shown in Table 13-1.

The hindrance of the section of the network which contains Y in parallel with the combination of Z and X' in series is written $Y(Z + X')$ according to the rules stated above. This grouping is, in turn, in series with contact X, so that the hindrance of the entire upper branch is $X + Y(Z + X')$. The lower branch of the network contains only the make contact W, which is effectively in parallel with the upper branch. The total hindrance of the circuit is, therefore, the product of W and the hindrance of the upper branch, or $W[X + Y(Z + X')]$, as shown in Table 13-1.

13-4 Laws and Identities of Boolean Algebra

The associative, distributive, and commutative laws also apply to the Boolean algebra, so that Boolean expressions can be manipulated in a

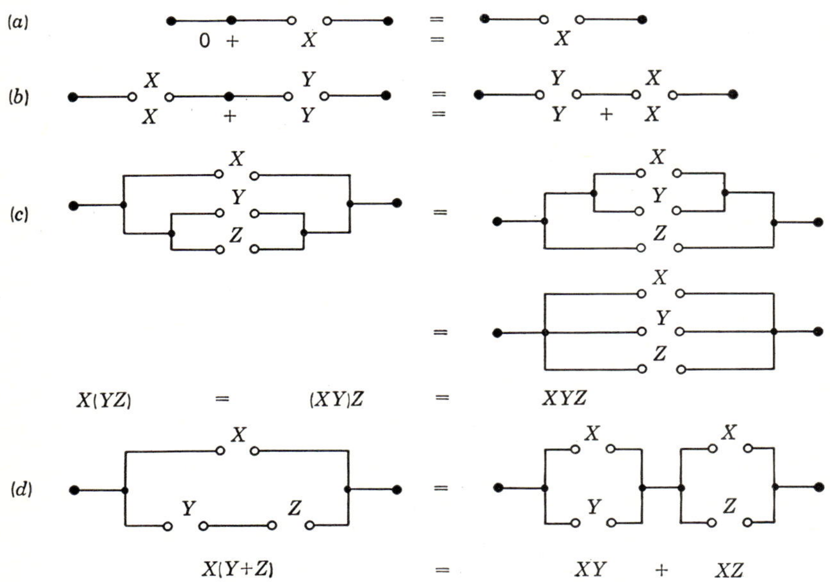

FIG. 13-1 Interpretation of some Boolean algebraic identities.

manner very similar to ordinary algebraic expressions. The following identities exist in the Boolean algebra:

$$\begin{aligned} 0 + X &= X \\ 0 \cdot X &= 0 \\ 1 \cdot X &= X \\ X + Y &= Y + X \\ XY &= YX \\ X + (Y + Z) &= (X + Y) + Z \\ X(YZ) &= (XY)Z \\ X(Y + Z) &= XY + XZ \end{aligned} \tag{13-1}$$

Interpretations of some of these identities in terms of switching circuits are shown in Fig. 13-1. In each case the hindrance function is written immediately below the circuit. It is important to note that in the Boolean algebra it is assumed that all contacts on a given relay, i.e., all contacts designated by the same letter, operate simultaneously. All of the identities in this section hold for both the Boolean algebra and normal algebraic expressions.

13-5 Identities Unique to Boolean Algebra

There are a number of rules unique to Boolean algebra which permit simplifications of expressions that are not possible by the rules of ordinary algebra. Some of these are

$$\begin{aligned} X = X + X &= X + X + X + X, \text{ etc.} \\ X = X \cdot X &= X \cdot X \cdot X, \text{ etc.} \\ X + 1 &= 1 \\ X + YZ &= (X + Y)(X + Z) \\ X \cdot X' &= 0 \\ X + X' &= 1 \\ (X + Y)' &= X'Y' \\ (XY)' &= X' + Y' \\ X + XY &= X \\ X(X + Y) &= X \end{aligned} \tag{13-2}$$

The circuit interpretation of some of these identities is shown in Fig. 13-2. These identities make the manipulation of Boolean expressions considerably simpler than that possible in ordinary algebra. Note that in the Boolean algebra there is no need for exponents, because according to Eq. (13-2), $X^n = X$. Also, there is no need for numerical coefficients, since $nX = X$. The identities in Eq. (13-2) can be proved by means of a so-called "truth table." Consider the last identity in Eq. (13-2), which states that $X(X + Y) = X$. The table is constructed in three columns as in Table 13-2. The first contains all of the possible values which X and Y can assume, which is a total of four combinations in this case.

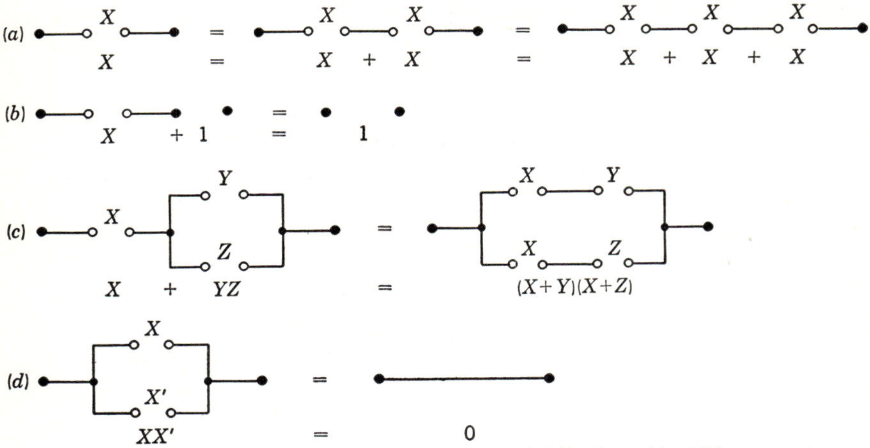

FIG. 13-2 Interpretation of some special Boolean identities.

The central column gives the corresponding value of $X(X + Y)$, and the column on the right gives the value of X under the conditions specified. From this table it can be seen that for every possible combination of the variables, the two sides of the identity are equal, and, therefore, the identity is a true statement under all conditions.

TABLE 13-2 PROOF OF IDENTITY $X(X + Y) = X$

Possible values for X and Y		Corresponding value of $X(X + Y)$	Value of X
1	1	$1(1 + 1) = 1$	1
1	0	$1(1 + 0) = 1$	1
0	1	$0(0 + 1) = 0$	0
0	0	$0(0 + 0) = 0$	0

13-6 Equivalent Circuits

Boolean algebra may be used to find many circuits equivalent in operating characteristics to a given circuit. To accomplish this, the hindrance of the original circuit is written and the expression is then manipulated according to the rules to yield different expressions. Each new expression represents a new circuit equivalent to the given one.

Boolean algebra is particularly useful in eliminating unnecessary contacts in complex switching circuits. Consider the circuit shown in Fig. 13-3*a*. The total hindrance H is the product of the hindrances of the three parallel branches, or

$$H = (X + U)(X + Y + V)(X + Y + Z + W) \tag{13-3}$$

This can be simplified as follows:

$$\begin{aligned} H &= (X + Y + Z + W)[X + (Y + V)U] \\ &= X + (Y + Z + W)(Y + V)U \\ &= X + [Y + (Z + W)V]U \end{aligned} \tag{13-4}$$

The circuit corresponding to the final expression in Eq. (13-4) is shown in Fig. 13-3*b*.

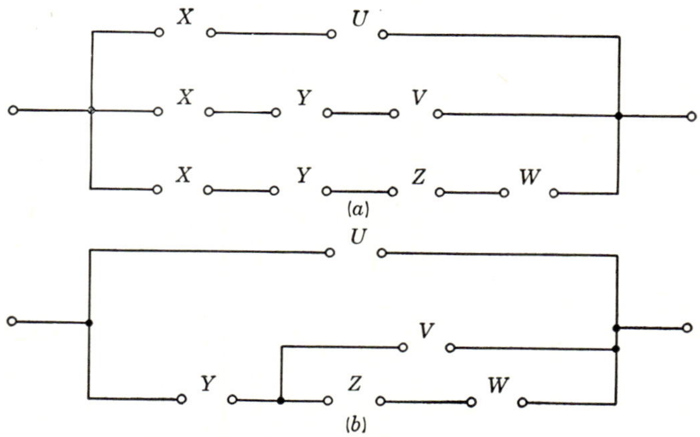

FIG. 13-3 Simplification of a network.

The hindrance function associated with a two-terminal network describes the network completely from the external point of view. By means of the hindrance function, it is possible to determine whether the circuit will be open or closed for any particular position of the relays. This is accomplished by given the variables corresponding to operated relays the value 0, which means that the make contacts will be closed and the break contacts open. In the same way, the variables corresponding to the unoperated relays are given the value 1, which means that the make contacts will remain open and the break contacts closed. For example, in the case of the last circuit shown in Table 13-1, the hindrance function was

$$H = W[X + Y(Z + X')] \tag{13-5}$$

Suppose that relays X and Y are operated and Z and W are not operated. Then

$$H = 1[0 + 0(1 + 1)] = 0 \tag{13-6}$$

and under these circumstances the circuit is closed.

Hindrance functions correspond explicitly to series-parallel types of circuits, because they are basically made up of sum and product operations. However, there is a hindrance function which represents the operating chracteristics for any network, series-parallel or not. The manner in which the hindrance for a non-series-parallel network is determined is illustrated by the case of the bridge network shown in Fig. 13-4. The hindrance of this circuit consists of the product of the hindrances

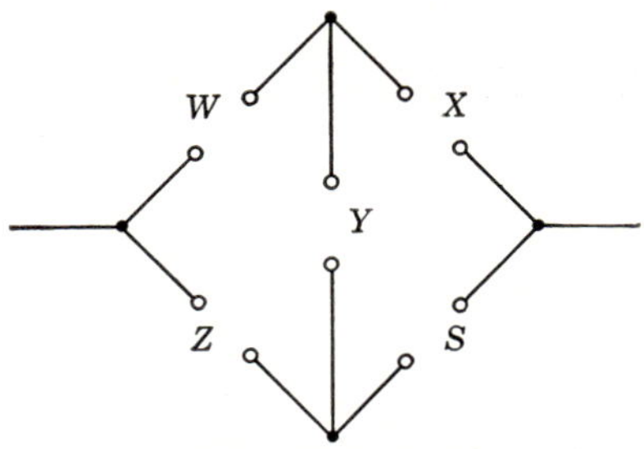

FIG. 13-4 Bridge circuit.

of all possible paths (four) between the input and output terminals. The four hindrances are, therefore, $(W + X)$, $(Z + S)$, $(W + Y + S)$, and $(Z + Y + X)$. The total hindrance of the bridge network is, therefore,

$$H = (W + X)(Z + S)(W + Y + S)(Z + Y + X) \qquad \text{(13-7)}$$

In simplifying networks, the appearance of repetitive factors is often a clue to the elimination of unnecessary contacts. It is usually helpful to multiply out all factors, because, quite often, many terms will automatically drop out in the process and the remainder may fall into a form which suggests additional simplification.

13-7 Synthesis of Switching Circuits

The theory of switching circuits may be divided into two major divisions, analysis and synthesis. The problem of analysis, determining the manner of operation of a given switching circuit, is comparatively simple. The inverse problem of finding a circuit satisfying certain given operating conditions, and in particular the *best* circuit, is, in general, more difficult and more important from the practical standpoint. A basic part of the general synthesis problem is the design of a two-terminal network with given operating characteristics. Shannon has considered this problem in detail in his paper on two-terminal switching circuits. Additional references on the more advanced phases of switching network design include the following:

T. L. Craven: Circuits and Logic, *Electronic Eng.*, **26** (317): 302–305, July, 1954.

F. E. Hohn: A Matrix Method for the Design of Relay Circuits, *I.R.E. Trans. on Circ. Theory*, **CT-2** (2): 154–161, June, 1955.

F. E. Hohn and L. R. Schissler: Boolean Matrices and the Design of Combinational Relay Switching Circuits, *Bell System Tech J.*, **34** (1): 177–202, January, 1955.

W. Keister, A. E. Ritchie, and S. H. Washburn: "The Design of Switching Circuits," D. Van Nostrand Company, Inc., New York, 1951.

Problems

Prob. 13-1 Evaluate the following expressions in terms of Boolean algebra:

(*a*) $0 \cdot 0$
(*b*) $1 + 1$
(*c*) $1 \cdot 1$
(*d*) $0 + 0$
(*e*) $1 \cdot 0$
(*f*) $0 \cdot 1$
(*g*) $0 + 1$
(*h*) $1 + 0$

Prob. 13-2 Simplify the expression $X(Y + XZ')(Z' + W)(Y' + XW)$.

Prob. 13-3 Employ a truth table to establish the validity of the identities given in Eq. (13-2).

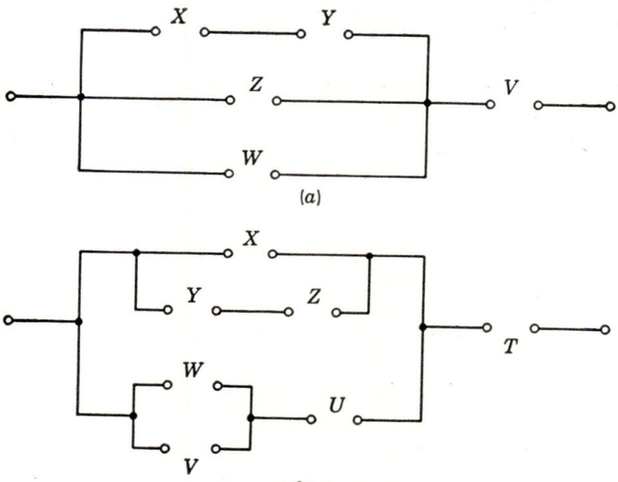

FIG. 13-5

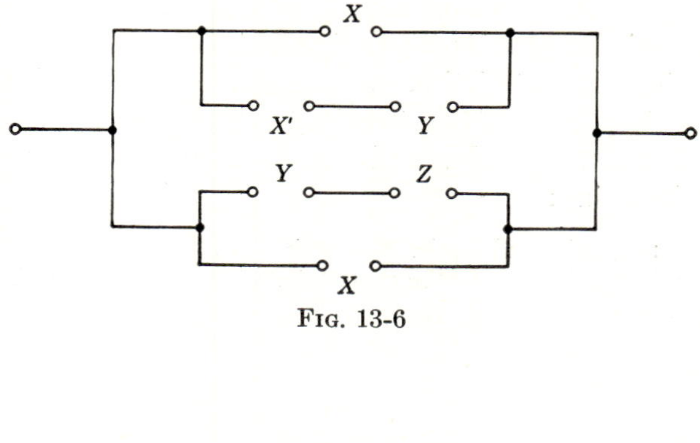

FIG. 13-6

Prob. 13-4 Write the hindrance of the circuits in Fig. 13-5*a* and *b*.
Prob. 13-5 Design a network such that $H = XY + [(Z + W)V]$.
Prob. 13-6 Draw an equivalent simplified network for Fig. 13-6.

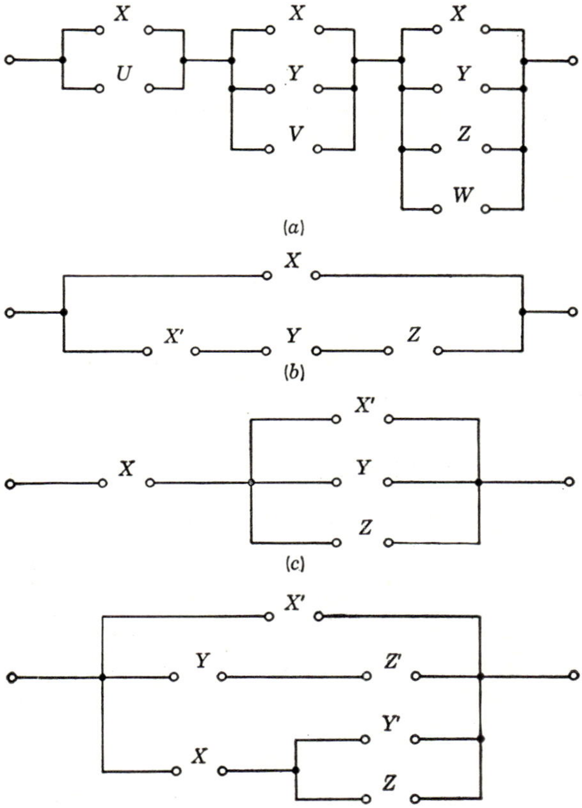

FIG. 13-7

Prob. 13-7 Simplify the networks shown in Fig. 13-7*a*, *b*, *c*, and *d*.
Prob. 13-8 Simplify the following expressions:

(*a*) $(X + Y + Z)(X' + Y' + Z')(X + Y' + Z')(X' + Y + Z)(X + Y + Z')(X' + Y + Z')$
(*b*) $XY(Z'W + X' + Y) + Y'(Z + W')(X + Y)$
(*c*) $(X + Y'Z)(X' + Y' + Z)$
(*d*) $(X + W')(Y'V) + X'Z'WV + Y'Z'VU$

Chapter 14

A STUDY PLAN IN MATHEMATICS FOR SPECIALISTS IN ELECTRONICS

A question often posed by the engineer specialist in the field of electronics is: On what phases of higher mathematics need I concentrate in order to keep abreast of technical developments in my field? The answer increases in complexity as the field of electronics advances. When the analysis of transient conditions in electric circuits became important to the engineer, the level of the terminal subject in his mathematical training was, of necessity, shifted upward from the elementary calculus to the study of differential equations. Thus, as the purview of electronics continues to expand, the range of topics with which a well-informed specialist in the field must be familiar constantly widens.

In earlier times, the mathematical needs of an engineer were well defined and rarely exceeded the elementary calculus level. Those mathematical topics which did not find ready application were relegated for graduate study by the pure mathematician. Nowadays, as the intricacy of technical problems enlarges, more and more subject matter which was formerly considered to lie within the realm of higher mathematics or pure mathematics has become commonplace in engineering, and no limits can be set on the mathematical needs of the engineer. As T. C. Fry has pointed out, "Higher mathematics means simply those branches of mathematics which have not yet found a wide field of application."[1]

The manner in which Boolean algebra has been appropriated from the realm of higher or pure mathematics and applied to problems in electronic engineering as described in Chap. 13 is not unusual. New topics are constantly being infused into the stream of electronic technical development. In the past, Bessel's functions were applied to the analysis of the frequency modulated carrier wave, Legendre polynomials to the analysis

[1] T. C. Fry, Industrial Mathematics, *Bell System Tech. J.*, **20** (3): 255–292, July, 1941.

of the field patterns of antennas, and elliptic integrals to the calculation of high-frequency inductance of coils. As early as 1931, Howitt[1] noted that, in the study of the electric network, one was likely to encounter (*a*) continued fractions, (*b*) Cauchy residue theory, (*c*) asymptotic series, (*d*) fractional and irrational derivatives and integrals, (*e*) group theory, (*f*) Fourier series and transforms, (*g*) integral equations, etc. To this list must surely be added the Laplace transform, Heaviside's operators, and tensor analysis, which was introduced by Gabriel Kron and which permits the statement of the physical laws which affect electronic circuits in terms which are independent of the system of coordinates used.[2]

Recently the Fourier integral has been applied to a wide variety of problems including network analysis, color television system analysis, and the evaluation of oscillator performance.[3]

Another mathematical term newly introduced to the electronic field is the spinor,[4] which can be described as a geometric object in three-dimensional space having a magnitude S and three Eulerian angles.[5] The theory of spinors has been previously known chiefly for its application to electron spin and relativistic quantum mechanics and was evolved by B. L. van der Waerden in 1929. The basic principles of the theory, however, have been known for many years, and the main forerunner of present-day spinor theory was Hamilton's quaternion theory (about 1843).[6] Van der Pol has reported applications of the theory of numbers

[1] N. Howitt, Group Theory and the Electric Circuit, *Phys. Rev.*, **37**: 1583–1595, June 15, 1931. See also W. Gaertner, The Group Theoretical Aspect of Linear Four-Pole Theory, *Conv. Rec. I.R.E.*, **2**: 36–43, 1954.

[2] See D. W. Epstein, Tensor Analysis and Its Application to Equivalent Circuits, *RCA Rev.*, **3**: 239–254, October, 1938; A. Kaufman, Applications of Tensor Theory to Linear Electronic Circuits, *Radio tech. dig.* (*France*), **6** (2): 67–76, **6** (3): 157–168, and **6** (4): 199–209, 1952; G. Ledig, Tensor Analysis of Transistor Feedback Circuits, *Arch. elekt. Ubertragung* **9**: 162–167, April, 1955.

[3] See E. A. Guillemin, The Fourier Integral—A Basic Introduction; C. H. Page, Applications of the Fourier Integral in Physical Science; W. R. Bennett, Application of the Fourier Integral in Circuit Theory and Circuit Problems; W. K. Linvill, R. E. Scott, and E. A. Guillemin, Evaluation of Fourier Transforms; T. Murakami and M. S. Corrington, Applications of the Fourier Integral in the Analysis of Color Television Systems; K. S. Miller and L. A. Zadeh, Generalization of the Fourier Integrals; G. Francini, Evaluation of Oscillatory Quality, *I.R.E. Trans. on Circuit Theory,* **CT-2**(**3**), September, 1955. See also W. L. Sullivan, Analysis of Systems with Known Transmission-Frequency Characteristics by Fourier Integrals, *Elec. Eng.*, **61** (5): 248–256, May, 1942.

[4] W. T. Payne, Elementary Spinor Theory, *Am. J. Phys*, **20** (5): 253–262, May, 1952.

[5] W. T. Payne, Spinor Theory of Four-terminal Networks, *J. Math. and Phys.*, **32** (1): 19–33, April, 1953.

[6] See Sir Edmund Whittaker, William Rowan Hamilton, *Sci. American*, **190** (5): 82–87, May, 1954.

Table 14-1 Frequency of Mathematical Applications in Technical Journals*

Periodical surveyed	Number of articles examined	Number of articles in which topic occurs: Diff. equations, Ordinary: Linear	Diff. equations, Ordinary: Vibrat. nonlin.	Diff. equations, Ordinary: Other	Diff. equations, Partial: Linear	Diff. equations, Partial: Other	Fourier series	Infinite series	Bessel functions	Gamma functions	Legendre functions	Vectors	Tensors	Matrix theory	Probability	Statistics	Curve fitting	Graphical solutions	Numerical solutions	Contour integration	Conformal transformation	Other	Laplace	Heaviside	Other	Integral equations	Fresnel integrals	Line integrals	Sine, cos integrals	Calculus of variations	Perturbation methods
Electronics																															
Proceedings of the Institute of Radio Engineers, vol. 38, 1950	80	23	2		16		8	17	8	5		8		3		1	1	4	26	3	2	32	7	1	6		1		1		2
Bell System Technical Journal, vol. 29, 1950	20	1		3	2	10	5	3	2	1	1	5	1	2	1							2	1			1			1		
Electrical Engineering, vol. 69, 1950	24	9	1		1	1	1	3				4		2	1	2	1	2	1				1	1		8					
Proceedings of the Institution of Electrical Engineers (London), vol. 96, 1949	116	13	1	2	5	3	3	10	4	1		7					1	3				1	3		1	10	2				
Electronics, vol. 23, 1950	120	2					2	2							1											2			1		
Total	360	48	4	5	24	14	19	35	14	7	1	24	1	7	3	3	3	9	27	3	2	35	12	2	7	21	3	—	3	—	2

PHYSICS

Physical Review, vol. 76, 1949	120	8		31		43		8		1		27	12	10	3	1				6		2	1		1	8			3	10	9
Zeitschrift für Physik, band 125, 126, 1949	68	4	1	9	2	10		11	4	1		6	2		1				2	2'		3	2		1	10		1	2	2	4
Journal of Applied Physics, July–Dec., vol. 20, 1949	151	21	8	3	13	9	16	16	22	3	5	29	10	7	6	3	1	3	5	1	1	5	13	1	4	10	3	3	2	3	4
Reviews of Modern Physics, Jan.–June, vol. 21, 1949	67	10	2		4	3	5	7	1	2		5	9	9	5	1	1	2					1			4		1	1		1
Transactions of the American Geophysical Union, vol. 30, 1949	77	5	2		11		2	1	2			3			2	8		3	3			1				5		1			
Total	483	48	13	43	30	65	23	43	29	7	5	70	33	26	17	13	2	8	10	9	1	11	17	1	6	37	3	6	8	15	18

AERONAUTICS

Journal of Aeronautical Science, vol. 17, 1950	74	41	10	3	24	6	6	6	8		2	4	1	5				3	11	4	4	5	6	1		21	1	4	7	3	7

* E. W. Anderson, Mathematics in Electrical Graduate Education, *Elec. Eng.*, **70** (6): 508, June, 1951.

to electronic problems[1] and describes the use of the sawtooth function $\mathrm{Sa}(x)$ which can be used to synthesize all types of discontinuous wave forms without the occurrence of the Gibbs phenomenon or overshoot at steep wave fronts as in the Fourier synthesis.

An estimate of the extent of employment of mathematics in electronics can be gained from an examination of an applied-mathematics frequency chart for the years 1949 to 1950 from *Electrical Engineering* (Table 14-1). Of the three fields of application which appear in the table—electronics, physics, and aeronautics—it is apparent that the greatest mathematical density, i.e., mathematical applications per paper, is encountered in aeronautics. For example, 41 of 74 papers in the aeronautical journal examined contained applications of linear differential equations, a density unequaled by other fields.

Of greatest interest to the electronics and communications engineer, however, is the distribution of mathematical applications in his own field. From Table 14-1, the 12 topics which occur most frequently are ranked as follows:

1. Linear differential equations
2. Infinite series
3. Numerical solutions
4. Vector analysis
5. Linear partial differential equations
6. Integral equations
7. Fourier series
8. Bessel functions
9. Laplace transforms
10. Graphical solutions
11. Matrix theory
12. Gamma functions

In terms of subdivisions of mathematics, as indicated by the double-lined sections of Table 14-1, the subject ranking is as follows:

1. Differential equations
2. Series
3. Complex variables
4. Graphical and numerical solutions
5. Vectors and tensors
6. Special functions
7. Operational calculus (Laplace transforms)
8. Integral equations
9. Matrix theory
10. Statistics
11. Special integrals
12. Perturbation methods

It can be seen that these subdivisions conform closely to courses in mathematics presented in colleges, universities, and technical institutes, and should obviously be the targets of the advanced student who intends to develop a proficiency in those fields of mathematics which are important in electronics.

[1] B. Van der Pol, Radio Technology and the Theory of Numbers, *J. Franklin Inst.*, **255** (6): 475–495, June, 1953.

Appendix

TABLE A-1 COMMON LOGARITHMS

N	0	1	2	3	4	5	6	7	8	9
10	0000	0043	0086	0128	0170	0212	0253	0294	0334	0374
11	0414	0453	0492	0531	0569	0607	0645	0682	0719	0755
12	0792	0828	0864	0899	0934	0969	1004	1038	1072	1106
13	1139	1173	1206	1239	1271	1303	1335	1367	1399	1430
14	1461	1492	1523	1553	1584	1614	1644	1673	1703	1732
15	1761	1790	1818	1847	1875	1903	1931	1959	1987	2014
16	2041	2068	2095	2122	2148	2175	2201	2227	2253	2279
17	2304	2330	2355	2380	2405	2430	2455	2480	2504	2529
18	2553	2577	2601	2625	2648	2672	2695	2718	2742	2765
19	2788	2810	2833	2856	2878	2900	2923	2945	2967	2989
20	3010	3032	3054	3075	3096	3118	3139	3160	3181	3201
21	3222	3243	3263	3284	3304	3324	3345	3365	3385	3404
22	3424	3444	3464	3483	3502	3522	3541	3560	3579	3598
23	3617	3636	3655	3674	3692	3711	3729	3747	3766	3784
24	3802	3820	3838	3856	3874	3892	3909	3927	3945	3962
25	3979	3997	4014	4031	4048	4065	4082	4099	4116	4133
26	4150	4166	4183	4200	4216	4232	4249	4265	4281	4298
27	4314	4330	4346	4362	4378	4393	4409	4425	4440	4456
28	4472	4487	4502	4518	4533	4548	4564	4579	4594	4609
29	4624	4639	4654	4669	4683	4698	4713	4728	4742	4757
30	4771	4786	4800	4814	4829	4843	4857	4871	4886	4900
31	4914	4928	4942	4955	4969	4983	4997	5011	5024	5038
32	5051	5065	5079	5092	5105	5119	5132	5145	5159	5172
33	5185	5198	5211	5224	5237	5250	5263	5276	5289	5302
34	5315	5328	5340	5353	5366	5378	5391	5403	5416	5428
35	5441	5453	5465	5478	5490	5502	5514	5527	5539	5551
36	5563	5575	5587	5599	5611	5623	5635	5647	5658	5670
37	5682	5694	5705	5717	5729	5740	5752	5763	5775	5786
38	5798	5809	5821	5832	5843	5855	5866	5877	5888	5899
39	5911	5922	5933	5944	5955	5966	5977	5988	5999	6010
40	6021	6031	6042	6053	6064	6075	6085	6096	6107	6117
41	6128	6138	6149	6160	6170	6180	6191	6201	6212	6222
42	6232	6243	6253	6263	6274	6284	6294	6304	6314	6325
43	6335	6345	6355	6365	6375	6385	6395	6405	6415	6425
44	6435	6444	6454	6464	6474	6484	6493	6503	6513	6522
45	6532	6542	6551	6561	6571	6580	6590	6599	6609	6618
46	6628	6637	6646	6656	6665	6675	6684	6693	6702	6712
47	6721	6730	6739	6749	6758	6767	6776	6785	6794	6803
48	6812	6821	6830	6839	6848	6857	6866	6875	6884	6893
49	6902	6911	6920	6928	6937	6946	6955	6964	6972	6981
50	6990	6998	7007	7016	7024	7033	7042	7050	7059	7067
51	7076	7084	7093	7101	7110	7118	7126	7135	7143	7152
52	7160	7168	7177	7185	7193	7202	7210	7218	7226	7235
53	7243	7251	7259	7267	7275	7284	7292	7300	7308	7316
54	7324	7332	7340	7348	7356	7364	7372	7380	7388	7396
N	0	1	2	3	4	5	6	7	8	9

TABLE A-1 COMMON LOGARITHMS (*Continued*)

N	0	1	2	3	4	5	6	7	8	9
55	7404	7412	7419	7427	7435	7443	7451	7459	7466	7474
56	7482	7490	7497	7505	7513	7520	7528	7536	7543	7551
57	7559	7566	7574	7582	7589	7597	7604	7612	7619	7627
58	7634	7642	7649	7657	7664	7672	7679	7686	7694	7701
59	7709	7716	7723	7731	7738	7745	7752	7760	7767	7774
60	7782	7789	7796	7803	7810	7818	7825	7832	7839	7846
61	7853	7860	7868	7875	7882	7889	7896	7903	7910	7917
62	7924	7931	7938	7945	7952	7959	7966	7973	7980	7987
63	7993	8000	8007	8014	8021	8028	8035	8041	8048	8055
64	8062	8069	8075	8082	8089	8096	8102	8109	8116	8122
65	8129	8136	8142	8149	8156	8162	8169	8176	8182	8189
66	8195	8202	8209	8215	8222	8228	8235	8241	8248	8254
67	8261	8267	8274	8280	8287	8293	8299	8306	8312	8319
68	8325	8331	8338	8344	8351	8357	8363	8370	8376	8382
69	8388	8395	8401	8407	8414	8420	8426	8432	8439	8445
70	8451	8457	8463	8470	8476	8482	8488	8494	8500	8506
71	8513	8519	8525	8531	8537	8543	8549	8555	8561	8567
72	8573	8579	8585	8591	8597	8603	8609	8615	8621	8627
73	8633	8639	8645	8651	8657	8663	8669	8675	8681	8686
74	8692	8698	8704	8710	8716	8722	8727	8733	8739	8745
75	8751	8756	8762	8768	8774	8779	8785	8791	8797	8802
76	8808	8814	8820	8825	8831	8837	8842	8848	8854	8859
77	8865	8871	8876	8882	8887	8893	8899	8904	8910	8915
78	8921	8927	8932	8938	8943	8949	8954	8960	8965	8971
79	8976	8982	8987	8993	8998	9004	9009	9015	9020	9025
80	9031	9036	9042	9047	9053	9058	9063	9069	9074	9079
81	9085	9090	9096	9101	9106	9112	9117	9122	9128	9133
82	9138	9143	9149	9154	9159	9165	9170	9175	9180	9186
83	9191	9196	9201	9206	9212	9217	9222	9227	9232	9238
84	9243	9248	9253	9258	9263	9269	9274	9279	9284	9289
85	9294	9299	9304	9309	9315	9320	9325	9330	9335	9340
86	9345	9350	9355	9360	9365	9370	9375	9380	9385	9390
87	9395	9400	9405	9410	9415	9420	9425	9430	9435	9440
88	9445	9450	9455	9460	9465	9469	9474	9479	9484	9489
89	9494	9499	9504	9509	9513	9518	9523	9528	9533	9538
90	9542	9547	9552	9557	9562	9566	9571	9576	9581	9586
91	9590	9595	9600	9605	9609	9614	9619	9624	9628	9633
92	9638	9643	9647	9652	9657	9661	9666	9671	9675	9680
93	9685	9689	9694	9699	9703	9708	9713	9717	9722	9727
94	9731	9736	9741	9745	9750	9754	9759	9763	9768	9773
95	9777	9782	9786	9791	9795	9800	9805	9809	9814	9818
96	9823	9827	9832	9836	9841	9845	9850	9854	9859	9863
97	9868	9872	9877	9881	9886	9890	9894	9899	9903	9908
98	9912	9917	9921	9926	9930	9934	9939	9943	9948	9952
99	9956	9961	9965	9969	9974	9978	9983	9987	9991	9996
N	0	1	2	3	4	5	6	7	8	9

TABLE A-2 NATURAL LOGARITHMS

0.0 to 9.9

N	0	1	2	3	4	5	6	7	8	9
0*	.....	7.697	8.391	8.796	9.084	9.307	9.489	9.643	9.777	9.895
1	0.000	0.095	0.182	0.262	0.336	0.405	0.470	0.531	0.588	0.642
2	0.693	0.742	0.788	0.833	0.875	0.916	0.956	0.993	1.030	1.065
3	1.099	1.131	1.163	1.194	1.224	1.253	1.281	1.308	1.335	1.361
4	1.386	1.411	1.435	1.459	1.482	1.504	1.526	1.548	1.569	1.589
5	1.609	1.629	1.649	1.668	1.686	1.705	1.723	1.740	1.758	1.775
6	1.792	1.808	1.825	1.841	1.856	1.872	1.887	1.902	1.917	1.932
7	1.946	1.960	1.974	1.988	2.001	2.015	2.028	2.041	2.054	2.067
8	2.079	2.092	2.104	2.116	2.128	2.140	2.152	2.163	2.175	2.186
9	2.197	2.208	2.219	2.230	2.241	2.251	2.262	2.272	2.282	2.293

10 to 99

N	0	1	2	3	4	5	6	7	8	9
1	2.303	2.398	2.485	2.565	2.639	2.708	2.773	2.833	2.890	2.944
2	2.996	3.045	3.091	3.135	3.178	3.219	3.258	3.296	3.332	3.367
3	3.401	3.434	3.466	3.497	3.526	3.555	3.584	3.611	3.638	3.664
4	3.689	3.714	3.738	3.761	3.784	3.807	3.829	3.850	3.871	3.892
5	3.912	3.932	3.951	3.970	3.989	4.007	4.025	4.043	4.060	4.078
6	4.094	4.111	4.127	4.143	4.159	4.174	4.190	4.205	4.220	4.234
7	4.248	4.263	4.277	4.290	4.304	4.317	4.331	4.344	4.357	4.369
8	4.382	4.394	4.407	4.419	4.431	4.443	4.454	4.466	4.477	4.489
9	4.500	4.511	4.522	4.533	4.543	4.554	4.564	4.575	4.585	4.595

100 to 209†

N	0	1	2	3	4	5	6	7	8	9
10	4.605	4.615	4.625	4.635	4.644	4.654	4.663	4.673	4.682	4.691
11	4.700	4.710	4.718	4.727	4.736	4.745	4.754	4.762	4.771	4.779
12	4.787	4.796	4.804	4.812	4.820	4.828	4.836	4.844	4.852	4.860
13	4.868	4.875	4.883	4.890	4.898	4.905	4.913	4.920	4.927	4.934
14	4.942	4.949	4.956	4.963	4.970	4.977	4.984	4.990	4.997	5.004
15	5.011	5.017	5.024	5.030	5.037	5.043	5.050	5.056	5.063	5.069
16	5.075	5.081	5.088	5.094	5.100	5.106	5.112	5.118	5.124	5.130
17	5.136	5.142	5.147	5.153	5.159	5.165	5.170	5.176	5.182	5.187
18	5.193	5.198	5.204	5.209	5.215	5.220	5.226	5.231	5.236	5.242
19	5.247	5.252	5.257	5.263	5.268	5.273	5.278	5.283	5.288	5.293
20	5.298	5.303	5.308	5.313	5.318	5.323	5.328	5.333	5.338	5.342

* In this line take tabular value −10.

† For numbers above 209 use either

$$\log_e 10^k N = \log_e N + k \log_e 10 = \log_e N + 2.303k$$

or

$$\log_e N = \log_e 10 \log_{10} N = 2.303 \log_{10} N$$

TABLE A-3 VALUES OF ϵ^x AND ϵ^{-x}

x	ϵ^x	ϵ^{-x}	x	ϵ^x	ϵ^{-x}
0.0	1.00	1.00	2.6	13.5	0.074
0.1	1.11	0.90	2.7	14.9	0.067
0.2	1.22	0.82	2.8	16.4	0.061
0.3	1.35	0.74	2.9	18.2	0.055
0.4	1.49	0.67	3.0	20.1	0.050
0.5	1.65	0.61	3.1	22.2	0.045
0.6	1.82	0.55	3.2	24.5	0.041
0.7	2.01	0.50	3.3	27.1	0.037
0.8	2.23	0.45	3.4	30.0	0.033
0.9	2.46	0.41	3.5	33.1	0.030
1.0	2.72	0.37	3.6	36.6	0.027
1.1	3.00	0.33	3.7	40.4	0.025
1.2	3.32	0.30	3.8	44.7	0.022
1.3	3.67	0.27	3.9	49.4	0.020
1.4	4.06	0.25	4.0	54.6	0.018
1.5	4.48	0.22	4.1	60.3	0.017
1.6	4.95	0.20	4.2	66.7	0.015
1.7	5.47	0.18	4.3	73.7	0.014
1.8	6.05	0.17	4.4	81.5	0.012
1.9	6.69	0.15	4.5	90.0	0.011
2.0	7.39	0.14	4.6	99.5	0.010
2.1	8.17	0.12	4.7	110.	0.0091
2.2	9.03	0.11	4.8	122.	0.0082
2.3	9.97	0.10	4.9	134.	0.0074
2.4	11.0	0.091	5.0	148.	0.0067
2.5	12.2	0.082			

TABLE A-4 TRIGONOMETRIC FUNCTIONS

Angles	Sines		Cosines		Tangents		Cotangents		Angles
	Nat.	Log.	Nat.	Log.	Nat.	Log.	Nat.	Log.	
0° 00′	.0000	∞	1.0000	**0**.0000	.0000	∞	∞	∞	90° 00′
10	.0029	**7**.4637	1.0000	0000	.0029	**7**.4637	343.77	**2**.5363	50
20	.0058	7648	1.0000	0000	.0058	7648	171.89	2352	40
30	.0087	9408	1.0000	0000	.0087	9409	114.59	0591	30
40	.0116	**8**.0658	.9999	0000	.0116	**8**.0658	85.940	**1**.9342	20
50	.0145	1627	.9999	0000	.0145	1627	68.750	8373	10
1° 00′	.0175	**8**.2419	.9998	**9**.9999	.0175	**8**.2419	57.290	**1**.7581	89° 00′
10	.0204	3088	.9998	9999	.0204	3089	49.104	6911	50
20	.0233	3668	.9997	9999	.0233	3669	42.964	6331	40
30	.0262	4179	.9997	9999	.0262	4181	38.188	5819	30
40	.0291	4637	.9996	9998	.0291	4638	34.368	5362	20
50	.0320	5050	.9995	9998	.0320	5053	31.242	4947	10
2° 00′	.0349	**8**.5428	.9994	**9**.9997	.0349	**8**.5431	28.636	**1**.4569	88° 00′
10	.0378	5776	.9993	9997	.0378	5779	26.432	4221	50
20	.0407	6097	.9992	9996	.0407	6101	24.542	3899	40
30	.0436	6397	.9990	9996	.0437	6401	22.904	3599	30
40	.0465	6677	.9989	9995	.0466	6682	21.470	3318	20
50	.0494	6940	.9988	9995	.0495	6945	20.206	3055	10
3° 00′	.0523	**8**.7188	.9986	**9**.9994	.0524	**8**.7194	19.081	**1**.2806	87° 00′
10	.0552	7423	.9985	9993	.0553	7429	18.075	2571	50
20	.0581	7645	.9983	9993	.6582	7652	17.169	2348	40
30	.0610	7857	.9981	9992	.0612	7865	16.350	2135	30
40	.0640	8059	.9980	9991	.0641	8067	15.605	1933	20
50	.0669	8251	.9978	9990	.0670	8261	14.924	1739	10
4° 00′	.0698	**8**.8436	.9976	**9**.9989	.0669	**8**.8446	14.301	**1**.1554	86° 00′
10	.0727	8613	.9974	9989	.0729	8624	13.727	1376	50
20	.0756	8783	.9971	9988	.0758	8795	13.197	1205	40
30	.0785	8946	.9969	9987	.0787	8960	12.706	1040	30
40	.0814	9104	.9967	9986	.0816	9118	12.251	0882	20
50	.0843	9256	.9964	9985	.0846	9272	11.826	0728	10
5° 00′	.0872	**8**.9403	.9962	**9**.9983	.0875	**8**.9420	11.430	**1**.0580	85° 00′
10	.0901	9545	.9959	9982	.0904	9563	11.059	0437	50
20	.0929	9682	.9957	9981	.0934	9701	10.712	0299	40
30	.0958	9816	.9954	9980	.0963	9836	10.385	0164	30
40	.0987	9945	.9951	9979	.0992	9966	10.078	0034	20
50	.1016	**9**.0070	.9948	9977	.1022	**9**.0093	9.7882	**0**.9907	10
6° 00′	.1045	**9**.0192	.9945	**9**.9976	.1051	**9**.0216	9.5144	**0**.9784	84° 00′
10	.1074	0311	.9942	9975	.1080	0336	9.2553	9664	50
20	.1103	0426	.9939	9973	.1110	0453	9.0098	9547	40
30	.1132	0539	.9936	9972	.1139	0567	8.7769	9433	30
40	.1161	0648	.9932	9971	.1169	0678	8.5555	9322	20
50	.1190	0755	.9929	9969	.1198	0786	8.3450	9214	10
7° 00′	.1219	**9**.0859	.9925	**9**.9968	.1228	**9**.0891	8.1443	**0**.9109	83° 00′
10	.1248	0961	.9922	9966	.1257	0995	7.9530	9005	50
20	.1276	1060	.9918	9964	.1287	1096	7.7704	8904	40
30	.1305	1157	.9914	9963	.1317	1194	7.5958	8806	30
40	.1334	1252	.9911	9961	.1346	1291	7.4287	8709	20
50	.1363	1345	.9907	9959	.1376	1385	7.2687	8615	10
8° 00′	.1392	**9**.1436	.9903	**9**.9958	.1405	**9**.1478	7.1154	**0**.8522	82° 00′
10	.1421	1525	.9899	9956	.1435	1569	6.9682	8431	50
20	.1449	1612	.9894	9954	.1465	1658	6.8269	8342	40
30	.1478	1697	.9890	9952	.1495	1745	6.6912	8255	30
40	.1507	1781	.9886	9950	.1524	1831	6.5606	8169	20
50	.1536	1863	.9881	9948	.1554	1915	6.4348	8085	10
9° 00′	.1564	**9**.1943	.9877	**9**.9946	.1584	**9**.1997	6.3138	**0**.8003	81° 00′
	Nat.	Log.	Nat.	Log.	Nat.	Log.	Nat.	Log.	
Angles	Cosines		Sines		Cotangents		Tangents		Angles

Table A-4 Trigonometric Functions (*Continued*)

Angles	Sines Nat.	Sines Log.	Cosines Nat.	Cosines Log.	Tangents Nat.	Tangents Log.	Cotangents Nat.	Cotangents Log.	Angles
9° 00′	.1564	9.1943	.9877	9.9946	.1584	9.1997	6.3138	0.8003	81° 00′
10	.1593	2022	.9872	9944	.1614	2078	6.1970	7922	50
20	.1622	2100	.9868	9942	.1644	2158	6.0844	7842	40
30	.1650	2176	.9863	9940	.1673	2236	5.9758	7764	30
40	.1679	2251	.9858	9938	.1703	2313	5.8708	7687	20
50	.1708	2324	.9853	9936	.1733	2389	5.7694	7611	10
10° 00′	.1736	9.2397	.9848	9.9934	.1763	9.2463	5.6713	0.7537	80° 00′
10	.1765	2468	.9843	9931	.1793	2536	5.5764	7464	50
20	.1794	2538	.9838	9929	.1823	2609	5.4845	7391	40
30	.1822	2606	.9833	9927	.1853	2680	5.3955	7320	30
40	.1851	2674	.9827	9924	.1883	2750	5.3093	7250	20
50	.1880	2740	.9822	9922	.1914	2819	5.2257	7181	10
11° 00′	.1908	9.2806	.9816	9.9919	.1944	9.2887	5.1446	0.7113	79° 00′
10	.1937	2870	.9811	9917	.1974	2953	5.0658	7047	50
20	.1965	2934	.9805	9914	.2004	3020	4.9894	6980	40
30	.1994	2997	.9799	9912	.2035	3085	4.9152	6915	30
40	.2022	3058	.9793	9909	.2065	3149	4.8430	6851	20
50	.2051	3119	.9787	9907	.2095	3212	4.7729	6788	10
12° 00′	.2079	9.3179	.9781	9.9904	.2126	9.3275	4.7046	0.6725	78° 00′
10	.2108	3238	.9775	9901	.2156	3336	4.6382	6664	50
20	.2136	3296	.9769	9899	.2186	3397	4.5736	6603	40
30	.2164	3353	.9763	9896	.2217	3458	4.5107	6542	30
40	.2193	3410	.9757	9893	.2247	3517	4.4494	6483	20
50	.2221	3466	.9750	9890	.2278	3576	4.3897	6424	10
13° 00′	.2250	9.3521	.9744	9.9887	.2309	9.3634	4.3315	0.6366	77° 00′
10	.2278	3575	.9737	9884	.2339	3691	4.2747	6309	50
20	.2306	3629	.9730	9881	.2370	3748	4.2193	6252	40
30	.2334	3682	.9724	9878	.2401	3804	4.1653	6196	30
40	.2363	3734	.9717	9875	.2432	3859	4.1126	6141	20
50	.2391	3786	.9710	9872	.2462	3914	4.0611	6086	10
14° 00′	.2419	9.3837	.9703	9.9869	.2493	9.3968	4.0108	0.6032	76° 00′
10	.2447	3887	.9696	9866	.2524	4021	3.9617	5979	50
20	.2476	3937	.9689	9863	.2555	4074	3.9136	5926	40
30	.2504	3986	.9681	9859	.2586	4127	3.8667	5873	30
40	.2532	4035	.9674	9856	.2617	4178	3.8208	5822	20
50	.2560	4083	.9667	9853	.2648	4230	3.7760	5770	10
15° 00′	.2588	9.4130	.9659	9.9849	.2679	9.4281	3.7321	0.5719	75° 00′
10	.2616	4177	.9652	9846	.2711	4331	3.6891	5669	50
20	.2644	4223	.9644	9843	.2742	4381	3.6470	5619	40
30	.2672	4269	.9636	9839	.2773	4430	3.6059	5570	30
40	.2700	4314	.9628	9836	.2805	4479	3.5656	5521	20
50	.2728	4359	.9621	9832	.2836	4527	3.5261	5473	10
16° 00′	.2756	9.4403	.9613	9.9828	.2867	9.4575	3.4874	0.5425	74° 00′
10	.2784	4447	.9605	9825	.2899	4622	3.4495	5378	50
20	.2812	4491	.9596	9821	.2931	4669	3.4124	5331	40
30	.2840	4533	.9588	9817	.2962	4716	3.3759	5284	30
40	.2868	4576	.9580	9814	.2994	4762	3.3402	5238	20
50	.2896	4618	.9572	9810	.3026	4808	3.3052	5192	10
17° 00′	.2924	9.4659	.9563	9.9806	.3057	9.4853	3.2709	0.5147	73° 00′
10	.2952	4700	.9555	9802	.3089	4898	3.2371	5102	50
20	.2979	4741	.9546	9798	.3121	4943	3.2041	5057	40
30	.3007	4781	.9537	9794	.3153	4987	3.1716	5013	30
40	.3035	4821	.9528	9790	.3185	5031	3.1397	4969	20
50	.3062	4861	.9520	9786	.3217	5075	3.1084	4925	10
18° 00′	.3090	9.4900	.9511	9.9782	.3249	9.5118	3.0777	0.4882	72° 00
	Nat.	Log.	Nat.	Log.	Nat.	Log.	Nat.	Log.	
Angles	Cosines		Sines		Cotangents		Tangents		Angles

TABLE A-4 TRIGONOMETRIC FUNCTIONS (*Continued*)

Angles	Sines Nat.	Sines Log.	Cosines Nat.	Cosines Log.	Tangents Nat.	Tangents Log.	Cotangents Nat.	Cotangents Log.	Angles
18° 00′	.3090	9.4900	.9511	9.9782	.3249	9.5118	3.0777	0.4882	72° 00′
10	.3118	4939	.9502	9778	.3281	5161	3.0475	4839	50
20	.3145	4977	.9492	9774	.3314	5203	3.0178	4797	40
30	.3173	5015	.9483	9770	.3346	5245	2.9887	4755	30
40	.3201	5052	.9474	9765	.3378	5287	2.9600	4713	20
50	.3228	5090	.9465	9761	.3411	5329	2.9319	4671	10
19° 00′	.3256	9.5126	.9455	9.9757	.3443	9.5370	2.9042	0.4630	71° 00′
10	.3283	5163	.9446	9752	.3476	5411	2.8770	4589	50
20	.3311	5199	.9436	9748	.3508	5451	2.8502	4549	40
30	.3338	5235	.9426	9743	.3541	5491	2.8239	4509	30
40	.3365	5270	.9417	9739	.3574	5531	2.7980	4469	20
50	.3393	5306	.9407	9734	.3607	5571	2.7725	4429	10
20° 00′	.3420	9.5341	.9397	9.9730	.3640	9.5611	2.7475	0.4389	70° 00′
10	.3448	5375	.9387	9725	.3673	5650	2.7228	4350	50
20	.3475	5409	.9377	9721	.3706	5689	2.6985	4311	40
30	.3502	5443	.9367	9716	.3739	5727	2.6746	4273	30
40	.3529	5477	.9356	9711	.3772	5766	2.6511	4234	20
50	.3557	5510	.9346	9706	.3805	5804	2.6279	4196	10
21° 00′	.3584	9.5543	.9336	9.9702	.3839	9.5842	2.6051	0.4158	69° 00′
10	.3611	5576	.9325	9697	.3872	5879	2.5826	4121	50
20	.3638	5609	.9315	9692	.3906	5917	2.5605	4083	40
30	.3665	5641	.9304	9687	.3939	5954	2.5386	4046	30
40	.3692	5673	.9293	9682	.3973	5991	2.5172	4009	20
50	.3719	5704	.9283	9677	.4006	6028	2.4960	3972	10
22° 00′	.3746	9.5736	.9272	9.9672	.4040	9.6064	2.4751	0.3936	68° 00′
10	.3773	5767	.9261	9667	.4074	6100	2.4545	3900	50
20	.3800	5798	.9250	9661	.4108	6136	2.4342	3864	40
30	.3827	5828	.9239	9656	.4142	6172	2.4142	3828	30
40	.3854	5859	.9228	9651	.4176	6208	2.3945	3792	20
50	.3881	5889	.9216	9646	.4210	6243	2.3750	3757	10
23° 00′	.3907	9.5919	.9205	9.9640	.4245	9.6279	2.3559	0.3721	67° 00′
10	.3934	5948	.9194	9635	.4279	6314	2.3369	3686	50
20	.3961	5978	.9182	9629	.4314	6348	2.3183	3652	40
30	.3987	6007	.9171	9624	.4348	6383	2.2998	3617	30
40	.4014	6036	.9159	9618	.4383	6417	2.2817	3583	20
50	.4041	6065	.9147	9613	.4417	6452	2.2637	3548	10
24° 00′	.4067	9.6093	.9135	9.9607	.4452	9.6486	2.2460	0.3514	66° 00′
10	.4094	6121	.9124	9602	.4487	6520	2.2286	3480	50
20	.4120	6149	.9112	9596	.4522	6553	2.2113	3447	40
30	.4147	6177	.9100	9590	.4557	6587	2.1943	3413	30
40	.4173	6205	.9088	9584	.4592	6620	2.1775	3380	20
50	.4200	6232	.9075	9579	.4628	6654	2.1609	3346	10
25° 00′	.4226	9.6259	.9063	9.9573	.4663	9.6687	2.1445	0.3313	65° 00′
10	.4253	6286	.9051	9567	.4699	6720	2.1283	3280	50
20	.4279	6313	.9038	9561	.4734	6752	2.1123	3248	40
30	.4305	6340	.9026	9555	.4770	6785	2.0965	3215	30
40	.4331	6366	.9013	9549	.4806	6817	2.0809	3183	20
50	.4358	6392	.9001	9543	.4841	6850	2.0655	3150	10
26° 00′	.4384	9.6418	.8988	9.9537	.4877	9.6882	2.0503	0.3118	64° 00′
10	.4410	6444	.8975	9530	.4913	6914	2.0353	3086	50
20	.4436	6470	.8962	9524	.4950	6946	2.0204	3054	40
30	.4462	6495	.8949	9518	.4986	6977	2.0057	3023	30
40	.4488	6521	.8936	9512	.5022	7009	1.9912	2991	20
50	.4514	6546	.8923	9505	.5059	7040	1.9768	2960	10
27° 00′	.4540	9.6570	.8910	9.9499	.5095	9.7072	1.9626	0.2928	63° 00′
	Nat.	Log.	Nat.	Log.	Nat.	Log.	Nat.	Log.	
Angles	Cosines		Sines		Cotangents		Tangents		Angles

TABLE A-4 TRIGONOMETRIC FUNCTIONS (*Continued*)

Angles	Sines		Cosines		Tangents		Cotangents		Angles
	Nat.	Log.	Nat.	Log.	Nat.	Log.	Nat.	Log.	
27° 00′	.4540	**9.**6570	.8910	**9.**9499	.5095	**9.**7072	1.9626	**0.**2928	**63° 00′**
10	.4566	6595	.8897	9492	.5132	7103	1.9486	2897	50
20	.4592	6620	.8884	9486	.5169	7134	1.9347	2866	40
30	.4617	6644	.8870	9479	.5206	7165	1.9210	2835	30
40	.4643	6668	.8857	9473	.5243	7196	1.9074	2804	20
50	.4669	6692	.8843	9466	.5280	7226	1.8940	2774	10
28° 00′	.4695	**9.**6716	.8829	**9.**9459	.5317	**9.**7257	1.8807	**0.**2743	**62° 00′**
10	.4720	6740	.8816	9453	.5354	7287	1.8676	2713	50
20	.4746	6763	.8802	9446	.5392	7317	1.8546	2683	40
30	.4772	6787	.8788	9439	.5430	7348	1.8418	2652	30
40	.4797	6810	.8774	9432	.5467	7378	1.8291	2622	20
50	.4823	6833	.8760	9425	.5505	7408	1.8165	2592	10
29° 00′	.4848	**9.**6856	.8746	**9.**9418	.5543	**9.**7438	1.8040	**0.**2562	**61° 00′**
10	.4874	6878	.8732	9411	.5581	7467	1.7917	2533	50
20	.4899	6901	.8718	9404	.5619	7497	1.7796	2503	40
30	.4924	6923	.8704	9397	.5658	7526	1.7675	2474	30
40	.4950	6946	.8689	9390	.5696	7556	1.7556	2444	20
50	.4975	6968	.8675	9383	.5735	7585	1.7437	2415	10
30° 00′	.5000	**9.**6990	.8660	**9.**9375	.5774	**9.**7614	1.7321	**0.**2386	**60° 00′**
10	.5025	7012	.8646	9368	.5812	7644	1.7205	2356	50
20	.5050	7033	.8631	9361	.5851	7673	1.7090	2327	40
30	.5075	7055	.8616	9353	.5890	7701	1.6977	2299	30
40	.5100	7076	.8601	9346	.5930	7730	1.6864	2270	20
50	.5125	7097	.8587	9338	.5969	7759	1.6753	2241	10
31° 00′	.5150	**9.**7118	.8572	**9.**9331	.6009	**9.**7788	1.6643	**0.**2212	**59° 00′**
10	.5175	7139	.8557	9323	.6048	7816	1.6534	2184	50
20	.5200	7160	.8542	9315	.6088	7845	1.6426	2155	40
30	.5225	7181	.8526	9308	.6128	7873	1.6319	2127	30
40	.5250	7201	.8511	9300	.6168	7902	1.6212	2098	20
50	.5275	7222	.8496	9292	.6208	7930	1.6107	2070	10
32° 00′	.5299	**9.**7242	.8480	**9.**9284	.6249	**9.**7958	1.6003	**0.**2042	**58° 00′**
10	.5324	7262	.8465	9276	.6289	7986	1.5900	2014	50
20	.5348	7282	.8450	9268	.6330	8014	1.5798	1986	40
30	.5373	7302	.8434	9260	.6371	8042	1.5697	1958	30
40	.5398	7322	.8418	9252	.6412	8070	1.5597	1930	20
50	.5422	7342	.8403	9244	.6453	8097	1.5497	1903	10
33° 00′	.5446	**9.**7361	.8387	**9.**9236	.6494	**9.**8125	1.5399	**0.**1875	**57° 00′**
10	.5471	7380	.8371	9228	.6536	8153	1.5301	1847	50
20	.5495	7400	.8355	9219	.6577	8180	1.5204	1820	40
30	.5519	7419	.8339	9211	.6619	8208	1.5108	1792	30
40	.5544	7438	.8323	9203	.6661	8235	1.5013	1765	20
50	.5568	7457	.8307	9194	.6703	8263	1.4919	1737	10
34° 00′	.5592	**9.**7476	.8290	**9.**9186	.6745	**9.**8290	1.4826	**0.**1710	**56° 00′**
10	.5616	7494	.8274	9177	.6787	8317	1.4733	1683	50
20	.5640	7513	.8258	9169	.6830	8344	1.4641	1656	40
30	.5664	7531	.8241	9160	.6873	8371	1.4550	1629	30
40	.5688	7550	.8225	9151	.6916	8398	1.4460	1602	20
50	.5712	7568	.8208	9142	.6959	8425	1.4370	1575	10
35° 00′	.5736	**9.**7586	.8192	**9.**9134	.7002	**9.**8452	1.4281	**0.**1548	**55° 00′**
10	.5760	7604	.8175	9125	.7046	8479	1.4193	1521	50
20	.5783	7622	.8158	9116	.7089	8506	1.4106	1494	40
30	.5807	7640	.8141	9107	.7133	8533	1.4019	1467	30
40	.5831	7657	.8124	9098	.7177	8559	1.3934	1441	20
50	.5854	7675	.8107	9089	.7221	8586	1.3848	1414	10
36° 00′	.5878	**9.**7692	.8090	**9.**9080	.7265	**9.**8613	1.3764	**0.**1387	**54° 00′**
	Nat.	Log.	Nat.	Log.	Nat.	Log.	Nat.	Log.	
Angles	Cosines		Sines		Cotangents		Tangents		Angles

TABLE A-4 TRIGONOMETRIC FUNCTIONS (*Continued*)

Angles	Sines		Cosines		Tangents		Cotangents		Angles
	Nat.	Log.	Nat.	Log.	Nat.	Log.	Nat.	Log.	
36° 00′	.5878	9.7692	.8090	9.9080	.7265	9.8613	1.3764	0.1387	54° 00′
10	.5901	7710	.8073	9070	.7310	8639	1.3680	1361	50
20	.5925	7727	.8056	9061	.7355	8666	1.3597	1334	40
30	.5948	7744	.8039	9052	.7400	8692	1.3514	1308	30
40	.5972	7761	.8021	9042	.7445	8718	1.3432	1282	20
50	.5995	7778	.8004	9033	.7490	8745	1.3351	1255	10
37° 00′	.6018	9.7795	.7986	9.9023	.7536	9.8771	1.3270	0.1229	53° 00′
10	.6041	7811	.7969	9014	.7581	8797	1.3190	1203	50
20	.6065	7828	.7951	9004	.7627	8824	1.3111	1176	40
30	.6088	7844	.7934	8995	.7673	8850	1.3032	1150	30
40	.6111	7861	.7916	8985	.7720	8876	1.2954	1124	20
50	.6134	7877	.7898	8975	.7766	8902	1.2876	1098	10
38° 00′	.6157	9.7893	.7880	9.8965	.7813	9.8928	1.2790	0.1072	52° 00′
10	.6180	7910	.7862	8955	.7860	8954	1.2723	1046	50
20	.6202	7926	.7844	8945	.7907	8980	1.2647	1020	40
30	.6225	7941	.7826	8935	.7954	9006	1.2572	0994	30
40	.6248	7957	.7808	8925	.8002	9032	1.2497	0968	20
50	.6271	7973	.7790	8915	.8050	9058	1.2423	0942	10
39° 00′	.6293	9.7989	.7771	9.8905	.8098	9.9084	1.2349	0.0916	51° 00′
10	.6316	8004	.7753	8895	.8146	9110	1.2276	0890	50
20	.6338	8020	.7735	8884	.8195	9135	1.2203	0865	40
30	.6361	8035	.7716	8874	.8243	9161	1.2131	0839	30
40	.6383	8050	.7698	8864	.8292	9187	1.2059	0813	20
50	.6406	8066	.7679	8853	.8342	9212	1.1988	0788	10
40° 00′	.6428	9.8081	.7660	9.8843	.8391	9.9238	1.1918	0.0762	50° 00′
10	.6450	8096	.7642	8832	.8441	9264	1.1847	0736	50
20	.6472	8111	.7623	8821	.8491	9289	1.1778	0711	40
30	.6494	8125	.7604	8810	.8541	9315	1.1708	0685	30
40	.6517	8140	.7585	8800	.8591	9341	1.1640	0659	20
50	.6539	8155	.7566	8789	.8642	9366	1.1571	0634	10
41° 00′	.6561	9.8169	.7547	9.8778	.8693	9.9392	1.1504	0.0608	49° 00′
10	.6583	8184	.7528	8767	.8744	9417	1.1436	0583	50
20	.6604	8198	.7509	8756	.8796	9443	1.1369	0557	40
30	.6626	8213	.7490	8745	.8847	9468	1.1303	0532	30
40	.6648	8227	.7470	8733	.8899	9494	1.1237	0506	20
50	.6670	8241	.7451	8722	.8952	9519	1.1171	0481	10
42° 00′	.6691	9.8255	.7431	9.8711	.9004	9.9544	1.1106	0.0456	48° 00′
10	.6713	8269	.7412	8699	.9057	9570	1.1041	0430	50
20	.6734	8283	.7392	8688	.9110	9595	1.0977	0405	40
30	.6756	8297	.7373	8676	.9163	9621	1.0913	0379	30
40	.6777	8311	.7353	8665	.9217	9646	1.0850	0354	20
50	.6799	8324	.7333	8653	.9271	9671	1.0786	0329	10
43° 00′	.6820	9.8338	.7314	9.8641	.9325	9.9697	1.0724	0.0303	47° 00′
10	.6841	8351	.7294	8629	.9380	9722	1.0661	0278	50
20	.6862	8365	.7274	8618	.9435	9747	1.0599	0253	40
30	.6884	8378	.7254	8606	.9490	9772	1.0538	0228	30
40	.6905	8391	.7234	8594	.9545	9798	1.0477	0202	20
50	.6926	8405	.7214	8582	.9601	9823	1.0416	0177	10
44° 00′	.6947	9.8418	.7193	9.8569	.9657	9.9848	1.0355	0.0152	46° 00′
10	.6967	8431	.7173	8557	.9713	9874	1.0295	0126	50
20	.6988	8444	.7153	8545	.9770	9899	1.0235	0101	40
30	.7009	8457	.7133	8532	.9827	9924	1.0176	0076	30
40	.7030	8469	.7112	8520	.9884	9949	1.0117	0051	20
50	.7050	8482	.7092	8507	.9942	9975	1.0058	0025	10
45° 00′	.7071	9.8495	.7071	9.8495	1.0000	0.0000	1.0000	0.0000	45° 00′
	Nat.	Log.	Nat.	Log.	Nat.	Log.	Nat.	Log.	
Angles	Cosines		Sines		Cotangents		Tangents		Angles

Table A-5 Integrals

The constant of integration is omitted in the following formulas:

Forms containing $ax + b$

1. $\int (ax+b)^m \, dx = \frac{(ax+b)^{m+1}}{a(m+1)} \qquad m \neq -1$

2. $\int \frac{dx}{ax+b} = \frac{1}{a} \log (ax+b)$

3. $\int \frac{x \, dx}{ax+b} = \frac{1}{a^2} [ax + b - b \log (ax+b)]$

4. $\int \frac{x \, dx}{(ax+b)^2} = \frac{1}{a^2} \left[\frac{b}{ax+b} + \log (ax+b) \right]$

5. $\int \frac{x^2 \, dx}{ax+b} = \frac{1}{a^3} \left[\frac{(ax+b)^2}{2} - 2b(ax+b) + b^2 \log (ax+b) \right]$

6. $\int \frac{x^2 \, dx}{(ax+b)^2} = \frac{1}{a^3} \left[ax + b - \frac{b^2}{ax+b} - 2b \log (ax+b) \right]$

7. $\int \frac{dx}{x(ax+b)} = \frac{1}{b} \log \frac{x}{ax+b}$

8. $\int \frac{dx}{x(ax+b)^2} = \frac{1}{b(ax+b)} + \frac{1}{b^2} \log \frac{x}{ax+b}$

9. $\int \frac{dx}{x^2(ax+b)} = -\frac{1}{bx} + \frac{a}{b^2} \log \frac{ax+b}{x}$

10. $\int \frac{x \, dx}{(ax+b)^3} = \frac{1}{a^2} \left[-\frac{1}{ax+b} + \frac{b}{2(ax+b)^2} \right]$

Forms containing $ax^2 + b$

11. $\int \frac{dx}{ax^2+b} = \frac{1}{\sqrt{ab}} \tan^{-1} \sqrt{\frac{a}{b}}\, x \qquad a > 0, \; b > 0$

12. $\int \frac{dx}{ax^2+b} = \frac{1}{2\sqrt{-ab}} \log \frac{\sqrt{b} + x\sqrt{-a}}{\sqrt{b} - x\sqrt{-a}} \qquad a < 0, \; b > 0$

13. $\int \frac{dx}{(ax^2+b)^2} = \frac{x}{2b(ax^2+b)} + \frac{1}{2b} \int \frac{dx}{ax^2+b}$

14. $\int \frac{dx}{x(ax^2+b)} = \frac{1}{2b} \log \frac{x^2}{ax^2+b}$

Forms containing $\sqrt{ax + b}$

15. $\int x\sqrt{ax+b} \, dx = \frac{2(3ax - 2b)\sqrt{(ax+b)^3}}{15a^2}$

16. $\int x^2\sqrt{ax+b} \, dx = \frac{2(15a^2x^2 - 12abx + 8b^2)\sqrt{(ax+b)^3}}{105a^3}$

17. $\int \frac{x \, dx}{\sqrt{ax+b}} = \frac{2(ax - 2b)\sqrt{ax+b}}{3a^2}$

18. $\int \frac{x^2 \, dx}{\sqrt{ax+b}} = \frac{2(3a^2x^2 - 4abx + 8b^2)\sqrt{ax+b}}{15a^3}$

TABLE A-5 INTEGRALS (*Continued*)

Forms containing $\sqrt{a^2 - x^2}$

19. $\int \frac{dx}{\sqrt{a^2 - x^2}} = \sin^{-1} \frac{x}{a}$

20. $\int \frac{x\,dx}{\sqrt{a^2 - x^2}} = -\sqrt{a^2 - x^2}$

21. $\int \frac{x^2\,dx}{\sqrt{a^2 - x^2}} = -\frac{x}{2}\sqrt{a^2 - x^2} + \frac{a^2}{2}\sin^{-1}\frac{x}{a}$

22. $\int \frac{x^3\,dx}{\sqrt{a^2 - x^2}} = \frac{1}{3}\sqrt{(a^2 - x^2)^3} - a^2\sqrt{a^2 - x^2}$

23. $\int \frac{dx}{x\sqrt{a^2 - x^2}} = -\frac{1}{a}\log\frac{a + \sqrt{a^2 - x^2}}{x}$

24. $\int \frac{dx}{x^2\sqrt{a^2 - x^2}} = -\frac{\sqrt{a^2 - x^2}}{a^2 x}$

25. $\int \frac{dx}{x^3\sqrt{a^2 - x^2}} = -\frac{\sqrt{a^2 - x^2}}{2a^2x^2} - \frac{1}{2a^3}\log\frac{a + \sqrt{a^2 - x^2}}{x}$

26. $\int \sqrt{a^2 - x^2}\,dx = \frac{x}{2}\sqrt{a^2 - x^2} + \frac{a^2}{2}\sin^{-1}\frac{x}{a}$

27. $\int x\sqrt{a^2 - x^2}\,dx = -\frac{1}{3}\sqrt{(a^2 - x^2)^3}$

28. $\int x^2\sqrt{a^2 - x^2}\,dx = -\frac{x}{4}\sqrt{(a^2 - x^2)^3} + \frac{a^2x}{8}\sqrt{a^2 - x^2} + \frac{a^4}{8}\sin^{-1}\frac{x}{a}$

29. $\int x^3\sqrt{a^2 - x^2}\,dx = \frac{1}{5}\sqrt{(a^2 - x^2)^5} - \frac{a^2}{3}\sqrt{(a^2 - x^2)^3}$

30. $\int \frac{\sqrt{a^2 - x^2}}{x}\,dx = \sqrt{a^2 - x^2} - a\log\frac{a + \sqrt{a^2 - x^2}}{x}$

31. $\int \frac{\sqrt{a^2 - x^2}}{x^2}\,dx = -\frac{\sqrt{a^2 - x^2}}{x} - \sin^{-1}\frac{x}{a}$

32. $\int \sqrt{(a^2 - x^2)^3}\,dx = \frac{x}{4}\sqrt{(a^2 - x^2)^3} + \frac{3a^2x}{8}\sqrt{a^2 - x^2} + \frac{3a^4}{8}\sin^{-1}\frac{x}{a}$

33. $\int \frac{dx}{\sqrt{(a^2 - x^2)^3}} = \frac{x}{a^2\sqrt{a^2 - x^2}}$

Forms containing $\sqrt{x^2 \pm a^2}$

34. $\int \frac{dx}{\sqrt{x^2 \pm a^2}} = \log(x + \sqrt{x^2 \pm a^2})$

35. $\int \frac{x\,dx}{\sqrt{x^2 \pm a^2}} = \sqrt{x^2 \pm a^2}$

36. $\int \frac{x^2\,dx}{\sqrt{x^2 \pm a^2}} = \frac{x}{2}\sqrt{x^2 \pm a^2} \mp \frac{a^2}{2}\log(x + \sqrt{x^2 \pm a^2})$

37. $\int \frac{x^3\,dx}{\sqrt{x^2 \pm a^2}} = \frac{1}{3}\sqrt{(x^2 \pm a^2)^3} \mp a^2\sqrt{x^2 \pm a^2}$

38. $\int \frac{dx}{x\sqrt{x^2 + a^2}} = -\frac{1}{a}\log\frac{a + \sqrt{x^2 + a^2}}{x}$

TABLE A-5 INTEGRALS (*Continued*)

39. $\int \frac{dx}{x\sqrt{x^2 - a^2}} = \frac{1}{a}\cos^{-1}\frac{a}{x}$

40. $\int \frac{dx}{x^2\sqrt{x^2 \pm a^2}} = \mp \frac{\sqrt{x^2 \pm a^2}}{a^2 x}$

41. $\int \sqrt{x^2 \pm a^2}\,dx = \frac{x}{2}\sqrt{x^2 \pm a^2} \pm \frac{a^2}{2}\log\,(x + \sqrt{x^2 \pm a^2})$

42. $\int x\sqrt{x^2 \pm a^2}\,dx = \frac{1}{3}\sqrt{(x^2 \pm a^2)^3}$

43. $\int x^2\sqrt{x^2 \pm a^2}\,dx = \frac{x}{4}\sqrt{(x^2 \pm a^2)^3} \mp \frac{a^2 x}{8}\sqrt{x^2 \pm a^2} - \frac{a^4}{8}\log\,(x + \sqrt{x^2 \pm a^2})$

44. $\int x^3\sqrt{x^2 \pm a^2}\,dx = \frac{1}{5}\sqrt{(x^2 \pm a^2)^5} \mp \frac{a^2}{3}\sqrt{(x^2 \pm a^2)^3}$

45. $\int \frac{\sqrt{x^2 + a^2}}{x}\,dx = \sqrt{x^2 + a^2} - a\log\frac{a + \sqrt{x^2 + a^2}}{x}$

46. $\int \frac{\sqrt{x^2 - a^2}}{x}\,dx = \sqrt{x^2 - a^2} - a\cos^{-1}\frac{a}{x}$

47. $\int \sqrt{(x^2 \pm a^2)^3}\,dx = \frac{x}{4}\sqrt{(x^2 \pm a^2)^3} \pm \frac{3a^2 x}{8}\sqrt{x^2 \pm a^2} + \frac{3a^4}{8}\log\,(x + \sqrt{x^2 \pm a^2})$

48. $\int \frac{dx}{\sqrt{(x^2 \pm a^2)^3}} = \frac{\pm x}{a^2\sqrt{x^2 \pm a^2}}$

Forms containing $ax^2 + bx + c$, $\sqrt{ax^2 + bx + c}$

49. $\int \frac{dx}{ax^2 + bx + c} = \frac{1}{\sqrt{b^2 - 4ac}}\log\frac{2ax + b - \sqrt{b^2 - 4ac}}{2ax + b + \sqrt{b^2 - 4ac}} \qquad b^2 - 4ac > 0$

50. $\int \frac{dx}{ax^2 + bx + c} = \frac{2}{\sqrt{4ac - b^2}}\tan^{-1}\frac{2ax + b}{\sqrt{4ac - b^2}} \qquad b^2 - 4ac < 0$

51. $\int \frac{dx}{(ax^2 + bx + c)^n} = \frac{2ax + b}{(n - 1)(4ac - b^2)(ax^2 + bx + c)^{n-1}} + \frac{2(2n - 3)a}{(n - 1)(4ac - b^2)}\int \frac{dx}{(ax^2 + bx + c)^{n-1}} \qquad n \neq 1$

52. $\int \frac{x\,dx}{ax^2 + bx + c} = \frac{1}{2a}\log\,(ax^2 + bx + c) - \frac{b}{2a}\int \frac{dx}{ax^2 + bx + c}$

53. $\int \frac{x^2\,dx}{ax^2 + bx + c} = \frac{x}{a} - \frac{b}{2a^2}\log\,(ax^2 + bx + c) + \frac{b^2 - 2ac}{2a^2}\int \frac{dx}{ax^2 + bx + c}$

54. $\int \frac{x\,dx}{(ax^2 + bx + c)^n} = \frac{2c + bx}{(n - 1)(b^2 - 4ac)(ax^2 + bx + c)^{n-1}} + \frac{b(2n - 3)}{(n - 1)(b^2 - 4ac)}\int \frac{dx}{(ax^2 + bx + c)^{n-1}} \qquad n \neq 1$

55. $\int \frac{dx}{\sqrt{ax^2 + bx + c}} = \frac{1}{\sqrt{a}}\log\left(\sqrt{ax^2 + bx + c} + \frac{2ax + b}{2\sqrt{a}}\right) \qquad a > 0$

56. $\int \frac{dx}{\sqrt{ax^2 + bx + c}} = \frac{1}{\sqrt{-a}}\sin^{-1}\left(\frac{-2ax - b}{\sqrt{b^2 - 4ac}}\right) \qquad a < 0$

57. $\int \frac{x\,dx}{\sqrt{ax^2 + bx + c}} = \frac{1}{a}\sqrt{ax^2 + bx + c} - \frac{b}{2a}\int \frac{dx}{\sqrt{ax^2 + bx + c}}$

TABLE A-5 INTEGRALS (*Continued*)

58. $\int \sqrt{ax^2+bx+c}\,dx = \frac{(2ax+b)\sqrt{ax^2+bx+c}}{4a} + \frac{4ac-b^2}{8a}\int \frac{dx}{\sqrt{ax^2+bx+c}}$

59. $\int x\sqrt{ax^2+bx+c}\,dx = \frac{(ax^2+bx+c)^{3/2}}{3a} - \frac{b}{2a}\int \sqrt{ax^2+bx+c}\,dx$

60. $\int \frac{dx}{x\sqrt{ax^2+bx+c}} = -\frac{1}{\sqrt{c}}\log\left(\frac{\sqrt{ax^2+bx+c}+\sqrt{c}}{x} + \frac{b}{2\sqrt{c}}\right) \qquad c > 0$

61. $\int \frac{dx}{x\sqrt{ax^2+bx+c}} = \frac{1}{\sqrt{-c}}\sin^{-1}\frac{bx+2c}{x\sqrt{b^2-4ac}} \qquad c < 0$

62. $\int \frac{dx}{(ax^2+bx+c)^{3/2}} = -\frac{2(2ax+b)}{(b^2-4ac)\sqrt{ax^2+bx+c}}$

Trigonometric forms

63. $\int \sin x\,dx = -\cos x$

64. $\int \sin^2 x\,dx = \frac{x}{2} - \frac{1}{4}\sin 2x$

65. $\int \sin^3 x\,dx = -\cos x + \frac{1}{3}\cos^3 x$

66. $\int \sin^n x\,dx = -\frac{1}{n}\sin^{n-1} x\cos x + \frac{n-1}{n}\int \sin^{n-2} x\,dx$

67. $\int \frac{dx}{\sin^n x} = -\frac{\cos x}{(n-1)\sin^{n-1} x} + \frac{n-2}{n-1}\int \frac{dx}{\sin^{n-2} x} \qquad n \neq 1$

68. $\int \cos x\,dx = \sin x$

69. $\int \cos^2 x\,dx = \frac{x}{2} + \frac{1}{4}\sin 2x$

70. $\int \cos^3 x\,dx = \sin x - \frac{1}{3}\sin^3 x$

71. $\int \cos^n x\,dx = \frac{1}{n}\cos^{n-1} x\sin x + \frac{n-1}{n}\int \cos^{n-2} x\,dx$

72. $\int \frac{dx}{\cos^n x} = \frac{\sin x}{(n-1)\cos^{n-1} x} + \frac{n-2}{n-1}\int \frac{dx}{\cos^{n-2} x} \qquad n \neq 1$

73. $\int \sin^n x\cos x\,dx = \frac{1}{n+1}\sin^{n+1} x$

74. $\int \cos^n x\sin x\,dx = -\frac{1}{n+1}\cos^{n+1} x$

75. $\int \sin^m x\cos^n x\,dx = -\frac{1}{m+n}\sin^{m-1} x\cos^{n+1} x + \frac{m-1}{m+n}\int \sin^{m-2} x\cos^n x\,dx$

76. $\int \sin^m x\cos^n x\,dx = \frac{1}{m+n}\sin^{m+1} x\cos^{n-1} x + \frac{n-1}{m+n}\int \sin^m x\cos^{n-2} x\,dx$

77. $\int \tan x\,dx = -\log\cos x$

78. $\int \tan^2 x\,dx = \tan x - x$

79. $\int \tan^n x\,dx = \frac{1}{n-1}\tan^{n-1} x - \int \tan^{n-2} x\,dx$

80. $\int \cot x\,dx = \log\sin x$

81. $\int \cot^2 x\,dx = -\cot x - x$

TABLE A-5 INTEGRALS (*Continued*)

82. $\int \cot^n x\,dx = -\frac{1}{n-1}\cot^{n-1} x - \int \cot^{n-2} x\,dx$

83. $\int \sec x\,dx = \log (\sec x + \tan x = \log \tan \left(\frac{x}{2} + \frac{\pi}{4}\right)$

84. $\int \sec^2 x\,dx = \tan x$

85. $\int \sec^n x\,dx = \int \frac{dx}{\cos^n x}$

86. $\int \csc x\,dx = \log (\csc x - \cot x) = \log \tan \frac{x}{2}$

87. $\int \csc^2 x\,dx = -\cot x$

88. $\int \csc^n x\,dx = \int \frac{dx}{\sin^n x}$

89. $\int x \sin x\,dx = \sin x - x \cos x$

90. $\int x \cos x\,dx = \cos x + x \sin x$

91. $\int x^n \sin x\,dx = -x^n \cos x + n\int x^{n-1} \cos x\,dx$

92. $\int x^n \cos x\,dx = x^n \sin x - n\int x^{n-1} \sin x\,dx$

93. $\int \sin mx \sin nx\,dx = \frac{\sin (m-n)x}{2(m-n)} - \frac{\sin (m+n)x}{2(m+n)}$

94. $\int \sin mx \cos nx\,dx = -\frac{\cos (m-n)x}{2(m-n)} - \frac{\cos (m+n)x}{2(m+n)}$

95. $\int \cos mx \cos nx\,dx = \frac{\sin (m-n)x}{2(m-n)} + \frac{\sin (m+n)x}{2(m+n)}$

96. $\int \sin^{-1} x\,dx = x \sin^{-1} x + \sqrt{1-x^2}$

97. $\int \cos^{-1} x\,dx = x \cos^{-1} x - \sqrt{1-x^2}$

98. $\int \tan^{-1} x\,dx = x \tan^{-1} x - \frac{1}{2}\log (1+x^2)$

99. $\int \cot^{-1} x\,dx = x \cot^{-1} x + \frac{1}{2}\log (1+x^2)$

Exponential and logarithmic forms

100. $\int e^{ax}\,dx = \frac{1}{a} e^{ax}$

101. $\int b^{ax}\,dx = \frac{b^{ax}}{a \log b}$

102. $\int xe^{ax}\,dx = \frac{e^{ax}}{a^2}(ax-1)$

103. $\int x^n e^{ax}\,dx = \frac{x^n e^{ax}}{a} - \frac{n}{a}\int x^{n-1}e^{ax}\,dx$

104. $\int \log x\,dx = x \log x - x$

105. $\int x^n \log x\,dx = x^{n+1}\left[\frac{\log x}{n+1} - \frac{1}{(n+1)^2}\right]$

106. $\int \log^n x\,dx = x \log^n x - n\int \log^{n-1} x\,dx$

107. $\int e^{ax} \sin bx\,dx = \frac{e^{ax}}{a^2+b^2}(a \sin bx - b \cos bx)$

108. $\int e^{ax} \cos bx\,dx = \frac{e^{ax}}{a^2+b^2}(a \cos bx + b \sin bx)$

Author Index

Subject Index